Lecture Notes in Networks and Systems

Volume 701

The series "Lecture Notes in Networks and Systems" publishes the latest developments in Networks and Systems—quickly, informally and with high quality. Original research reported in proceedings and post-proceedings represents the core of LNNS.

Volumes published in LNNS embrace all aspects and subfields of, as well as new challenges in, Networks and Systems.

The series contains proceedings and edited volumes in systems and networks, spanning the areas of Cyber-Physical Systems, Autonomous Systems, Sensor Networks, Control Systems, Energy Systems, Automotive Systems, Biological Systems, Vehicular Networking and Connected Vehicles, Aerospace Systems, Automation, Manufacturing, Smart Grids, Nonlinear Systems, Power Systems, Robotics, Social Systems, Economic Systems and other. Of particular value to both the contributors and the readership are the short publication timeframe and the world-wide distribution and exposure which enable both a wide and rapid dissemination of research output.

The series covers the theory, applications, and perspectives on the state of the art and future developments relevant to systems and networks, decision making, control, complex processes and related areas, as embedded in the fields of interdisciplinary and applied sciences, engineering, computer science, physics, economics, social, and life sciences, as well as the paradigms and methodologies behind them.

Indexed by SCOPUS, INSPEC, WTI Frankfurt eG, zbMATH, SCImago.

All books published in the series are submitted for consideration in Web of Science.

For proposals from Asia please contact Aninda Bose (aninda.bose@springer.com).

Ivana Salopek Čubrić · Goran Čubrić ·
Kristian Jambrošić · Tanja Jurčević Lulić ·
Davor Sumpor
Editors

Proceedings of the 9th International Ergonomics Conference

ERGONOMICS 2022

Springer

Editors
Ivana Salopek Čubrić
Faculty of Textile Technology
University of Zagreb
Zagreb, Croatia

Goran Čubrić
Faculty of Textile Technology
University of Zagreb
Zagreb, Croatia

Kristian Jambrošić
Faculty of Electrical Engineering
and Computing
University of Zagreb
Zagreb, Croatia

Tanja Jurčević Lulić
Faculty of Mechanical Engineering
and Naval Architecture
University of Zagreb
Zagreb, Croatia

Davor Sumpor
Faculty of Transport and Traffic Sciences
University of Zagreb
Zagreb, Croatia

ISSN 2367-3370 ISSN 2367-3389 (electronic)
Lecture Notes in Networks and Systems
ISBN 978-3-031-33985-1 ISBN 978-3-031-33986-8 (eBook)
https://doi.org/10.1007/978-3-031-33986-8

This Springer imprint is published by the registered company Springer Nature Switzerland AG
The registered company address is: Gewerbestrasse 11, 6330 Cham, Switzerland

Preface

The Croatian Ergonomics Society (CrES) has organized the *9th International Ergonomics Conference—ERGONOMICS 2022* in Zagreb, Croatia. We are proud that this conference is part of the series "Ergonomics" organized by CrES since 2001 with the aim to promote ergonomics worldwide. The fact that we are in Zagreb in December, when the world-famous Zagreb Advent season is back in full glory, makes this venue and the conference even more special.

The organization of the conference is traditionally a joint project of the three faculties—University of Zagreb Faculty of Transport and Traffic Sciences (FPZ), University of Zagreb Faculty of Mechanical Engineering and Naval Architecture (FSB), and University of Zagreb Faculty of Textile Technology (TTF). As in previous years, the conference is endorsed by the International Ergonomics Association (IEA), the Federation of European Ergonomics Societies (FEES), and the Acoustical Society of Croatia (ASC).

Over the years, this conference has brought together enthusiasts, experts, and scientists from Croatia and from all over the world. Our big Ergonomics family has grown over the years. We always eagerly look forward to meeting old friends and making new acquaintances here. At this year's edition of the conference, participants came from 21 countries, namely Bosnia and Herzegovina, Brazil, Croatia, France, Germany, Greece, Hungary, India, Israel, Italy, Japan, Malaysia, Poland, Portugal, Romania, Russia, Slovakia, Slovenia, South Korea, Ukraine, and the United States of America.

All submissions were peer-reviewed, and 47 abstracts were accepted for publication and presentation at the conference. The post-conference proceedings consists of double peer-reviewed full texts of the accepted abstracts and will be published by the Springer Publishing Co. as "Proceedings of the 9th International Ergonomics Conference—ERGONOMICS 2022" in the series "Lecture Notes in Networks and Systems".

As President of CrES, and as conference President, I would like to express my deep appreciation to the members of the conference committees involved in this challenging organization (unfortunately, still in the pandemic era), as well as to all the supporting organizations that all together have once again made this conference in the "Ergonomics" series possible and a success. My sincere thanks go to all the participants for their valuable contribution, which makes the picture valuable and real.

We hope that all participants had a very pleasant stay in Zagreb and will return home with positive impressions and new scientific ideas. As in the last two decades, we cordially invite you to participate in our anniversary 10th conference and to spread the word to your colleagues so that our Ergonomics family continues to grow.

Sincerely,

Zagreb, Croatia
December 2022

Prof. Ivana Salopek Čubrić, PhD
President of the Croatian Ergonomics Society,
General Chair of the Ergonomics 2022 Conference

Organization

9th International Ergonomics Conference—*ERGONOMICS 2022*

The conference is **organized by**:

CROATIAN ERGONOMICS SOCIETY

The conference is **endorsed by**:

Conference Endorsed by the IEA International Ergonomics Association

Endorsed by FEES Federation of European Ergonomics Societies

Hrvatsko akustičko društvo HAD Acoustical Society of Croatia

The conference is **co-organized by**:

FSB University of Zagreb, Faculty of Mechanical Engineering and Naval Architecture

TTF University of Zagreb, Faculty of Textile Technology

University of Zagreb, Faculty of Transport and Traffic Sciences

The conference is organized **under the auspices of**:

Ministry of Labour, Pension System, Family and Social Policy of the Republic of Croatia

Croatian Academy of Engineering

International Scientific Committee

Kristian Jambrošić, Croatia (Scientific Committee President)
Shamsul Bahri Mohd Tamrin, Malaysia
Tino Bucak, Croatia
Denis Coelho, Sweden
Ivana Salopek Čubrić, Croatia
Apurba Das, India
Bernard Dugué, France
Jose Orlando Gomez, Brazil
Szabó Gyula, Hungary
Ray Yair Lifshitz, Israel
Tanja Jurčević Lulić, Croatia
Abhijit Majumdar, India
Budimir Mijović, Croatia
Diana Milčić, Croatia
Beata Mrugalska, Poland
Alessandro Naddeo, Italia
Georgius Priniotakis, Greece
Nebojša Rašović, Bosnia and Herzegovina
Uwe Reischl, USA
Davor Sumpor, Croatia
Adisa Vučina, Bosnia and Herzegovina
Eric Min-Yang Wang, Taiwan
Emilija Zdraveva, Croatia

Organizing Committee

Ivana Salopek Čubrić, Croatia (Organizing Committee President, General Chair)
Tino Bucak, Croatia
Goran Čubrić, Croatia
Anca Draghici, Romania
Szabó Gyula, Hungary
Jurica Ivošević, Croatia
Tanja Jurčević Lulić, Croatia
Jasna Leder Horina, Croatia
Dorotea Kovačević, Croatia
Diana Milčić, Croatia
Beata Mrugalska, Poland
Elma Mulaomerović, Taiwan
Ivanka Nikolić, Croatia
Davor Sumpor, Croatia
Irena Šabarić, Croatia
Sandro Tokić, Croatia

Program Committee

Davor Sumpor, Croatia (Program Committee President)
Goran Čubrić, Croatia
Kristian Jambrošić, Croatia
Tanja Jurčević Lulić, Croatia
Ivana Salopek Čubrić, Croatia
Aleksandr Volosiuk, Russia
Eric Min-Yang Wang, Taiwan

Technical Committee

Jasna Leder Horina, Croatia (Technical Committee President)
Božena Jurčić, Croatia
Ivanka Nikolić, Croatia
Irena Šabarić, Croatia
Sandro Tokić, Croatia

Reviewers

Shamsul Bahri Mohd Tamrin
Blaženka Brlobašić Šajatović
Tino Bucak
Denis Coelho
Goran Čubrić
Danijela Domljan
Anca Draghici
Bernard Dugué
Darina Dupláková
Goran Đukić
Alin Gaureanu
Szabó Gyula
Hrvoje Haramina
Kristian Jambrošić
Tanja Jurčević Lulić
Evgeniy Lavrov
Jasna Leder Horina
Renata Hrženjak
Abhijit Majumdar
Budimir Mijović
Diana Milčić
Melita Milenković
Beata Mrugalska
Alessandro Naddeo
Tihomir Opetuk
Georgius Priniotakis
Nebojša Rašović
Hassan Sadeghi Naeini
Ivana Salopek Čubrić
Davor Sumpor
Irena Šabarić
Adisa Vučina
Ray Yair Lifshitz
Emilija Zdraveva
Mislav Stjepan Žebec

Acknowledgments

The 8th International Ergonomics Conference—ERGONOMICS 2020 was held from December 2–5, 2020, at the University Campus Borongaj "ZUK Borongaj", in Zagreb, the capital of Croatia.

Ergonomics in Croatia has existed as a scientific and professional discipline for several decades. Let's remember that it all started in 1974 when Prof. Emeritus Dragutin Taboršak, PhD, the founder and the first president of the Croatian Ergonomics Society (CrES), established CrES in Croatia. The achievements as well as recent and relevant ideas in the field of ergonomics have been continuously discussed and exchanged with the international scientific community at the conferences in the series "Ergonomics", along with other means of communication and networking.

The conferences in the series "Ergonomics" have been organized by the (CES) from 2001 as part of CES objectives to promote ergonomics and exchange knowledge and experience with the scientific and professional community from Croatia and the world. The conference traditionally brings together enthusiasts, experts, and scientists from Croatia and from all over the world by bringing their know-how to the table, establishing valuable contacts for future cooperation.

This time, the conference ERGONOMICS 2020 is a joint project with our co-organizing partners:

- CAES (Chinese Association of Ergonomics Societies),
- CES (Chinese Ergonomics Society),
- ErgoWork (Romanian Society on Ergonomics and Workplace Management),
- EST (Ergonomics Society of Taiwan),
- HKES (Hong Kong Ergonomics Society),
- MET (Hungarian Ergonomics Society),
- FPZ (Faculty of Transport and Traffic Sciences), University of Zagreb, Croatia
- FSB (Faculty of Mechanical Engineering and Naval Architecture), University of Zagreb, Croatia
- TTF (Faculty of Textile Technology), University of Zagreb, Croatia.

Furthermore, the conference has been this time endorsed by

- IEA (International Ergonomics Association),
- FEES (Federation of European Ergonomics Societies),
- ASC (Acoustical Society of Croatia).

The conference program this time included more types of oral presentations of papers from the following GROUPS OF TOPICS (not limited):

Aesthetics and Ergonomics;
Biomechanics and Modeling in Ergonomics;
Cognitive Ergonomics;
Education and Trainings in Work Safety and Ergonomics;
Ergonomics for People with Disabilities and Aging Population;
Ergonomics in Product and Process Design;
Ergonomic Regulations, Standards, and Guidelines;
Healthcare Ergonomics;
Physical Ergonomics and Human Factors;
Human Comfort;
Safety and Risk Ergonomics;
Psychoacoustic Ergonomics;
Social and Occupational Ergonomics;
Traffic and Transport Ergonomics.

The Organizing Committee (OC) of the ERGONOMICS 2020 Conference received more than 45 contributions within a diverse range of conference topics. All submissions have been peer-reviewed by the International Scientific Committee and referees from abroad and Croatia, regardless of what type of oral presentation has been chosen (live oral presentation on the spot, real-time online presentation, pre-recorded lecture).

Furthermore, the post-conference Proceedings titled "Proceedings of the 8th International Ergonomics Conference—ERGONOMICS 2020" in printed form (Hardcover ISBN: 978-3-030-66936-2) and online form (eBook ISBN: 978-3-030-66937-9) with full texts of all twenty-five accepted and reviewed papers as well as two full texts invited lectures was published by the Springer Publishing Co., in the series "Advances in Intelligent Systems and Computing" available in printed form (Series ISSN: 2194-5357) and online form (Series E-ISSN: 2194-5365).

The forty-five participants came from the following countries: Austria, Croatia, Brazil, Bulgaria, Bosnia and Herzegovina, Greece, Hungary, India, Italy, Poland, Portugal, Romania, Russian Federation, Ukraine, Slovenia, Spain, and Taiwan ROC.

During the conference opening ceremony, Davor Sumpor (President of CrES and Conference Chair) invited Tomislav Mlinarić (FPZ Dean) and Jose Orlando Gomes (IEA Vice-President) to address the participants with a few words of welcome.

In the introductory part of the conference sessions, the following invited speakers contributed with their lectures: Jose Orlando Gomes (Brazil), Eric Min-Yang Wang (Taiwan ROC), Abhijit Majumdar (India), Nejc Šarabon (Slovenia), Beata Mrugalska (Poland), and Ivana Salopek Čubrić (Croatia).

Due to the global and local situation caused by the COVID-19 virus pandemic, in the circumstances of a situation called "New normal", the conference was held as a mixed conference which included the following types of oral presentations: live oral presentation on the spot, real-time online presentation, and pre-recorded lecture (only voice as part of Power Point presentation).

The conference provided a great opportunity for all the participants and all stakeholders from Croatia and abroad to contribute to the advances in ergonomics in Croatia once again despite the situation called "New normal".

I would like to express deep appreciation to all co-organizing partners, patrons and sponsors, reviewers and all the members of all the conference bodies, authors of the papers and conference participants, who have all together enabled and helped to make this conference in the "Ergonomics" series successful once again.

Finally, I am pleased to note that even in the circumstances of the pandemic situation called "New normal", CrES has been receiving active support of our constant and reliable international partners, who we are very proud of, and for which we are very grateful.

We are excited to meet you again in Croatia at the Ergonomics 2022 Conference, with the belief it will be held in person, at the same location in Zagreb.

On behalf of the Organizing Committee of Ergonomics 2020.

Report on Humanitarian Action for the Earthquake-Affected Areas in the Republic of Croatia

In the wake of a devastating earthquake that affected Croatia on December 29, 2020, the humanitarian action of the Croatian Ergonomics Society for the earthquake-affected areas in the Republic of Croatia (31.12.2020–27.03.2021) which was started by the decision of the Executive Board of the Croatian Ergonomics Society dated 31.12.2020 was finally successfully completed on 27.3.2021 when the recipient of the donation, Štefan Rožanković from Greda, Greda195, Sisak, and his family (a small child of 20 months with two older children aged 14 and 20) received a donated mini kitchen with accompanying equipment, which was installed in the previous donated new housing container on 9.2.2021 manufactured by Tehnix doo, in accordance with the decision of the Executive Board of the Croatian Ergonomics Society dated 8.2.2021.

The house of Štefan Rožanković's family was severely damaged in the first earthquakes, and subsequent earthquakes completely destroyed it (as a result, on February 16, 2021, the orange sticker was replaced with a red sticker, and the remains of the house were planned to be demolished).

With the mediation of members of the Railroad Engineer Trade Union of Croatia (SSH), the company Komunalac Vrbovec joined the humanitarian action and donated the service of transporting a housing container from the factory Tehnix doo in Donji Kraljevec to the recipient's location in Greda near Sisak.

For the purchase of the housing container, money is donated from (in alphabetical order): Izidor Alfirević (Croatia), M. Dugue Bernard (France), Tino Bucak (Croatia), David Caple (Republic of South Africa), Denis Alves Coelho (Portugal), Mammas Contantinos (Greece), Czech Ergonomics Association - Premedis Foundation (Czech Republic), Jose Orlando Gomes (Brazil), International Ergonomics Association, Eduard Ivanjko (Croatia), Juraj Ivanjko (Croatia), Marijan Jakovljević (Croatia), Kristian Jambrošić (Croatia), Tanja Jurčević Lulić (Croatia), Diana Milčić (Croatia), Davor Sumpor (Croatia), Irena Šabarić (Croatia), Emilija Zdraveva (Croatia), Jasenka Pibernik (Croatia), Ivana Salopek Čubrić (Croatia), Pero Škorput (Croatia), and Adrian Wagner (Austria).

During January and February 2021, in the area of Sisak, Petrinja, and Moščenica, a van of the University of Zagreb Faculty of Transport Sciences (FPZ) twice delivered humanitarian aid to earthquake victims (with an emphasis on families with several small children).

Humanitarian aid was collected by the members of the Croatian Ergonomic Society in an action supported by several partners, individuals, and organizations as follows:

- teachers, staff, and students of the University of Zagreb, Faculty of Transport and Traffic Sciences (FPZ),
- members of the Croatian Acoustic Society,
- parents of students of the foreign language school, Heren Doron Špansko from Zagreb,
- members of the Railroad Engineer Trade Union of Croatia (SSH),
- other citizens of the Republic of Croatia.
- On behalf of the Croatian Ergonomics Society.

Davor Sumpor
CrES Vice-President

Contents

Future Relevance of Identified Organizational Measures for Coping with the COVID-19 Pandemic: A Delphi Study

Caroline Adam, Klaus Bengler, Sophie Berger, Christopher Brandl, Birte Emmermann, Mirlinda Hajdari, Verena Nitsch, Gritt Ott, Sebastian Pütz, and Martin Schmauder

Abstract The COVID-19 pandemic engendered massive restrictions in all areas of our social and work lives. Companies and organizations have had to make extensive adjustments to their working conditions and processes. Fifty-two interviews at 33 companies and organizations were conducted to gain insights to help companies

C. Adam (✉) · K. Bengler · B. Emmermann
Chair of Ergonomics, Technical University of Munich, Munich, Germany
e-mail: caroline.adam@tum.de

K. Bengler
e-mail: bengler@tum.de

B. Emmermann
e-mail: birte.emmermann@tum.de

S. Berger · G. Ott · M. Schmauder
CIMTT Centre of Production Engineering and Management, Technische Universität Dresden, Dresden, Germany
e-mail: sophie.berger@tu-dresden.de

G. Ott
e-mail: gritt.ott@tu-dresden.de

M. Schmauder
e-mail: martin.schmauder@tu-dresden.degritt.ott

C. Brandl · M. Hajdari · V. Nitsch · S. Pütz
Institute of Industrial Engineering and Ergonomics, RWTH Aachen University, Aachen, Germany
e-mail: c.brandl@iaw.rwth-aachen.de

M. Hajdari
e-mail: m.hajdari@iaw.rwth-aachen.de

V. Nitsch
e-mail: v.nitsch@iaw.rwth-aachen.de

S. Pütz
e-mail: mgritt.ott@tu-dresden.de

I. Salopek Čubrić et al. (eds.), *Proceedings of the 9th International Ergonomics Conference*, Lecture Notes in Networks and Systems 701,
https://doi.org/10.1007/978-3-031-33986-8_1

and organizations to cope with the pandemic. The applicability and future relevance of these findings were ascertained in a practice-oriented and systematic manner in a Delphi study. For this purpose, 17 statements were formulated, which were evaluated by 21 experts from different areas based on their suitability for coping with the COVID-19 pandemic as well as their suitability for working outside a crisis situation. The assessment dimensions considered included general usefulness and company-specific relevance. The results show that not all of measures introduced in response to the pandemic can be transferred on a one-to-one basis to a period of time outside a crisis situation. Based on the results, however, it can be assumed that the COVID-19 pandemic has triggered a long-term change process in companies and organizations and that some of the measures introduced in the short term can also contribute to increasing value creation and improving working conditions in the future.

Keywords COVID-19 · Pandemic · Delphi-Study · Measures · Change Process

1 Introduction

Within a short period of time, the COVID-19 virus spread rapidly to become a worldwide pandemic with serious consequences that impacted nearly all areas of life. The development of the pandemic triggered a series of regulations and measures all around the world all aimed at containing the virus. For example, the federal government in Germany established distance and hygiene rules for the population and took measures to reduce social contacts in private and business settings to a minimum, resulting in temporarily closed schools, daycare centers, universities, pubs and stores, canceled cultural events and vacations away from home [1–4]. Both the first nationwide *lockdown* (from March to mid-June 2020) and the second *lockdown* (from mid-December 2020 to May 2021) exacerbated these measures and forced companies and organizations to restructure their operations. In various industries, companies immediately had employees work from home to minimize interpersonal contact among colleagues and to enable them to continue working. Some companies even had to close down completely or introduce short-time working [2, 3]. In many companies, the situation has settled down after about two years of the pandemic and a *new normal* has developed. Companies and organizations introduced new ways of working and measures that enable them to continue working despite the restrictions [5, 6]. The aim of the study described in this article is to determine which of the measures introduced during the pandemic and identified in [5] are suitable for coping with the pandemic and similar crisis situations as well as for a future that is not affected by a pandemic.

2 Theoretical Background

The changes in the world of work in Germany that are increasingly perceivable, particularly in connection with the COVID-19 pandemic, have already been the subject of numerous studies and publications.

According to a survey of around 700 employees, the number of employees working from home rose from 12.5% in 2016 to around 35% in 2020 [7]. In addition, a large majority of respondents (81%) would also like to continue working remotely, at least in part, during the post-pandemic period to benefit from increased self-determination through flexibilization of working hours and location as well as the associated improvement in work-life balance [7]. Despite these positive aspects for employees resulting from the increase in working from home, a survey of industrial companies in Germany conducted by the Federal Ministry for Economic Affairs and Energy (BMWi) found that many employees develop a feeling that they have a higher workload and more time pressure, as well as fatigue due to video conferencing [8]. After more than two years of the pandemic, the equipment of the workplace at home often remains provisional. There is an increasing demand for ergonomic desks, chairs and external monitors [8].

A study by [9] examines working from home in various fields of social work. In a survey of 2,064 participants, it was found that the option to work from home was primarily available to managers and public sector employees. Work output deteriorated due to the fact that only about half of those working from home received the necessary equipment for their home office and that contact with superiors, colleagues and cooperation partners changed [9].

In a preliminary study to the study presented here, [5] identified measures and work design solutions for dealing with the pandemic in the three sectors of the economy. For this, managers and employees of 33 companies and organizations were interviewed. In particular, an increasing flexibilization of work location and working hours as well as the introduction of numerous hygiene measures were observed, such as the obligation to wear masks, to disinfect surfaces and hands, or a spatial separation of workplaces through rearrangement or through the use of strategically placed flexible Perspex panels. In addition, an acceleration in the digital transformation was reported. This includes the digitization of work processes and the introduction or expansion of digital working resources. Other key findings relate to the impact on leadership culture, which particularly includes establishing a culture of trust instead of a culture of presence. Finally, adjustments to communication processes and to operational handling of crisis-related challenges were analyzed [5].

3 Method

The aforementioned measures [5] were used to identify best practices to help companies and organizations manage the pandemic. Then, 17 projections for potential measures (Table 1) were formulated based on the identified measures and were evaluated using a Delphi study. In the course of the two-round Delphi study, experts were first asked to rate and comment on the presented measures and second, to discuss their assessment asynchronously with the other experts and if necessary, to adapt their own ratings. The Delphi study was conducted online using the SurveyLet software (calibrum) for a period of three weeks in October and November 2021. Anonymity was ensured by assigning each respondent an ID, which enabled the experts to state their opinion without perceiving social pressure or influence from other participants [10].

Sample
A total of 21 participants were included in the survey, 71% of whom were male and 29% female. The average age was 43.24 years ($SD = 10.55$). Not all experts completed the survey in both rounds, resulting in different sample sizes. Experts from the following fields were recruited: Twelve experts were from the *business/company* field, seven from the field of *science/research* and two from the field of *employer representation of interests.* No participant could be obtained from the field of *employee representation of interests.*

Procedure
At the beginning of the first round, each expert was informed about the study's purpose and process. Subsequently, general information about the person was requested, after which the experts evaluated the 17 measures (Table 1) with respect to two evaluation dimensions and scenarios respectively.

Respondents were familiarized with the two scenarios. The first scenario (*Work during the COVID19 pandemic,* S1) describes the situation at the peak of an infection

Table 1. 17 identified measures. For the full titles see the supplemental material

No	Short title	No	Short title
1	Extended hygiene protection	10	Trust in leadership
2	Permanent crisis team	11	Avoiding team splits
3	Regular status updates	12	Transparency of accessibility
4	Flexible working time arrangements	13	Recommendations for the home office
5	Flexible choice of place of work	14	Avoidance of work intensification
6	New workplace concepts	15	Hybrid meetings
7	Non-locally based employees	16	Reduction of business trips
8	Working in fixed teams	17	Capacities for data protection
9	Further training during short-time work		

wave with a Germany-wide lockdown and corresponding restrictions on private and working life. The second scenario (*Work outside a crisis situation,* S2) describes the situation of a normal working day without crisis-related restrictions. Each measure was evaluated for these scenarios in two dimensions: its general usefulness (GU) as well as the relevance for the experts' individual workplace – company-specific relevance (CR). The measures were assessed quantitatively on a six-point Likert scale and qualitatively with a written explanation of the assessment by each expert. After the experts had completed the first round of the survey, the Delphi study was adapted. In the second round of the survey, the experts were given response distributions, descriptive statistics, and a list with the comments for each scenario and dimension. Based on this, each participant was able to read and question the assessments of the other experts, and if necessary, re-evaluate their own initial assessment (Table 1).

4 Results

The expert assessments showed high stability between the two rounds of the Delphi study. Thus, results are reported for participant final ratings provided in the second round. Figure 1 presents the full quantitative results for the dimension GU for both scenarios (S1 and S2). Quantitative results for the CR ratings and a full discussion of open responses are provided in supplementary materials.

The common Delphi method practice was followed to investigate whether the experts reached a consensus in their assessments. An interquartile range (*IQR*) of ≤ 1 of the ratings was defined as the threshold for a consensus (threshold adjusted to the length of the scale; see e.g., [11–13]). For S1, a consensus was indicated for 10 of the 17 measures regarding GU and for 6 measures regarding CR. In the context of S2, the ratings for five measures yielded a consensus for GU and CR respectively.

Further, it was examined whether the experts' ratings differed systematically between scenarios and dimensions. To determine this, the average ratings for all measures per scenario and per dimension was calculated and the non-parametric Van der Waerden test (cf. [14]) was applied. The test showed a significant difference between scenarios, $\chi^2(1) = 8.21$, $p < 0.01$, with more positive ratings for S1 ($M = 3.75$) than for S2 ($M = 3.25$). In addition, there was a significant, albeit small, difference between the ratings regarding GU ($M = 3.52$) and CR ($M = 3.45$), $\chi^2(1) = 4.03$, $p < 0.05$. There was no significant interaction between the two factors $\chi^2(1) = 1.67$, $p = 0.197$.

In general, most of the identified measures were assessed as useful/relevant by the experts for both scenarios. In S1, the introduction of extended hygiene concepts and the reduction of business trips, followed by the establishment of frequent status updates, a permanent crisis unit, and more flexible arrangements for employee working times and location were rated highest. These measures were almost exclusively assessed positively by the experts, whereby open responses highlighted them

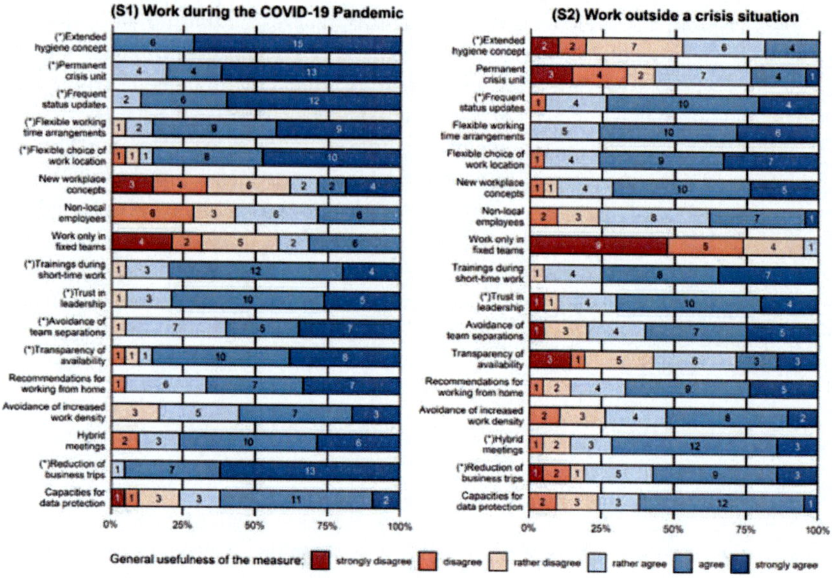

Fig. 1 Distributions of expert responses for all 17 measures evaluating the general usefulness of the respective measure for (S1) work during the COVID-19 pandemic and (S2) a work outside a crisis. situation. (*)- before measures labels indicate that an expert consensus was reached ($IQR \leq 1$)

as most effective for lowering the infection rate within the company and making organizational processes more flexible for dealing with the legal and social requirements related to crisis management.

Making working hours and the place of work more flexible as well as continuing with frequent status updates were also among the measures rated as most suitable for S2. Corresponding with allowing employees to switch between office and work from home more flexibly, experts see benefits in the development of new workplace concepts that enable a closer match between the provided working environment and the requirements of tasks that continue to be performed at the primary workplace. Moreover, experts consider the use of future phases of short-time work for employee training to be a valuable opportunity.

The measure that was assessed most negatively for both scenarios was the division of the workforce into fixed teams. Even though this measure can contribute to minimize infections, experts emphasized the associated risks of impeding flexibility as well as collaboration between employees. For S1, they were also skeptical about changing workplace concepts, since they focused primarily on enabling employees to work from home, and about hiring non-local candidates, since this would need to be integrated into a long-term recruitment strategy. As to be expected, extended hygiene concepts and permanent crisis units were assessed to be the least suitable measure for S2. That said, even these measures were evaluated positively by about half of the experts.

5 Discussion

Overall, the measures assessed in this Delphi study were mostly viewed positively by the experts. This supports the validity of the measure identified and formulated by [5]. As such, strategies that have enabled companies to still be operable to a certain extent while facing a global pandemic will likely play an influential and positive role in the future of work even after the COVID-19 pandemic has passed. In accordance with the findings of [7], the results suggest that, in particular, an increase in employees working from home will persist in the post-pandemic work environment. However, not all measures can be transferred directly to other future work scenarios. This is indicated by the significantly higher rating of the measures in the context of the COVID-19 pandemic than of the post-pandemic work environment. The experts view certain measures to be inadequate for a time outside a pandemic, e.g., extensive hygiene measures and working in fixed teams. This effect could additionally be caused by inter-individual differences in the experts' prediction of what the future work environment will look like. Thus, the measures hold stronger applicability for the context they were identified in.

Furthermore, a significant difference was found in the experts' assessment of the measures between their general usefulness and their company-specific relevance. The measures were viewed more critically when the experts took their companies' specific circumstances and individual practices into consideration. For example, multiple experts report that, while generally welcoming a company's support of their employees in creating a work-life-balance, they object to the measure considering their individual work tasks and position in the company.

Results showed high stability and little discussion or changes in ratings between the first and second round of the Delphi study. This effect can be caused by multiple factors: First, the measures in themselves are very agreeable. Since they are based on elaborate data collection and analysis [5], the high consensus between the experts—especially in the context of the pandemic—is understandable. Second, the statements describing the measures are formulated in a non-critical, appeasing manner. More polarizing statements might have led to a livelier discussion. Third, the experts showed homogenous characteristics. The extension of the sample pool to a more heterogeneous representation (e.g., other industry sectors and personnel from various hierarchy levels within the companies) could potentially bring forth a higher dissensus as more complexity would be added. Fourth, even though a software was used that specializes in the conduction of Delphi studies, its user interface reportedly did not encourage a re-examination of the measures and thus, discussion and adaptions of the ratings in the second round of the Delphi study were hindered. Lastly, participation in the first round of the study was elaborate and demanded extensive mental and time capacities of the experts. A decrease of the study's complexity, e.g., by reducing the number of measures or scenarios, might have allowed for better circumstances in which a discussion between the experts could take place.

The Delphi study was conducted using a small sample size of experts from industry who predominantly work in the tertiary sector. Furthermore, the participants are located in the German states of Bavaria, North Rhine-Westphalia, and Saxony, since the collected measures were also identified in these regions. Therefore, the results are limited in their transferability with regard to the characteristics of the sample.

6 Outlook

The results of this Delphi study indicate that especially the trends of digitalization and more flexible work arrangements, such as working from home or training during short-time work, will play an important role in the future world of work. The COVID-19 pandemic has pushed companies to implement drastic changes to their working structure within a short period of time. Even in the future, these now-established measures can create value. However, the specific circumstances of the employee and their tasks need to be considered and the measure adapted accordingly.

The Delphi method is a means to predict future developments based on experts' present estimations and experience. In the context of this study, this allows the distinction between, on the one hand, measures that emerged due to the pandemic and which provide a potential value for the post-pandemic work environment, and, on the other hand, those measures whose value is likely restricted to times of the pandemic. However, one cannot evaluate for the future but merely collect subjective assessments of individuals of the present. Therefore, further research is needed to establish whether measures will have indeed proven valuable at the specific point in time. It is indispensable to investigate the long-term effects of the pandemic on the future of work and, with changing circumstances, to continuously adapt recommendations, both for companies and organizations as well as for legal regulations.

Supplemental Material

The qualitative results of the Delphi study can be found via: https://osf.io/gxm8j/

Acknowledgements This article describes research conducted as part of the research project COVID-19 Lessons Learned (Grant Number 02L18A700), which is funded by the German Federal Ministry of Education and Research (BMBF).

References

1. Bünning, M., Hipp, L., Munnes, S.: Erwerbsarbeit in Zeiten von Corona. WZB Ergebnisbericht. WZB, Berlin (2020)
2. Frodermann, C., et al.: Online-befragung von beschäftigten: wie corona den arbeitsalltag verändert hat. IAB-Kurzbericht (13/2020). IAB, Nürnberg (2020)
3. Imöhl, S., Ivanov, A.: Corona in Deutschland in der aktuellen Zusammenfassung. https://www.handelsblatt.com/politik/deutschland/covid-19-corona-in-deutschland-in-der-aktuellen-zusammenfassung/25584942.html?ticket=ST-8276847-VDwgNafEyhMzUFVxMamK-ap2. Accessed 21 Feb 2022

4. Möhring, K., et al.: Die Mannheimer Corona-Studie: Schwerpunktbericht zu Erwerbstätigkeit und Kinderbetreuung. Mannheim (2020)
5. Adam, C., Bengler, K., Brandl, C., Nitsch, V., Ott, G., Pütz, S., Schmauder, M.: Maßnahmen und Lösungen zur Arbeitsgestaltung für den Umgang mit der COVID-19 Pandemie: Eine systematische Analyse der Arbeit im Primär–, Sekundär- und Tertiärsektor in Deutschland. Z. Arb. Wiss. **75**, 527–541 (2021)
6. Bengler, K., et al.: COVID-19LL: a systematic approach to identify best practices and lessons learned in german economic sectors. In: Black N.L., Neumann W.P., Noy I. (eds.) Proceedings of the 21st Congress of the International Ergonomics Association 2021, vol. 1, pp. 515–522. Springer, Cham (2021)
7. Kunze, F., Hampel, K., Zimmermann, S.: Homeoffice in der corona-krise: eine nachhaltige transformation der Arbeitswelt? Policy paper (2), Konstanz (2020)
8. BMWi: Das neue Normal? Die Arbeitswelt nach der Corona-Pandemie: Bundesministerium für Wirtschaft und Energie. Berlin (2021)
9. Meyer, N., Alsago, E.: Soziale Arbeit am Limit? Sozial Extra **45**(3), 210–218 (2021)
10. Gnatzy, T., Warth, J., von der Gracht, H., Darkow, I.-L.: Validating an innovative real-time Delphi approach - a methodological comparison between real-time and conventional Delphi studies. Technol. Forecast. Soc. Chang. **78**(9), 1681–1694 (2011)
11. Scheibe, M., Skutsch, M., Schofer, J.: Experiments in Delphi methodology. In: Linstone, H.A., Turoff, M. (eds.) The Delphi Method — Techniques and Applications, pp. 262–287. Addison-Wesley, Reading (1975)
12. von der Gracht, H.A.: Consensus measurement in Delphi studies: review and implications for future quality assurance. Technol. Forecast. Soc. Chang. **79**(8), 1525–1536 (2012)
13. von der Gracht, H.A., Darkow, I.L.: Scenarios for the logistics services industry: a Delphi-based analysis for 2025. Int. J. Prod. Econ. **127**(1), 46–59 (2010)
14. Luepsen, H.: Comparison of nonparametric analysis of variance methods: a vote for van der Waerden. Commun. Stat. Simul. Comput. **47**(9), 2547–2576 (2018)

Reachability Simulation of Car Dashboard Commands: A Comparison Between Delmia™ v5 and Unreal Engine™ v4

Francesco Adinolfi, Verdiana Anna Faustini, Andrea Terracciano, Anil Yalcin, Rosaria Califano, Nicola Cappetti, and Alessandro Naddeo

Abstract The aim of this study is to provide an example of a methodology to simulate human–machine interaction in human centric design approach for performing ergonomics and (dis)comfort analyses. It consists of gathering data from the real world, creating a virtual model of the environment and a digital human model, and finally simulating interactions with artefacts and human body-parts motions in different software applications. Firstly, motion capture has been carried out using a low-cost motion-capture system with markers and cameras, then data acquired have been processed using Python© and MATLAB© codes to extract useful information about the movements. This information has been processed to recreate the movements in a virtual environment using DELMIA™ and Unreal Engine™. Both methods proved their reliability in testing reachability, but the comparison showed that Unreal Engine™ appears much more realistic in manikin and movements' simulations than DELMIA™.

Keywords Reachability · DELMIA · Unreal Engine · Car Seat · Virtual Environment · Ergonomics · Comfort

1 Introduction

In recent years, virtual environments (VEs) are increasingly being studied as a complementary approach for product development, for running simulations and for learning tasks in safety. By these technologies, it is in fact possible to avoid physical mockups in order to reduce costs and product development time and to recognize any

---

F. Adinolfi · V. A. Faustini · A. Terracciano · A. Yalcin · R. Califano · N. Cappetti · A. Naddeo (✉)
Department of Industrial Engineering, University of Salerno, Via Giovanni Paolo II, 132, 84084 Fisciano, Salerno, Italy
e-mail: anaddeo@unisa.it

I. Salopek Čubrić et al. (eds.), *Proceedings of the 9th International Ergonomics Conference*, Lecture Notes in Networks and Systems 701,
https://doi.org/10.1007/978-3-031-33986-8_2

dangerous conditions in the workplace before they occur. The automotive industry has also used new technologies for developing and testing parts through VE. In the early phases of automotive design, seating and dashboard commands can be virtually prototyped, and, using Digital Human Modelling (DHM) software, several kinds of interactions can be modelled to evaluate the ergonomics and comfort of designed solutions. Several studies demonstrated that DHM approaches are preferable in virtual reachability and usability tests as well as in macro-ergonomics evaluations [1]. In this work the reachability of some commands of the dashboard of a vehicle was analyzed, as it was shown that the reachability strongly influences the perception of general comfort. In recent years human factors involved in car design have become one of the main areas of focus, thanks to Virtual prototyping (VP) and DHM, in fact, researchers are able to perform several kinds of simulations to assess the required performance of products, particularly in the field of human factors and ergonomics [2].

2 Acquisition of Data

Motion capture systems allow to capture human body posture and motion in a realistic way using synthetic or fiducial markers. One of the most popular approaches is the use of binary square fiducial markers [3]. The main benefit of these markers is that a single marker provides enough correspondences (its four corners) to obtain the pose. Also, the inner binary codification makes them especially robust, allowing the possibility of applying error detection and correction techniques [4]. Since it is not possible to perform motion capture for a wide population with all the possible anthropometric parameters, a realistic way would be capturing data of motion for sitting position of occupants by sampling of subjects: these subjects will belong to the normal distribution of the chosen population for specified confidence bounds and can be used as reference to conduct biomechanics and occupational health analysis [5]. The idea was to create an environment in which one can perform some experiments for checking and verifying the software performances in simulating the real movements of users. To provide an example of a driver reachability assessment, the tester performed several movements using a simulation station. The tests were carried out in the Virtual Reality Laboratory (VR Lab) at the University of Salerno using a physical mock-up of a car seat with fully reconfigurable seat, steering wheel, gearbox and movable Infotainment (Fig. 1a). To track body movements during recordings, ArUco markers have been applied on the tester's garments. An ArUco marker is a synthetic square marker composed by a wide black border and an inner binary matrix which determines its identifier (id) (Fig. 1b).

The black border facilitates its fast detection in the image and the binary codification allows its identification and the application of error detection and correction techniques [6]. In order to obtain consistent data from the tracking of the markers, cameras must be calibrated. Calibrating a camera involves measuring the internal camera parameters and the nonlinearities resulting from lens distortion [7]. To correctly

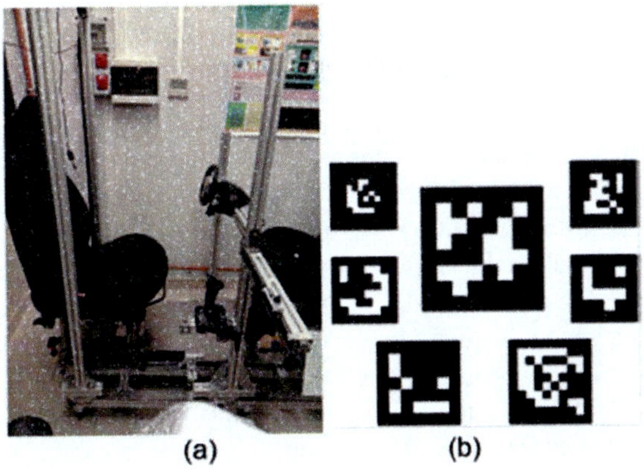

Fig. 1 (a) Experimental setup (b) Example of ARUCO marker

acquire all aspects of the movements in the space, three cameras have been used from three different points of view: front, side and top. Two markers have been attached at each important point (human body joints and segments) in order to make the acquisition of the data more reliable. In fact, markers can't be recognized by the software when they are partially covered, when they are too much inclined or when the light reflects on them, so having redundant markers is useful because, if one of them is unrecognizable, another can be used.

Then the videos, obtained from the cameras, have been processed using a Python script in order to acquire the position and orientation of the markers, while the initial frames have been analysed by Kinovea software, that has been used to measure the angles in the starting position, as explained in the next paragraph.

Anthropometric data of the tester are in Table 1.

Table 1 Anthropometric measurements of the tester.

Variables	Measures [mm]
Stature	1600
Axilla height	1240
Waist height	1020
Crotch height (standing)	800
Sleeve outseam	508
Acromion-radiale lenght	285
Radiale-stylion length	230

3 Setting the Starting Position with Kinovea and Video Processing by Python

To acquire data on the three-dimensional point coordinates, a photogrammetric analysis of the initial frames of the three cameras such as in Fig. 2, was performed using the Kinovea software, which, after manually applying vectors on the image, allowed to calculate the angles between the markers [8]. These values were taken to set the starting position of the manikin on DELMIA [1]. A Python script was used to process the images acquired by the cameras in order to obtain positions and rotations of each marker, thus the movements performed by the tester [9]. The code generates as output a ".txt" file containing eight columns (Fig. 3), which report: a) the number of the frame of the video; b) the identifier of the recognized marker; c) the coordinates of the center of the marker along the three axes; d) the inclination of the marker around the three axes.

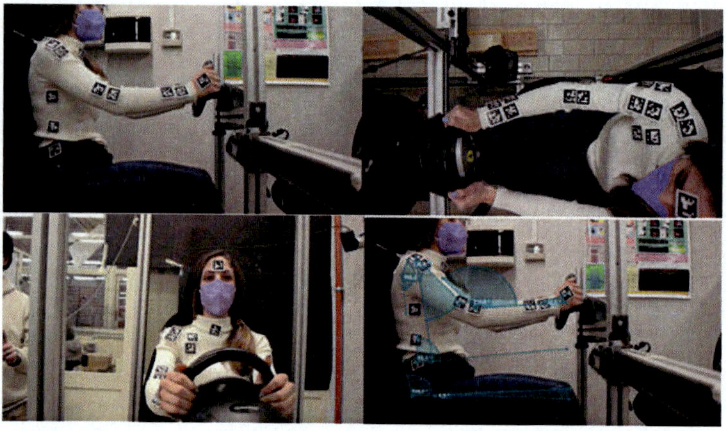

Fig. 2 Three views of cameras for Kinovea setting

Fig. 3 Results of Python elaboration

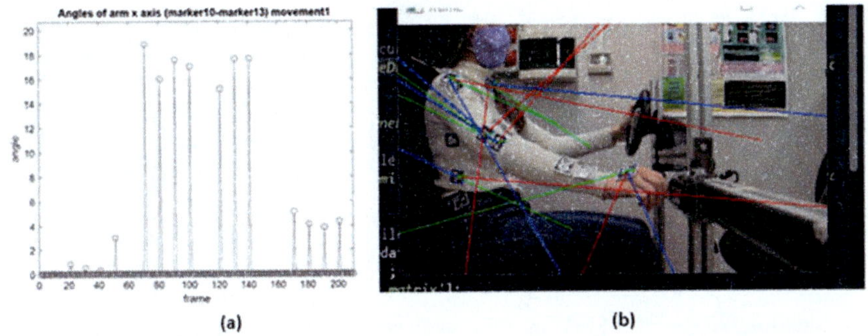

Fig. 4 (a) Results of MATLAB elaboration (b). View of the side camera and data acquisition

With such data it was possible to completely track the instantaneous position and orientation of all the markers acquired by the individual cameras at any time.

4 MATLAB Processing

A MATLAB script was used to process the data obtained from Python to evaluate the rotations of the various parts of the body. The code takes as input the file ".txt" and two couples of markers for a part of the body (e.g.: markers of shoulder and elbow to identify the upper arm) and it calculates the rotation of that part of the body for each frame [10]. Then the results are represented with a graph and saved as a ".pdf" file. This procedure has been performed for each part of the body involved in the movement.

The graph in Fig. 4a shows the rotations of the arm, identified by the relative angular displacements between markers 10 and 13, during the first movement along the X axis (which correspond to the side camera – Fig. 4b). Data obtained from the MATLAB script are also saved in ".xls" files.

5 Reproduction of the Simulation in Delmia Environment

In order to reproduce the simulation station in a virtual environment, DELMIA software has initially been used. The seating buck components were measured and modelled.

The parts have been modelled and then assembled in one Product file. After the creation of the virtual environment on DELMIA, it was also necessary to create a manikin that could reproduce the actions performed by the tester (see Fig. 5). It's important to take the right measurements of the occupant to have consistent results. Anthropometry is the science that defines physical measures of a person's size, form,

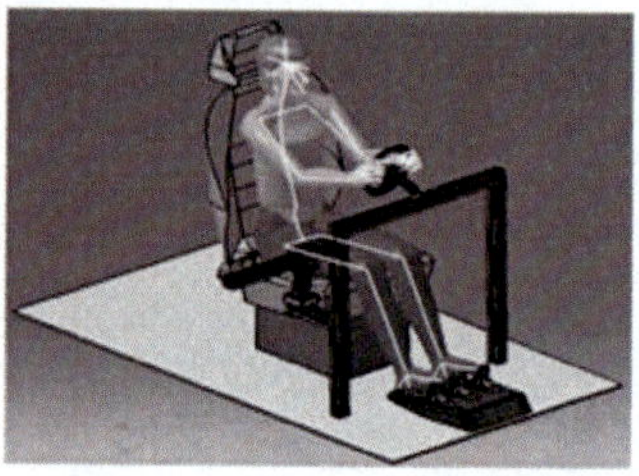

Fig. 5. 3D Model of manikin in DELMIA VE

and functional capacities [11]. In order to create one that faithfully represented the person who performed the tests, accurate measurements of the tester were taken and entered into the DELMIA software through the Human Builder module. Comparing the measurements taken with the current data available, the manikin represents the 34th percentile of women in the South European Database [12]. This manikin was inserted into the virtual environment and its starting position specified according to the values obtained with Kinovea and, after that, the simulation of the movements during the following five different tasks has been performed: Movement 1: reaching the Infotainment position 1; Movement 2: reaching the gear stick; Movement 3: turning the wheel; Movement 4: reaching the Infotainment position 2; Movement 5: reaching the Infotainment position 3.

Carrying out the 5 tasks, different positions of the Infotainment were taken into consideration. In the nearest position (n. 1) the left edge of the Infotainment is 4 cm to the left of the frame, while in position 2 it is 12 cm to the right of the frame and in the last configuration it is 29 cm far from the frame.

Movement 1: Tester reaches the Infotainment position 1, turns the volume wheel, and comes back to the steering wheel.
Movement 2: Tester reaches the gear stick, changes the gear, and comes back to the steering wheel.
Movement 3: Tester turns the steering wheel approximately 45 degrees clockwise and comes back to the starting position.
Movement 4: Tester reaches the Infotainment position 2, turns the volume wheel, and comes back to the steering wheel.
Movement 5: Tester reaches the Infotainment position 3, turns the volume wheel and comes back to the steering wheel.

6 Reproduction of the Simulation Station

For the use in Blender and Unreal, we created a manikin in MakeHuman software reproducing anthropometric measures of the tester. Open-source software systems such as MakeHuman and Blender have become powerful tools in creating and animating Digital Human Modelling (DHM) beyond the use in video and computer gaming; in combination, they provide the basis for the application of dynamic DHM in scientific and engineering virtual environments [13]. MakeHuman is a computer graphics software designed for prototyping a photorealistic humanoid. MakeHuman mesh is used in industrial design, to verify the anthropometry of a project, and in virtual reality research, to quickly produce avatars from measures or camera views [14]; it is capable to associate a parameterized base mesh for human representation with a master rig (skeleton made by joints and segments) including geometries for topology, clothes, tongue, eyes, eyelashes, eyebrows, teeth, and hair" [13]. A fundamental step in the manikin's creation is the choice of the skeleton: in our case the Game Engine skeleton proposed by the software, which is the most suitable for Blender and Unreal Engine applications, has been chosen because has approximately the same number of joints and segments of the DELMIA manikin.

Blender is a free and open-source 3D computer graphics software toolset used for creating animated films, visual effects, art, 3D printed models, motion graphics, interactive 3D applications, virtual reality, and computer games [14]. The manikin has been exported from MakeHuman as a ".FBX" (FilmBox) file to successively import it in Blender. The 3D model of the physical environment of the test was imported in the same file with STEP extension directly from DELMIA thanks to a Blender add-on, which allows importing ".stp" files.

The manikin was later placed in the right seating position on the car seat (see Fig. 6 - left). After that, the same movements captured in the acquisition phase were reproduced in the Animation environment. The process performed to create an animation in Blender was to move the bones to reproduce the starting, ending and some intermediate positions of the movements made by the tester during the acquisition. Then these poses were associated with a certain keyframe in the action editor and the software automatically made a smooth transition between them. Then textures were added to the parts.

Unreal Engine is a game engine developed by Epic Games. Initially developed for PC first-person shooters, it has since been used in a variety of genres of three-dimensional (3D) games and has seen adoption by other industries, most notably the film and television industry [15]. The created files were imported in Unreal environment in ". fbx" format, including the manikin with its animations. Using the Level Sequencer, the sequence of the five movements, that we previously simulated

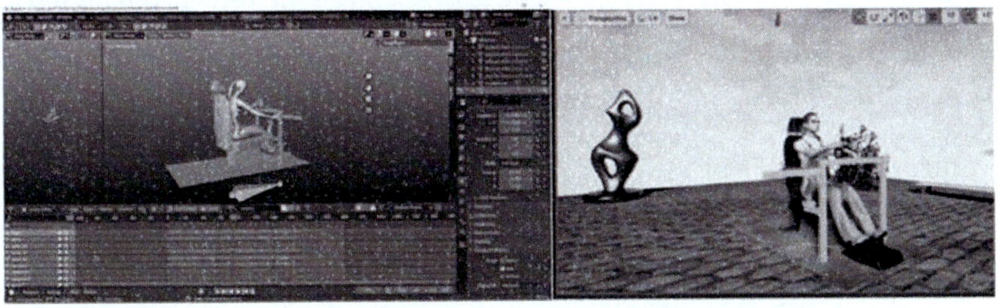

Fig. 6 Blender workspace (left) and Unreal Engine Virtual Environment (right)

on Blender, was imported and videos from different perspectives have been recorded (see Fig. 6 - right).

7 Discussion and Conclusion

This study involved the acquisition of movement performed by a driver of a vehicle and the recreation and simulation of those movements in a virtual environment. The focus of the project was put on the comparison of two different software: Delmia V5 and Unreal Engine 4. Both software proved to be powerful and useful tools to create virtual environments and to perform simulations. Time of development of scenarios and of implementation of manikins within the software is about the same for those who have good skills in DELMIA modelling and in Unreal programming. The comparison showed that Unreal Engine has more features than Delmia and also that the manikin and the movements are more realistic, furthermore Unreal Engine allows to add textures to the objects and provides a lot of pre-made entities such objects, textures and so on. This result is in line with our expectations: since Unreal Engine was originally developed for video games, it makes it easy to recreate actions and realize the interaction between environment and characters. Nevertheless, even if Unreal is able to reproduce in realistic way the movements for performing body posture analyses, Delmia has several tools for performing Ergonomic Analyses, like OCRA [16] or RULA [17], directly inside the software environment an give an answer to designers' needs.

To conclude, from the study of literature and thanks to the recent development of VR, it is possible to affirm that virtual reality software will obtain a fundamental place in the industrial field.

Future developments of this work may include an ergonomic assessment of head, neck, trunk and upper limbs by analyzing the rotations of the joints involved in the different actions according to international legislations and standards. This could be useful for example in the design phase to establish the maximum distance of the volume command on the dashboard based on reachability capabilities of the car driver. The procedure shown in this study could be easily generalized with different percentiles and anthropometric characteristics of people and different human activities.

References

1. Naddeo, A., Cappetti, N., Ippolito, O.: Dashboard reachability and usability tests: a cheap and effective method for drivers' comfort rating. In: SAE Technical Paper, vol. 1 (2014). https://doi.org/10.4271/2014-01-0455
2. Naddeo, A., Cannavacciuolo, O., Cialeo, G., De Stefano, P., Goglia, I., Russo, R.: the effect of reachability on global comfort perception: the case of front-seat car passengers. In: SAE Technical Paper, vol. 2018, pp. 1–8 (2018). https://doi.org/10.4271/2018-01-1320

3. Sarmadi, H., Muñoz-Salinas, R., Olivares-Mendez, M.A., Medina-Carnicer, R.: Detection of binary square fiducial markers using an event camera. In: IEEE Access, vol. 9, pp. 27813–27826 (2021). https://doi.org/10.1109/ACCESS.2021.3058423
4. https://docs.opencv.org/3.4/d9/d6a/group__aruco.html. Access 06th Nov 2022
5. Tao, Q., Kang, J., Sun, W., Li, Z., Huo, X.: Digital evaluation of sitting posture comfort in human-vehicle system under industry 4.0 framework. Chinese J. Mech. Eng. **29**(6), 1096–1103 (2016). https://doi.org/10.3901/CJME.2016.0718.082
6. https://docs.opencv.org/4.x/d5/dae/tutorial_aruco_detection.html
7. Schmalstieg, D., Höllerer, T.: Augmented Reality: Principles and Practice. Addison-Wesley, ISBN 9780321883575 (2016)
8. Mousavi, S.A., Shahbazi, M., Shirzad, E.: Comparison of performance and movement kinematics of virtual reality practice with real-world practice in the dart-throwing skill. Asian J. Sports Med. **13**(2) (2022). https://doi.org/10.5812/asjsm-123574
9. Fiorillo, I., Piro, S., Anjani, S., Smulders, M., Song, Y., Naddeo, A., Vink, P.: Future vehicles: the effect of seat configuration on posture and quality of conversation. In: Ergonomics, vol. 62, no. 11, pp. 1400–1414 (2019). https://doi.org/10.1080/00140139.2019.1651904
10. Fiorillo, I., Nasti, M., Naddeo, A.: Design for comfort and social interaction in future vehicles: a study on the leg space between facing-seats configuration. In: International Journal of Industrial Ergonomics, vol. 83 (2021). https://doi.org/10.1016/j.ergon
11. https://www.cdc.gov/niosh/topics/anthropometry/default.html
12. https://www.gigacalculator.com/calculators/height-percentile-calculator.php. Access 06th Nov 2022
13. Gunther, P., Scataglini, S.: Open-source software to create a kinematic model in digital human modelling. DHM Posturography (2019). https://doi.org/10.1016/B978-0-12-816713-7.00017-9
14. http://www.makehumancommunity.org/. Access 06th Nov 2022
15. Erdei, T.I., Krakó, R., Husi, G.: Design of a digital twin training centre for an industrial robot arm. Appl. Sci. (Switzerland) **12**(17) (2022). https://doi.org/10.3390/app12178862
16. Occhipinti, E., Colombini, D.: The OCRA method: updating of reference values and prediction models of occurrence of work-related musculo-skeletal diseases of the upper limbs (UL-WMSDs) in working populations exposed to repetitive movements and exertions of the upper limbs. In: Medicina Del Lavoro, vol. 95, no 4, pp.305–319 (2004)
17. McAtamney, L., Nigel Corlett, E.: RULA: a survey method for the investigation of work-related upper limb disorders. In: Applied Ergonomics, vol. 24, no. 2, pp. 91–99 (1993). https://doi.org/10.1016/0003-6870(93)90080-S

Starting a Chainsaw: A Postural Load Assessment

M. Bačić, Z. Pandur, M. Landekić, M. Šušnjar, M. Bakarić, and M. Šporčić

Abstract A gasoline-powered chainsaw is still an irreplaceable tool in forest harvesting operations. Operating a chainsaw requires training and skill since it is a potentially dangerous tool if not operated according to the safety guidelines. Safely starting a chainsaw is often overlooked in practice which can cause serious physical injuries. To safely start a chainsaw, the operator has two options; starting from a ground position or starting a chainsaw secured between the operator's legs. The suggested options can lead to awkward body postures and consequently increased postural load. More so if the operator is starting a chainsaw multiple times a day. A sample of 67 workers on felling and processing was recorded with a video camera. Frames that include the chainsaw starting procedure were isolated, extracted, and assessed in Ergofellow 3.0 using REBA (rapid entire body assessment) tool. Results show that almost 50% of chainsaw operators ignored safety instructions and preferred "drop starting" a chainsaw. Obtained REBA scores show significant differences between three methods of starting a chainsaw.

Keywords Chainsaw · Starting a chainsaw · Postural load · REBA

M. Bačić (✉) · Z. Pandur · M. Landekić · M. Šušnjar · M. Bakarić · M. Šporčić
Faculty of Forestry and Wood Technology, University of Zagreb, Zagreb, Croatia
e-mail: mbacic1@sumfak.hr

Z. Pandur
e-mail: zpandur@sumfak.hr

M. Landekić
e-mail: mlandekic@sumfak.hr

M. Šušnjar
e-mail: msusnjar@sumfak.hr

M. Bakarić
e-mail: mbakaric@sumfak.hr

M. Šporčić
e-mail: msporcic@sumfak.hr

I. Salopek Čubrić et al. (eds.), *Proceedings of the 9th International Ergonomics Conference*, Lecture Notes in Networks and Systems 701,
https://doi.org/10.1007/978-3-031-33986-8_3

1 Introduction

The forest worker uses a chainsaw with both hands, holding it by the front handle and controlling it with the rear handle. In logging operations, workers use professional chainsaws with a total weight of over 6 kg. Long-term work with a chainsaw weighing 6 to 8 kg causes a high static load on the muscles of the upper limbs due to holding the chainsaw itself, and the dynamic load caused by handling the chainsaw [1]. Their work requires a great deal of physical strength and the adoption of forced, awkward postures [2–4]. Due to the forced postures, the generation and application of significant forces, and exposure to vibrations, logging is associated with a high risk of musculoskeletal disorders (MSDs). Recurring musculoskeletal pain might cause reflex posture changes, leading to distortion of body coordination and stability, thus increasing the risk of accidents [5].

Starting a chainsaw is an important step in chainsaw handling. OSHA (Occupational Safety and Health Administration) proposes two safe methods of starting a chainsaw [6]. The first method is starting a chainsaw firmly secured between the knees while holding a front handle with a left hand, arm straight, and a pulling starter grip with the right hand until it starts (Fig. 1).

The second method is starting a chainsaw from the ground by placing the left hand on the front handle while pressing it down, putting a right foot into the back handle, and pressing down to firmly secure the chainsaw on the ground and finally, pulling a starter grip with the right hand until it starts (Fig. 2).

When starting a chainsaw, the chain brake is to always be engaged. Some workers ignore safety measures and choose to start a chainsaw using the so-called "drop start" method. This means that they use the jerking motion of a chainsaw going down, held by the left hand on the front handle, and simultaneously pulling a starter grip with the right hand until it starts (Fig. 3). Drop starting a chainsaw can potentially cause

Fig. 1 "Between the knees" starting method

Fig. 2 "From the ground" starting method

injuries, especially if a chain brake is not engaged, because the chainsaw is not properly secured and can pivot around the front handle. Also, jerking a 6 to 8 kg chainsaw puts a worker's body under unnecessary physical and postural strain.

Working in logging, a chainsaw operator must start a chainsaw several times a day, and sometimes the starting procedure lasts several seconds, especially if the chainsaw is cold, putting an operator in an awkward position for a longer time.

The aim of this paper is to evaluate postural load during the three mentioned methods of starting a chainsaw using the REBA (rapid entire body assessment) tool with an assumption of significant difference between them.

Fig. 3 "Drop starting" method

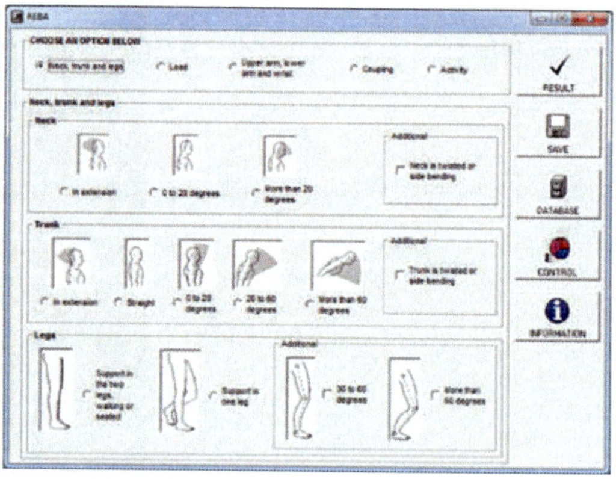

Fig. 4 Ergofellow 3.0 software - REBA tool

2 Materials and Methods

Total of 67 workers on logging were recorded using a GoPro Hero 8 videocamera. The obtained video material was viewed in the VLC media player. Frames that showed typical postures when starting a chainsaw were extracted using *print screen* function. One frame per worker was extracted and analyzed using REBA [7] tool in Ergofellow 3.0 software (Fig. 4). In REBA tool points are assigned in 5 categories (Fig. 4), and a final result is a total REBA score which is used to classifiy observed working posture in one of five levels of MSD risk. Minumum score is 1 – negligable risk, and maximum is 15 – very high risk.

Obtained data i.e. points per category and total score, were not normally distributed, and thus nonparametric statistical analyses were used. Statistical analyses were conducted in IBM SPSS Statistics software. Descriptive statistics were used to give a perspective on obtained total REBA scores for every method. To test statistical differences in assigned points per category and total REBA score between the three mentioned methods of starting a chainsaw, the Kruskal–Wallis test was used. Furthermore, to test statistical differences in assigned points per category and a total REBA score between the two methods of starting a chainsaw, a Mann–Whitney test was used. For practicality purposes, methods were marked as: A - "between the knees" method, B - "from the ground" method and C - "drop start" method.

3 Results and Discussion

Of the total 67 workers observed in this research, 32 used C method, 21 used A method and 14 used B method. Nearly 50% of workers used unsafe method of starting a chainsaw. While being the least safe method, mean and median of the REBA scores of "drop starting" were lowest of the three methods (Table 1).

Table 1 REBA scores - descriptive statistics

Method	Sample size	Minimum	Maximum	Mean	Median
A	21	3	7	5.33	6
B	14	4	10	6.64	7
C	32	3	8	4.5	4

Moreover, Kruskal–Wallis test ($p < 0.05$) confirmed significant difference between total REBA scores of the three methods, $H(2) = 19.49$, $p = 0.00$. A worker can "drop start" a chainsaw from an upright position, and there is almost no forced awkward body positions while doing it, meaning that big percentage of workers are ignoring the safety instructions and choosing more comfortable method. On the other hand, A and B methods require the worker to position himself to secure the chainsaw before starting, which includes a lot of trunk and leg bending.

Mann–Whitney test ($p < 0.05$) shows statisticly significant differences in several categories when testing one method versus another (Table 2, 3 and 4).

Table 2 shows significantly higher points were attributed to the Trunk category when using the B method rather than the A method. When starting a chainsaw from the ground, the worker must reach the front handle and starter grip with his hands meaning that he must bend the trunk (Fig. 1) significantly more in comparison to A method when the chainsaw is closer to the worker's upper extremities. There is also a significant difference between these two methods in total REBA scores.

In the category Legs_additional points, a significant difference was observed between the A and C methods (Table 3). To position the chainsaw firmly between the knees, and get in position to pull the starting grip, a worker must slightly squat (Fig. 2), which adds points to the category in question. While "drop starting" workers' legs

Table 2 Mann–Whitney test A vs. B method

Category	Method pair	Mean	Median	Z	p
Trunk	A	3.62	4	−4.08	<0.001
	B	4.79	5		
REBA score	A	5.33	6	−2.96	0.003
	B	6.67	7		

Table 3 Mann–Whitney test A vs. C method

Category	Method pair	Mean	Median	Z	p
Legs_additional points	A	0.66	1	−4.33	<0.001
	C	0.09	0.00		
REBA score	A	5.33	6	−2.43	0.015
	C	4.5	4		

Table 4 Mann–Whitney test B vs. C method

Category	Method pair	Mean	Median	Z	p
Neck	B	1.93	2	−2.36	0.018
	C	2.25	2		
Trunk	B	4.79	5	−4.84	<0.001
	C	3.41	3.5		
Legs_additional points	B	0.79	1	−4.64	<0.001
	C	0.09	0.00		
REBA score	B	6.64	7	−4.01	<0.001
	C	4.5	4		

are straight, and do not receive any points in this category. There is also a significant difference between these two methods in total REBA scores.

Scores in categories Neck, Trunk and Legs_additional points differ significantly between B and C method (Table 4). While B method received higher points in Trunk and Legs_additional points categories for obvious reasons, C method received higher points in Neck category. Although the worker's legs and trunk were straight during the C method, he had to pay more attention to an unproperly secured chainsaw, hence he had to look down and bend his neck (Fig. 3). REBA scores are also significantly different in this comparison.

4 Conclusion

The mean REBA score for all three methods is between 4 and 7, placing all methods at a medium level of MSD risk. However, there is a significant difference in REBA score between these methods and it can be concluded that the B method puts the worker in the highest postural strain, while the C method, which is considered unsafe, in the least postural strain.

There is also significant difference in respective categories of REBA tool when comparing two methods. In comparison of A an B methods – Trunk category; A and C methods - Legs_additional points category; B and C method – Neck, Trunk and Legs_additional points categories.

Acknowledgements The research was funded by the Croatian Science Foundation within the Project «Increasing the Competitiveness of Forestry Sector Through Development of Safety Culture (ForSaf2024)», project number IP-2020-02-7637.

References

1. Grzywiński, W., Wandycz, A., Tomczak, A., Jelonek, T.: The prevalence of self-reported musculoskeletal symptoms among loggers in Poland. Int. J. Ind. Ergon. **52**, 12–17 (2016)
2. Hagen, K.B.: Biomechanical analysis of spinal load in motor-manual cutting. J. Forest Eng. **2**(1), 39–41 (1990)
3. Harstela, P.: Work posture and strain of workers in nordic forest work: a selective review. Int. J. Ind. Ergon. **5**(3), 219–226 (1990)
4. Grzywiński, W.: The effect of selected factors on the type of working postures of chainsaw operator during felling. PhD thesis, Uniwersytet Przyrodniczy w Poznaniu, Poznan, Poland, pp. 1–99 (2011)
5. Sawastian, K., Grzywiński, W., Turowski, R.: Analysis of postural strain of loggers during timber harvesting in a spruce stand. Forestry Lett. **108**, 1–6 (2015)
6. https://www.osha.gov/sites/default/files/2018-12/fy10_sh-2086610_Chainsaw_Safety.pptx. Accessed 1 Sep 2022
7. Hignett, S., Mcatamney, L.: Rapid entire body assessment (REBA). Appl. Ergon. **31**(2), 201–205 (2000)

Predictive Safety Risk Assessment Methods Applicable in Aviation Ergonomics

Dajana Bartulović and Sanja Steiner

Abstract Ergonomics ensures the tools, procedures, and environment that employees use is appropriate to fit the employee's job requirements and personal capabilities. Ergonomic hazards are situations in workplace that can cause human failure. This can include repetition, awkward posture, forceful motion, stationary position, direct pressure, vibration, extreme temperature, noise, work stress, and other. In aviation, the SHELL model is a conceptual model that defines the relationship between aviation system environment and the human component. Overall goal is to improve human performance with respect to health, safety, and work environment. Performance measurement and risk assessment of any operator in aviation are fundamental to management planning and control activities. To achieve this purpose, safety risk assessment methods are developed and used. This paper provides the overview of safety risk assessment methods in ergonomics such as Systems-Theoretic Process Analysis (STPA) method, the Event Analysis of Systemic Teamwork Broken Links (EAST-BL) method, and the Network Hazard Analysis and Risk Management System (Net-HARMS) method, with the special emphasis on predictive safety risk assessment methods such as Lumbar Motion Monitor Method, the Occupational Repetitive Action (OCRA) Methods (OCRA Index and OCRA Checklist), Systematic Human Error Reduction and Prediction Approach (SHERPA), Task Analysis for Error Identification (TAFEI), the Multiple Resources Time-Sharing Model (MRTSM), the Situational Awareness Global Assessment Technique (SAGAT), and the Adaptive Neuro-Fuzzy Inference Systems (ANFIS). Paper outlines the advantages and applicability of predictive safety risk assessment methods in aviation ergonomics.

Keywords Predictive · Safety · Risk assessment methods · Aviation · Ergonomics

---

D. Bartulović (✉) · S. Steiner
Faculty of Transport and Traffic Sciences, Zagreb, Croatia
e-mail: dbartulovic@fpz.unizg.hr

I. Salopek Čubrić et al. (eds.), *Proceedings of the 9th International Ergonomics Conference*, Lecture Notes in Networks and Systems 701,
https://doi.org/10.1007/978-3-031-33986-8_4

1 Introduction

Ergonomics is a scientific discipline with task to investigate the human organism and behavior and provide data on the adaptability of objects with which a person comes into contact. Hence, ergonomics studies the anatomical, physiological, and other parameters of the human body, and ensures the tools, procedures, and environment that employees use is appropriate to fit the employee's job requirements and personal capabilities.

Ergonomic hazards are situations in workplace that can cause human failure. This can include repetition, awkward posture, forceful motion, stationary position, direct pressure, vibration, extreme temperature, noise, work stress, and other [1].

In aviation, as per [2], the SHELL model is a conceptual model that defines the relationship between aviation system environment and the human component. The SHELL model emphasizes the need to consider human factors as an integrated part of safety risk management. The SHELL model contains four main components: software (S) (procedures, training, support, etc.), hardware (H) (machines and equipment), environment (E) (the working environment in which the rest of the L–H-S system must function) and liveware (L) (other humans in the workplace). These components generate four main interfaces: Liveware-Hardware (L–H), Liveware-Software (L-S), Liveware-Liveware (L-L), and Liveware-Environment (L-E). The first Liveware-Hardware (L–H) interface refers to the relationship between the human and the physical attributes of equipment, machines, and facilities. This interface considers the ergonomics of operating the equipment by personnel, how safety information is displayed and how switches and operating levers are labelled and operated so they are logical and intuitive to operate. Overall goal is to improve human performance with respect to health, safety, and work environment.

Performance measurement and risk assessment of any operator in aviation are fundamental to management planning and control activities. To achieve this purpose, safety risk assessment methods are developed and used [2]. This paper provides the overview of safety risk assessment methods in ergonomics including Systems-Theoretic Process Analysis (STPA) method [3], the Event Analysis of Systemic Teamwork Broken Links (EAST-BL) method [4], and the Network Hazard Analysis and Risk Management System (Net-HARMS) method [5]. Special emphasis is on predictive safety risk assessment methods including Lumbar Motion Monitor method [6], the Occupational Repetitive Action (OCRA) methods (OCRA Index and OCRA Checklist) [7], Systematic Human Error Reduction and Prediction Approach (SHERPA) [8], Task Analysis for Error Identification (TAFEI) [1], the Multiple Resources Time-Sharing Model (MRTSM) [9], the Situational Awareness Global Assessment Technique (SAGAT) [10], and the Adaptive Neuro-Fuzzy Inference Systems (ANFIS) [11].

After the analysis of available safety risk assessment methods in ergonomics, the applicability of predictive safety risk assessment methods in aviation ergonomics, is outlined. Special attention is given to the advantages of Adaptive Neuro-Fuzzy

Inference Systems (ANFIS), and possibility of its application in the field of aviation ergonomics and safety risk management.

2 Overview of Safety Risk Assessment Methods in Ergonomics

This section gives an overview of safety risk assessment methods in ergonomics. Well-known safety risk assessment methods in ergonomics analyzed in this paper are the STPA method, the EAST-BL method, and the Net-HARMS method.

In order to explain STPA method it is necessary to explain the Systems-Theoretic Accident Model and Processes (STAMP) approach [12], as STPA methods requires development of a STAMP model the entire organizational activity structure. Such models include a wide range of control and feedback mechanisms that link different influential factors inside the system [3]. Such mechanisms of control and feedback provide source and exchange of information which reflect overall safety performance of the system. STPA method includes a wide range of new control and feedback mechanisms as opposed to older techniques (such as Fault Tree Analyses (FTA) or Hazard and Operability Study (HAZOP)). Hazardous states are usually the result of insufficient control in the system. The use of STPA method provides analysts with information in order to control hazardous states in the system more efficiently [3].

EAST-BL method represents an upgrade to the Event Analysis of Systemic Teamwork (EAST) method which evaluates the complex system's performance [13]. EAST method applies Hierarchical Task Analysis (HTA) and raw data gathered from direct observations to build three networks: a task network, a social network, and an information network. It combines task, social and information networks into one network. EAST-BL explains that accidents and adverse incidents are mostly caused by failure in the communicating information among systems. In the task network, "broken" links represent information that is not passed between tasks, whereas within the social network, the "broken" links represent information that is not passed between personnel. EAST-BL provides a systematic risk assessment approach to detect communication and information gaps in complex systems [4].

Net-HARMS method was established by analyzing existing risk assessment methods [4]. Net-HARMS introduces two crucial advances over existing risk assessment approaches. First advance is in the fact that it provides detection of risks across the entire system, as opposed to "sharp-end" risks only. Second advance lies in the fact that Net-HARMS provides detection of emerging risks that appear when various risks intersect each other. Net-HARMS incorporates HTA [14], task networks and the SHERPA methodology [8]. Net-HARMS, describes the goals, tasks, operations, and plans required across overall system, as well as it supports the detection of emerging risks. The Net-HARMS risk structure contains three categories and 10 risk modes [3, 5].

3 Analysis of Predictive Safety Risk Assessment Methods in Ergonomics

This section shows the analysis of predictive safety risk assessment methods in ergonomics including Lumbar Motion Monitor method, the Occupational Repetitive Action methods (Index and Checklist), Systematic Human Error Reduction and Prediction Approach, Task Analysis for Error Identification, the Multiple Resources Time-Sharing Model, the Situational Awareness Global Assessment Technique, and the Adaptive Neuro-Fuzzy Inference Systems.

The Lumbar Motion Monitor (LMM) method aims to assess so called Low Back Disorder (LBD) risk. The method assesses the dynamic components of LBD risk in occupational settings, e.g., manual tasks. The potentiometers are used to measure the instantaneous position of the spine, relative to the pelvis, in three-dimensional space. All gathered data are recorded using specially developed software, that calculates the velocity and acceleration of the spine for the motion of interest [1]. The LBD risk model was built from over 400 repetitive manual tasks [6]. Analysis detected five relevant factors including two "workplace" measures (the maximum external moment about the spine and the job's lift rate) and three "trunk-motion" parameters (maximum sagittal flexion position, maximum lateral velocity, and average twisting velocity). The LMM charts analyze each of these five factors and estimate the overall LBD risk [1].

In 1996, Occhipinti and Colombini developed the Occupational Repetitive Action (OCRA) methods to study employees' exposure to tasks that include upper-limb injury risks [7]. The OCRA index is used to predict the risk of upper extremity (UE) work-related musculoskeletal disorders (WMSDs). The OCRA checklist, based on the OCRA index, is simpler to apply and is generally recommended for the initial analysis of work positions that include repetitive tasks [7]. These methods assess four key risk factors based on their duration: repetitiveness, force, awkward posture and movements, and lack of proper recovery periods. Table 1 shows examples of the OCRA checklist score and the OCRA index used to measure exposure level [1, 7].

The SHERPA is a method introduced by Embrey, in 1986, as a method to predict human error with a possibility to propose potential solutions to detected errors [8]. The method is based upon a human error taxonomy, and originally, it emphasizes the psychological impact as an important part of the error [1]. Embrey recognized

Table 1 OCRA Checklist Score and OCRA Index

Checklist Score	Index	Exposure Level
≤7.5	2.2	no exposure
7.6–11	2.3–3.5	very low exposure
11.1–14.0	3.6–4.5	light exposure
14.1–22.5	4.6–9.0	medium exposure
≥22.6	≥9.1	high exposure

Source: Taken from Stanton, N. A., et al. (2005) [1]

two key issues. First one relates to the insufficient information on the external factors. Second one referred to overdependence made upon the analyst's judgment [1]. SHERPA analysis includes eight steps: HTA, task classification, Human Error Identification (HEI), consequence analysis, recovery analysis, ordinal probability analysis, criticality analysis, and remedy analysis [1].

Modeling the interaction between user and device gives with possibility to predict errors related to device exploitation. This technique is called TAFEI method. TAFEI is appropriate when addressing issues coming from the interaction between different system components (for example, human-device interface). [1]. TAFEI method includes three main phases: HTA, State-Space Diagrams (SSDs), and a transition matrix [1]

The technique called MRTSM model can predict the level of interference between two "time-shared" tasks, i.e., it can predict the loss in performance of one or both tasks carried out simultaneously, versus their single-task baseline measures. MRTSM model was developed upon multiple-resource theory by Navon, Gopher, and Wickens [9, 15], which can predict how well (or how poorly) a task will be performed when it is time-shared with another task [1]. The MRTSM model [9] involves the following five steps: code time-shared tasks, calculating total demand score, calculate resource-conflict score, calculate total interference score, and apportion interference score [1]. Table 2 shows the example of predicted interference values for two time-shared tasks resulting from the demand and conflict component [1].

Situation awareness (SA) defines the perception of the elements within time and space dimensions, the understanding of their meaning, and the projection of their state in the near future [1]. SAGAT is an alternative method for measuring SA, and it was introduced by Endsley, in 1995 [16]. SAGAT assesses situation awareness of individuals by directly questioning them regarding their current knowledge of their work environment and comparing their responses to the actual state of that work environment. Steps of this technique include developing SAGAT queries, selecting SAGAT queries, implementing the SAGAT, administering the SAGAT, and analyzing the data [10]. Examples of two SAGAT queries in a TRACON air traffic control simulation were presented by Endsley and Kiris, in 1995 [17].

To analyze the magnitude of identified safety risks, and accordingly evaluate them, Sadeghi and others proposed using the ANFIS method [11]. Fuzzy Inference System

Table 2 Predicted total interference values resulting from six task combinations

Tasks	Demand Component	Conflict Component	Total Interference
AA	$1 + 1 = 2$	$0.8 + 0 + 0 + 0 = 0.8$	2.8
BB	$1 + 1 = 2$	$0.8 + 1 + 0.6 + 0.6 = 3$	5.0
CC	$1.5 + 1.5 = 3$	$0.8 + 1 + 0.6 + 0.6 = 3$	6.0
AB	$1 + 1 = 2$	$0.8 + 0 + 0.6 + 0 = 1.4$	3.4
AC	$1 + 1.5 = 2.5$	$0.8 + 0 + 0.6 + 0 = 1.4$	3.9
BC	$1 + 1.5 = 2.5$	$0.8 + 1 + 0.6 + 0.6 = 3$	5.5

Source: Taken from Stanton, N. A., et al. (2005) [1]

Table 3 Sample of safety risk magnitudes and classifications on testing data sets using ANFIS

Safety Risks	Actual Risk Magnitude	Predicted Risk Magnitude	Error	Aggregated Risk Magnitude	Risk Magnitude Classification	Actions to be Taken	Ranks
SR13	0.202	0.125	0.077	0.112	Critical	Elimination	5
SR14	0.062	0.000	0.062	0.084	Negligible	Acceptance	28
SR15	0.048	0.048	0.000	0.039	Critical	Elimination	13
SR16	0.016	0.016	0.000	0.007	Minor	Mitigation	27
SR17	0.007	0.000	0.007	0.018	Major	Mitigation	23

Source: Taken from Sadeghi, H. et al. (2020) [11]

(FIS) is a decision-making methodology, which takes input values and creates output fuzzy values based on some logic rules [18]. Sometimes decisions need to be made based on real-world observations and some fuzzy logic rules (i.e., if–then rules) [19]. Technically, there are two types of FIS, such as Mamdani FIS and Takagi–Sugeno Fuzzy Model [11]. A variant of Takagi–Sugeno Fuzzy Model called ANFIS is presented to analyze safety risks [20]. The Artificial Neural Network (ANN) is a model that can learn features and patterns, from large amount of data. The coefficients in the network can be adjusted to minimize the output error [21]. In 1993, Jang [22], combined FIS with neural networks to overcome the shortcomings of FIS and introduced ANFIS. ANFIS is, in a nutshell, a neural network with five layers. The use of ANFIS includes six steps: data collection, fuzzy sets construction, dividing the dataset into training and test datasets, training the ANFIS model, evaluating the model and verifying the accuracy of the model. Variables or risk parameters are probability, severity, exposure, detectability (input variables), and risk magnitude (output variable) [11]. Statistical criteria such as Root Mean Square Error (RMSE) and Mean Absolute Percentage Error (MAPE) are sometimes used to confirm accuracy of ANFIS prediction [23]. Table 3 shows the example of safety risk magnitudes and classifications related to testing data sets using ANFIS model.

4 Conclusion

This paper aimed to present and analyze safety risk assessment methods in ergonomics, with special emphasis on predictive safety risk assessment methods, to assess the applicability of predictive safety risk assessment methods in aviation ergonomics.

Paper analyzed general safety risk assessment methods in ergonomics including STPA, EAST-BL, and Net-HARMS method. Special focus was on predictive safety risk assessment methods including LMM, OCRA, SHERPA, TAFEI, MRTSM, SAGAT and ANFIS method.

Aviation ergonomics is focused on the relationship between the human and the hardware component in aviation system, and it is a part of so-called SHELL model, specifically its interface called Liveware-Hardware or L-H interface. In aviation, safety risk assessment is conducted by ICAO prescribed safety risk management methodology, according to ICAO Safety Management Manual, by assessing safety risk index and safety risk tolerability of identified risks, using safety risk matrix which is composed of safety risk probability and safety risk severity.

After analyzing safety risk assessment methods presented in this paper, it can be observed that ANFIS is most appropriate and most compatible method, due to comparing its elements to elements of safety risk assessment used in aviation. For example, ANFIS method uses dataset which includes input variables (risk parameters) such as probability, severity, exposure, and detectability, and output variable such as risk magnitude. In aviation, this corresponds with input variables (risk parameters) called safety risk probability and safety risk severity, and output variable called safety risk index. The main reason why the ANFIS technique is used in risk assessment lies in the fact that the safety risk magnitude, or corresponding aviation safety risk index, as well as their related evaluation strategies, can be predicted. Hence, it can be concluded that ANFIS method could be applicable to safety risk management in aviation ergonomics.

Future research shall focus on verifying these conclusion by conducting a case study on the sample aviation organization, using ANFIS method and actual safety data, to predict and assess ergonomic aviation safety-related risks.

References

1. Stanton, N.A., Hedge, A., Brookhuis, K., Salas, E., Hendrick, H.: Handbook of Human Factors and Ergonomics Methods. CRC Press, London (2005)
2. International Civil Aviation Organization (ICAO): Safety Management Manual (SMM) [Doc 9859], 4th edn. International Civil Aviation Organization, Montreal (2018)
3. Hulme, A., et al.: Systems thinking-based risk assessment methods applied to sports performance: a comparison of STPA, EAST-BL, and Net-HARMS in the context of elite women's road cycling. Appl. Ergon. **91**(1), 103297 (2021). https://doi.org/10.1016/j.apergo.2020.103297
4. Dallat, C., Salmon, P.M., Goode, N.: Identifying risks and emergent risks across sociotechnical systems: the NETworked hazard analysis and risk management system (NET-HARMS). Theor. Issues Ergon. Sci. **19**(4), 456–482 (2018). https://doi.org/10.1080/1463922X.2017.1381197
5. Thatcher, A., Nayak, R., Waterson, P.: Human factors and ergonomics systems-based tools for understanding and addressing global problems of the twenty-first century. Ergonomics **63**(3), 367–387 (2020). https://doi.org/10.1080/00140139.2019.1646925
6. Marras, W.S., et al.: The role of dynamic three dimensional trunk motion in occupationally related low back disorders: the effects of workplace factors, trunk position and trunk motion characteristics on injury. Spine **18**(5), 617–628 (1993). https://doi.org/10.1097/00007632-199304000-00015
7. Colombini, D., Occhipinti, E., Grieco, A.: Risk Assessment and Management of Repetitive Movements and Exertions of Upper Limbs. Elsevier Science, New York (2002)
8. Embrey, D.E.: SHERPA: a systematic human error reduction and prediction approach. In: Proceedings of the International Topical Meeting on Advances in Human Factors in Nuclear Power Systems, Knoxville, TN (1986)

9. Wickens, C.D.: Multiple resources and performance prediction. Theor. Issues Ergon. Sci. **3**(2), 159–177 (2002). https://doi.org/10.1080/14639220210123806
10. Endsley, M.R.: Direct measurement of situation awareness: validity and use of SAGAT. In: Situation awareness analysis and measurement. Lawrence Erlbaum Associates, Mahwah (2000)
11. Sadeghi, H., Mohandes, S.R., Hosseini, M.R., Banihashemi, S., Mahdiyar, A., Abdullah, A.: Developing an ensemble predictive safety risk assessment model: case of Malaysian construction projects. Int. J. Environ. Res. Public Health **17**(22), 1–25 (2020). https://doi.org/10.3390/ijerph17228395
12. Leveson, N.: A new accident model for engineering safer systems. Saf. Sci. **42**(4), 237–270 (2004). https://doi.org/10.1016/S0925-7535(03)00047-X
13. Stanton, N.A., Baber, C., Harris, D.: Modelling Command and Control: Event Analysis of Systemic Teamwork. CRC Press, Boca Raton (2008)
14. Shepherd, A.: HTA as a framework for task analysis. Ergonomics **41**(11), 1537–1552 (1998). https://doi.org/10.1080/001401398186063
15. Navon, D., Gopher, D.: On the economy of the human processing systems. Psychol. Rev. **86**(3), 254–255 (1979). https://doi.org/10.1037/0033-295X.86.3.214
16. Endsley, M.R.: Measurement of situation awareness in dynamic systems. J. Hum. Factors Ergon. Soc. **37**(1), 65–84 (1995). https://doi.org/10.1518/001872095779049499
17. Endsley, M.R., Kiris, E.O.: Situation Awareness Global Assessment Technique (SAGAT) TRACON Air Traffic Control Version User's Guide. Texas Tech University Press, Lubbock (1995)
18. Mahdiyar, A., Mohandes, S.R., Durdyev, S., Tabatabaee, S., Ismail, S.: Barriers to green roof installation: an integrated fuzzy-based MCDM approach. J. Clean. Prod. 269(1), 122365 (2020). https://doi.org/10.1016/j.jclepro.2020.122365122365
19. Jahangiri, M., Jamshidi Solukloei, H.R., Kamalinia, M.: A neuro-fuzzy risk prediction methodology for falling from scaffold. Saf. Sci. **117**(1), 88–99 (2019). https://doi.org/10.1016/j.ssci.2019.04.009
20. Chang, P.C., Wu, J.L., Lin, J.J.: A Takagi-Sugeno fuzzy model combined with a support vector regression for stock trading forecasting. Appl. Soft Comput. **38**(1), 831–842 (2016). https://doi.org/10.1016/j.asoc.2015.10.030
21. Naderpour, H., Mirrashid, M.: Moment capacity estimation of spirally reinforced concrete columns using ANFIS. Complex Intell. Syst. **6**(1), 97–107 (2020). https://doi.org/10.1007/s40747-019-00118-2
22. Jang, J. S. R.: ANFIS: adaptive network-based fuzzy inference system. In: IEEE Transactions on Systems, Man, and Cybernetics, vol. 23, no. 3, pp. 665–685 (1993). https://doi.org/10.1109/21.256541
23. Mohandes, S.R., Zhang, X., Mahdiyar, A.: A comprehensive review on the application of artificial neural networks in building energy analysis. Neurocomputing **340**(1), 55–75 (2019). https://doi.org/10.1016/j.neucom.2019.02.040

Application and Advantage of 3D Body Scanning in the Ergonomic Design of a Sitting Workplace

Blaženka Brlobašić Šajatović and Slavenka Petrak

Abstract The paper deals with researching the advantages of anthropometric measurements using a 3D body scanner. The working principle of the VITUS Smart laser 3D scanner is described, on which measurements were made for the purposes of this research according to ISO 20685–1:2018 standards. Anthropometric measurements obtained by applying this measurement are more accurate, precise and repeatable. The paper selected the anthropometric measures that are used in the ergonomic design of the workplace for the computer, and highlighted the advantages but also the disadvantages of measuring with a 3D scanner.

Keywords Anthropometry · Ergonomics · 3D scanner

1 Introduction

Anthropometry is a method of anthropology, it deals with the measurement of the human body and the relationship between the dimensions, that is, the proportions of individual parts of the body, and it is of great importance in the field of clothing construction. The goal of anthropometry is to quantitatively characterize the morphological and physiological features of the human body, which differ within and between certain populations, based on body measurements. [1].

Anthropometric measurements of the body for the needs of the clothing industry are carried out using the conventional method and application of non-contact computer determination of body measurements, using a 3D body scanner. There are certain disadvantages and advantages of 3D body scanning compared to conventional measurements. [1, 2]. The importance of anthropometric measurements is extremely great in the application of the textile and clothing industry, the automotive industry, in architecture, interior design, medicine, in art, sports, workplace ergonomics, etc.

B. B. Šajatović (✉) · S. Petrak
Faculty of Textile Technology University of Zagreb, Zagreb, Croatia
e-mail: blazenka.brlobasic@ttf.hr

I. Salopek Čubrić et al. (eds.), *Proceedings of the 9th International Ergonomics Conference*, Lecture Notes in Networks and Systems 701,
https://doi.org/10.1007/978-3-031-33986-8_5

Significant development in anthropometric research has also contributed to the development of computing in the field of developing 3D humanoid models. The aim of the work is to show the applicability of anthropometric measurements in the ergonomic design of a sitting workplace working at a computer, using a 3D scanner according to ISO 20685–1:2018.

The international standard ISO 20685–1:2018 was developed to ensure the comparability of body measurements determined by the standards ISO 7250 (basic human body measurements of technological design) and ISO 8559 (garment construction and anthropometric surveys-body dimensions) determined using different 3D body scanners [3].

The aim of ergonomics is to adapt the working environment to the worker, design workplaces, increase workplace safety and reduce health problems, and improve work efficiency. The application of ergonomics achieves a positive effect on the worker and the place where the worker performs tasks [4, 5] When the working conditions are adapted to the employee, the likelihood of fatigue and illness is lower, and health, motivation and productivity are more likely to improve. According to previous research on anthropometric measurements that are applied in the field of ergonomics of a sitting workplace working at a computer, it was noticed that the measurements are performed using standardized measurements of anthropometric variables according to the International Biological Program (IBP) [6, 7]. Most of the scientific research deals with the requirements to be met by computer equipment, requirements related to space, lighting, glare andglare, noise, microclimatic conditions, radiation and how the working environment is harmonized with ergonomic principles. Scientific research into the analysis of a workplace with a computer from an ergonomic point of view and research into the problem of long-term sitting while working, emphasize the health aspects of working with a computer [8, 9].

Most of the previous researchs are based on conventional body measurement. The aim of this paper is to investigate the advantages and disadvantages of 3D body scanning for the purpose of ergonomically designing a sitting workplace.

1.1 Conventional Anthropometric Measurement

At the beginning of the anthropometric measurement, it is important to determine the exact position of individual anthropometric points on the human body. Anthropometric points are divided into "fixed" and "virtual". Fixed anthropometric points are always located on the same part of the body. They are clearly visible and their position can be easily determined. Virtual anthropometric points are points whose position changes with respect to the position of the body. Whenever possible, the position of the anthropometric points is marked on the right side of the body. Only after placing the body in a standard position and marking the position of individual anthropometric points, one moves on to anthropometric measurement, so that the parts of the measuring instruments are placed on the subject's body, on individual anthropometric points. The measurement error using the conventional measurement method results from the error of the meter or measuring instruments. By extracting the standards ISO 3635, ISO 8559 and ISO 9407, the foundations were laid for the unique definition of the measurements of the human body for the needs of the clothing and footwear industry, and thus for the way of performing anthropometric measurements using the conventional method [2].

1.2 Non-contact Measurement of Body Measurements

In addition to conventional measuring instruments, 3D body scanners are used to measure body measurements. A 3D scanner analyzes an object in the real world or environment to gather shape information. 3D body scanners are systems that enable obtaining information about the shape and surface of the human body. Lasers or structured white light and CCD cameras (Charge-Coupled Device) are used for recording. Three-dimensional scanners scan the subject or object in three dimensions, and in most cases they are based on laser technology. The light source, i.e. laser and cameras are placed on the same stand, and their distance is known. The computer measures the difference between the electrical signals received from the two video cameras and thus determines the distance from the object, that is, a certain point in space. The main advantages of laser triangulation technology for 3D scanning are high scanning resolution and accuracy. As a disadvantage, sensitivity to the properties of the surface being scanned can be mentioned. Very shiny and dark and transparent surfaces are particularly problematic. Also, during the scanning, the person must be very still, considering that the scanning time is up to approx. 10 s, depending on the type of scanner [1, 10].

1.2.1 Laser 3D scanner VITUS Smart

The scanning system uses triangulation sensors that allow scanning the object in one pass, and are directed towards the object from different directions. By moving the sensor from the top dead center to the bottom, the object is scanned, during which the laser passes over the object, and each camera records a part of the body. Triangulation 3D laser scanners are active scanners that use laser light. A laser beam is projected onto an object and a camera is used to locate the laser point. Depending on the distance of the laser impact on the surface of the object, the laser point appears in different places in the field of view of the camera. Object scanning is performed in an area of 120 × 80 cm and 210 cm high. After scanning the subject, the obtained data is stored in the computer, followed by data processing. Software package ScanWorx V 2.7.2. enables the analysis of body models, and the performance of automatic or interactive determination of body measurements. Using the 3D VITUS Smart scanner and the accompanying compute program, 153 body measurements are determined [11] (Fig. 1).

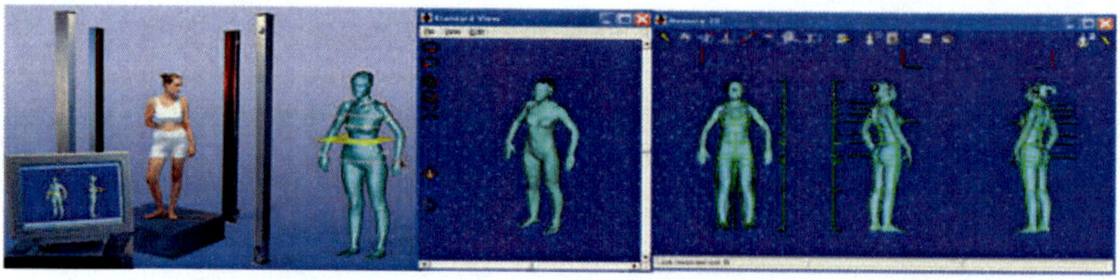

Fig. 1 Presentation of the results of computer determination of body measurements [12]

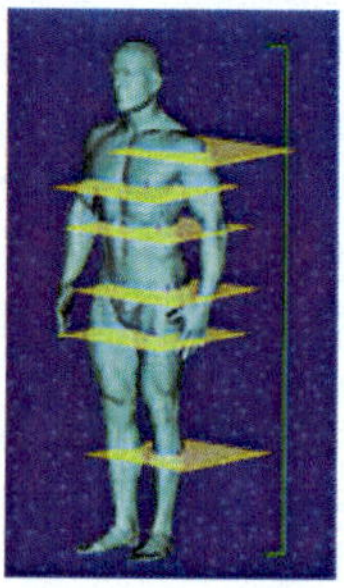

Fig. 2 Characteristic circumferences on the body determined by cutting the model using planes [1]

By analyzing the computer model of the subject's body obtained based on the results of a 3D body scan, information was obtained about body posture, body symmetry and the shape of individual body parts. The rotation of the body model gives the possibility to view all the details of the body. By aligning with a moving vertical line on the body model, the posture of the body is determined. Some parts of the body are mutually balanced. Asymmetric parts of the body can be determined by analyzing the symmetry of the body in relation to the central axis of the body. Analyzing the cross-sections by moving the plane on the body, the measurements of the circumference of individual parts of the body are read, which enables an insight into the shapes of the cross-sections (Fig. 2) [1, 11].

2 Anthropometric Measures in Application for Ergonomic Design of a Sitting Workplace

The position of the subject during the scan is important to obtain reliable data that can be used in an anthropometric database. Basic body positions according to the ISO 20685–1;2018 standard during scanning are: standing position A Standing position B Standing position C and Sitting position D (Fig. 3) [13].

For the purposes of ergonomic design of a sitting workplace, work at a computer, use the following variables: 1) standing height, 2.) sitting height, 3.) eye height, 4.) shoulder height,5.) elbow height, 6.) arm length, 7.) distance of the knee from the back, 8.) sitting length of the thigh (length of the upper leg), 9.) length of the lower

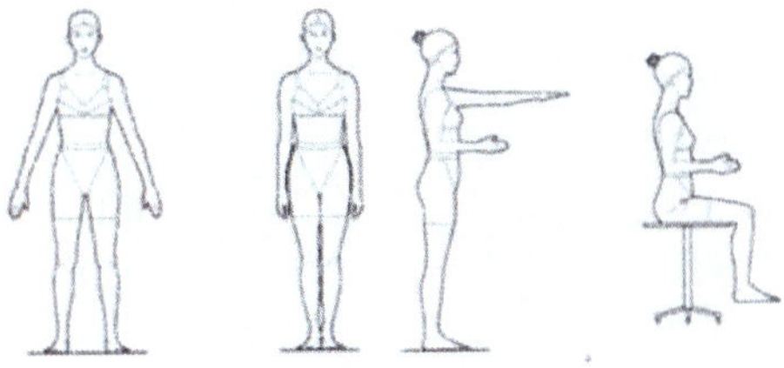

Fig. 3 Standing and sitting positions of subjects in which 3D scanning is performed according to ISO 20685–1:2018 [13]

leg, 10.) length of the foot, 11) length of the hand 12.) height of the sitting thigh. The body measurements shown were taken on a male body (Fig. 4). Of the 153 body measurements that are obtained by body scanning, only those measurements that are necessary for the ergonomics of a sitting workplace are shown. There is a possibility of selecting additional body measurements if the need for it is shown during the ergonomic design of the workplace (Table 1).

The body measurements shown were taken at the Faculty of Textile Technology, University of Zagreb, in the Department of Clothing Technology, on the VITUS Smart 3D body scanner, which is equipped with a modular system with 4 vertical consoles that house cameras and a laser light source (Fig. 4.). The measurements shown in Fig. 4 were determined by interactive measurement on a 3D body model, given that the computer program enables automatic body measurement only in the basic, standing position of the body. In this sense, the precision of the person carrying out the measurement is required, so that the measuring points that need to be added on the model are correctly and precisely positioned, and in order to determine the exact values of the necessary measurements. Precise anthropometric measurement of the body using a 3D body scanner enables a better adaptation of the body to a neutral sitting position, which implies a balanced head position, gaze directed straight ahead, relaxed shoulders, upper arms close to the body, free elbow, straight wrist, forearms and lower legs parallel to the floor, the back is slightly tilted back and rests its entire length on the back of the chair, sufficient space for the legs is provided, the angle of the knee joint is slightly greater than 90 degrees, the bend of the knee is slightly above the edge of the seat, the feet are placed on the floor or on the footrest (Fig. 5).

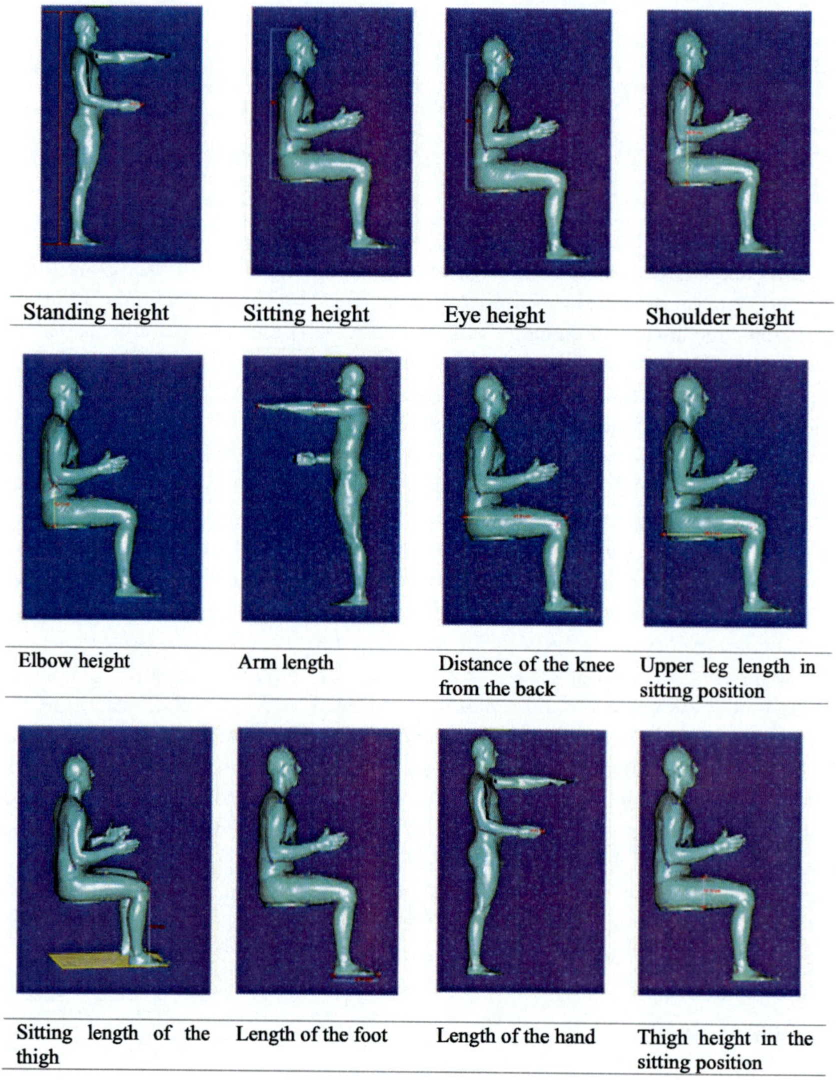

Fig. 4 Body measurements for the needs of ergonomic design of a sitting workplace working at a computer

Table 1 Values of body measurements

Body measurement	Standing height	Sitting height	Eye height	Shoulder height	Elbow height	Arm length
	189,3 cm	98,0 cm	87,3 cm	63,9 cm	26,3 cm	87,2 cm
Body measurement	Distance of the knee from the back	Upper leg length in sitting position	Sitting length of the thigh	Length of the foot	Length of the hand	Thigh height in the sitting position
	61,8 cm	50,3 cm	56,6 cm	28,1 cm	18,8 cm	17,9 cm

Advantage	Noncontact measurement
	Measurement accuracy
	Information about body posture
	Information about the shape and surface of the human body
	A large number of anthropometric measures
	High measurement speed
Disadvantage	VITUS Smart - not intended for measurements outside the laboratory
	As equipment, it is more expensive than conventional anthropometric instruments
	Comparability of measurement techniques between scanners since there are differences in the way individual body measurements are taken
	Lack of data on certain body surfaces due to shadows

Fig. 5 Advantages and disadvantages of anthropometric measurements using 3D body scanners

3 Conclusion

By scanning the body with a 3D scanner, it is possible to obtain 153 anthropometric measurements. The paper presents selected measures applicable to the ergonomic design of a sitting workplace for working at a computer. By emphasizing the advantages of anthropometric measurement with a 3D scanner such as: noncontact measurements, measurements accuracy, information about body posture, inforamtion about the shape and surface of the human body, a large number of anthropometric measures, hight measurement speed. The contribution of this measurement is emphasized, for a better and more precise anthropometric alignment of the sitting workplace working at the computer with the size of the body and its position during work. When ergonomically designing a sitting workplace, it is possible to select additional measures from the computer model of the subject's body obtained on the basis of the results of a 3D scan of the model's body if the need arises. The mentioned measures enable precise anthropometric alignment of the body in order to adjust the workplace where the worker works with the computer, according to the required requirements, such as: distance of the screen from the eyes, appropriate size of the keyboard, position of the keyboard in relation to other elements of the workplace, dimensions of the table, height of the seat of the work chair (adjustable), back of the chair (adjustable), correct height and length of the footrest.

References

1. Brlobašić Šajatović, B.: Investigation of body proportions of trained and untrained population as factor in the processes of garment construction, PhD Thesis, University of Zagreb, Croatia (2014)
2. Ujević, D., at al.: Theoretical aspects and application of Croatian anthropometric system, University of Zagreb, Faculty of Textile Technology, Croatia (2009)
3. Petrak, S., Mahnić, M., Ujević, D.: Research of 3D body models computer adjustment based on anthropometric data determined by laser 3D scanner. In: Proceedings of the 3rd International Conference on 3D Body Scanning Technologies, Hometrica Consulting, Ascona, Lugano, pp. 115–126, Switzerland (2012)
4. Kirin, S.: Rad radnika za računalom; 3. Međunarodni stručno-znanstveni skup; Veleučilište u Karlovcu, Karlovac (2010)
5. Kovačvić, T.: ergonomsko uređenje radnog mjesta. http://www.iusinfo.hr/DailyContent/..%5CDocuments%5CErgonomija%20radnog%20prostora.pdf. Accessed 20 Sep 2022
6. Tusic, K: Ergonomija uredskog prostora i opreme. http://www.iusinfo.hr/DailyContent/..%5CDocuments%5CErgonomija%20radnog%20prostora.pdf. Accessed 20 Sep 2022
7. Bozić, M: Ergonomija radnog mjesta za računalom – suvremeni pristup: http://web.zpr.fer.hr/ergonomija/2000/mustapic/erg_misevi.htm. Accessed 20 Sep 2022
8. Kardum Z.: Rad na siguran način s računalom. http://issuu.com/hdusluge/docs/priru__nik_za_rad_na_siguran_na__in. 25 Apr 25 2015.
9. Sigurnost i zaštita zdravlja pri radu s računalom; Ergonomija radnog prostora. http://www.iusinfo.hr/DailyContent/..%5CDocuments%5CErgonomija%20radnog%20prostora.pdf; 31.5.2015.
10. http://issuu.com/hdusluge/docs/priru__nik_za_rad_na_siguran_na__in. Accessed 20 Sep 2022
11. Petrak, S., Mahnić, M., Rogale, D.: Impact of male body posture and shape on design and garment fit. Fibres Textiles Eastern Europe **6**(114), 150–158 (2015)
12. Petrak, S., Rogale, D.: Uspostava i razvoj Laboratorija za računalnu konstrukciju odjeće na Tekstilno-tehnološkom fakultetu Sveučilišta u Zagrebu, Godišnjak 2021. Akademije tehničkih znanosti Hrvatske, Doprinos članova Akademije razvoju sustava znanstveno-nastavnih institucija, Akademija tehničkih znanosti Hrvatske, pp. 441–458, Zagreb (2022)
13. ISO 20685–1: 2018 3D 3-D scanning methodologies for internationally compatible anthropometric databases — Part 1: Evaluation protocol for body dimensions extracted from 3-D body scan (2018)

Pressure Distribution When Sitting on a Hard Surface Without Cushioning – A Case Study

Marko Čeredar, Tanja Jurčević Lulić, Jasna Leder Horina, and Danijela Domljan

Abstract Today's children and young people spend most of their time sitting. Also, a large part of that time refers to the time spent in an educational institution, where the furniture is not adapted to the anthropometric and physiological characteristics of pupils and students, meaning it is only suitable for a very small number of them. When sitting, most pupils and students find the sitting surface too high or too low. From a biomechanical point of view, one of the indicators of comfort when sitting is the distribution of pressure on the contact surface of the seat. Since chairs in educational institutions generally do not have a cushioned seating surface, in this study an investigation was carried out on the distribution of contact pressures on a hard, uncushioned sitting surface in different ways of sitting (different angles of torso inclination forward and backward, crossed legs) and different heights of the sitting surface ($h_S = 28$ cm, 47 cm, 68 cm). The measurement was performed on one subject - a 23-year-old student ($h = 178$ cm, $G = 685$ N). For the sake of comparability, the results are expressed as percentage values in relation to the results obtained when sitting at a suitable height of the chair with an upright torso, which are marked with 100%. The distribution of pressures and sizes of contact surfaces are shown and explained. Asymmetry of the left and right side of the body was observed. Future research will be conducted on a significant sample of male and female subjects.

Keywords Sitting · Uncushioned Seating · Body posture · Pressure Distribution

M. Čeredar (✉) · T. Jurčević Lulić
Faculty of Mechanical Engineering and Naval Architecture, University of Zagreb, Zagreb, Croatia
e-mail: mc217709@stud.fsb.hr

T. Jurčević Lulić
e-mail: tjurcev@fsb.hr

J. Leder Horina
University North, Varaždin, Croatia
e-mail: jlederhorina@unin.hr

D. Domljan
Faculty of Forestry and Wood Technology, University of Zagreb, Zagreb, Croatia
e-mail: ddomljan@sumfak.hr

I. Salopek Čubrić et al. (eds.), *Proceedings of the 9th International Ergonomics Conference*, Lecture Notes in Networks and Systems 701,
https://doi.org/10.1007/978-3-031-33986-8_6

1 Introduction

Every year, the amount of time people spend sitting increases, leaving great consequences on people's health. A special problem is that children and young people - pupils in primary and secondary schools and students, spend most of their time in a sitting position; at the same time, they spend a large part of their time in educational institutions with inadequate furniture. The design of school chairs is not suitable for most students; the dimensions of the chair are not appropriate, above all the height of the sitting surface, the support for the spine is not adequate, and the sitting surface is not cushioned. According to the literature [1], about 75% of the body's weight is taken by the sitting surface, and the rest of the weight is distributed on the contact surfaces of the feet and the base, the back and backrest, and the arms and armrests, if any. Comfort when sitting is a psychophysical subjective experience of an individual that can be described as a lack of unpleasant pressure and pain, and it depends on the load acting on the individual and on the duration of sitting [2], and it does not have to be, if something is comfortable, that it is also the most effective in terms of reducing whole body stress [3]. From a biomechanical point of view, one of the indicators of comfort when sitting is the distribution of pressure on the contact surfaces. Contact pressure affects physiological and biomechanical processes, primarily blood vessels and circulation, nerve fibers and transmission of nerve impulses. Decreased blood circulation results in an insufficient supply of organs with oxygen, thus affecting the concentration of pupils and students in class. Also, the weight of the upper body is transferred to the pelvis, which transfers the load to the contact surface of the seat via soft tissue. With non-cushioned seats, deformation of the soft tissue occurs, and with padded seats, the seat is deformed, which is better because there is no reduction in circulation. Sitting on uncushioned surfaces is less favorable, and this is precisely the type of chairs in educational institutions. The aim of this paper is to investigate the distribution of pressures in different ways of sitting on an uncushioned surface.

2 Methods and Subjects

One male respondent participated in the research - a student aged 23, height $h = 178$ cm, weight $G = 685$ N, body mass index BMI $= 22.1$. The subject sat on a wooden uncushioned surface without a backrest, where the height of the sitting surface could be changed by placing pads under the feet. The tests were conducted at the Department of Biomechanics and Ergonomics of the Faculty of Mechanical Engineering and Naval Architecture in Zagreb. Measurements were made on a portable platform of the company Tekscan^{TM}, model MobileMat (Fig. 1), which was placed on a sitting surface. The platform has dimensions of $63.6 \times 55.9 \times 4.2$ cm, the measurement frequency is 100 Hz, and it works on a trailer of capacitive sensors that are arranged 1 sensor/cm^2. The software package FootMat Research was used for measurements and analysis of pressure, forces and contact surfaces.

Fig. 1 MobileMat platform by Tekscan™ [4]

Measurements were made for different sitting postures, including sitting with a vertical torso (angle 90°), sitting with the torso tilted forward at an angle to the sitting surface of 60° and 45°, leaning back at an angle of 120° and sitting with legs crossed. All measurements were made for a chair without a backrest: in the case where the legs are in contact with the ground ($h_S = 47$ cm), in the case of a chair that is too low ($h_S = 28$ cm) when the upper legs are lifted from the seat, and in the case of a high chair ($h_S = 68$ cm) where the feet are not in contact with the ground. The sitting positions are shown in Fig. 2. The depth of the sitting surface was 44 cm. The sitting height was regulated using footrests. All measurement variants are presented in Table 1.

The subject sat in each position for 2 min, but due to the static position of the body, the recording lasted only 5 s, the recording frequency was 5 Hz, and for all recorded *frames* the average pressure in each sensor was given and this pressure is shown as a result of the measurement.

In Figs. 3, 6, 9 and 12, the image on the left shows the case of a high chair, the image in the middle shows a normal height chair, and the image on the right shows a low chair. Figures 4, 7, 10 and 13 show the 2D contact surface with marked zones

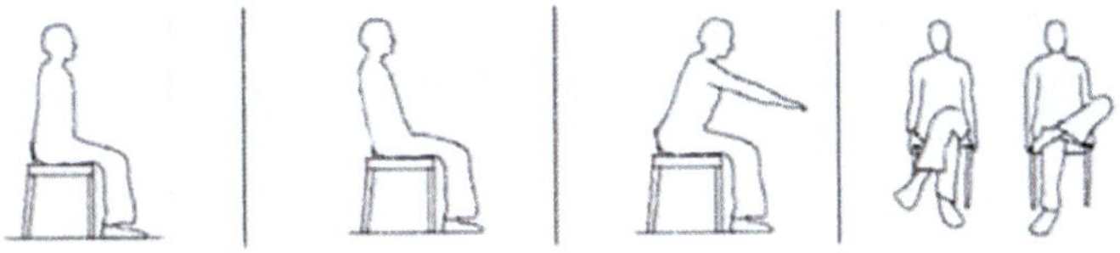

Fig. 2 Sitting positions - trunk upright, trunk leaning backwards, trunk leaning forward, legs crossed

Table 1 Display of measurements

trunk tilt angle	90°	60°	45°	120°
low chair	+	+	+	+
normal chair	+	+	+	+
high chair	+	+	+	+
leg crossed over thigh	+	–	–	–
leg crossed over the knee	+	–	–	–

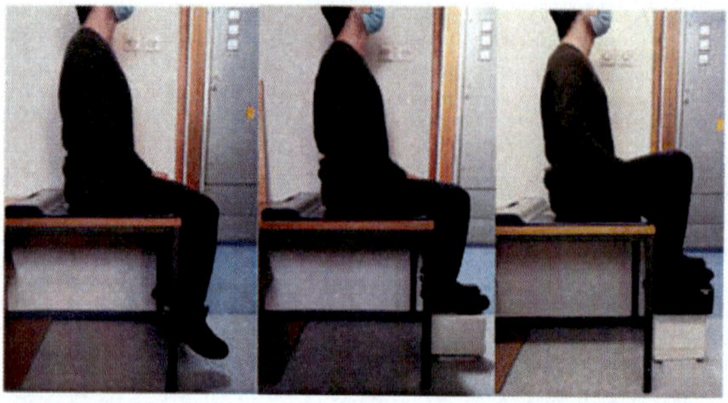

Fig. 3 Sitting with a vertical trunk at different heights of the sitting surface

of the highest pressure at various angles of trunk inclination, while Figs. 5, 8, 11 and 14 show a 3D representation of the pressure distribution at different angles of trunk inclination. In the representation of the 2D contact surface with marked zones of highest pressure, the red rectangle represents the pressure under the left contact surface, and the green one under the right contact surface. The origin of the coordinate system is located in the projection of the center of gravity on the platform.

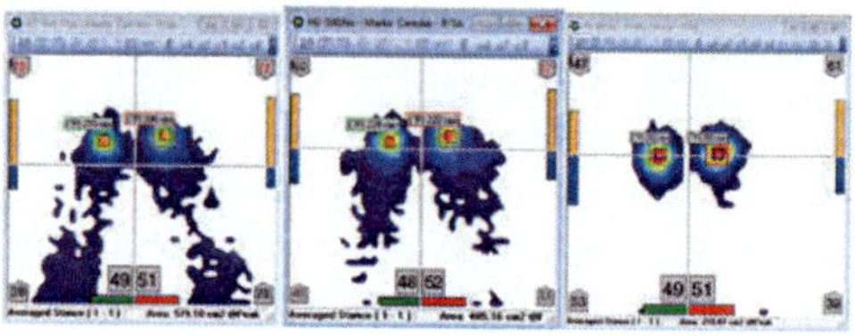

Fig. 4. 2D representation of the contact surfaces with pressure distribution (trunk at an angle of 90°)

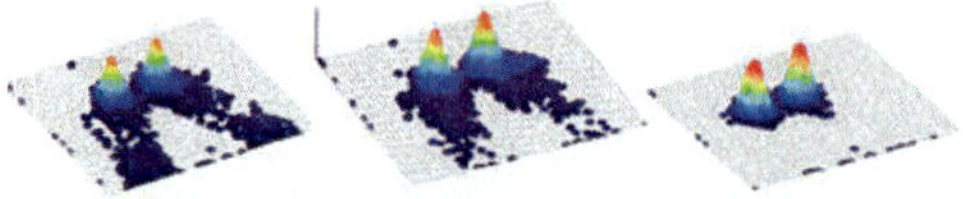

Fig. 5. 3D representation of the pressure distribution (trunk at an angle of 90°)

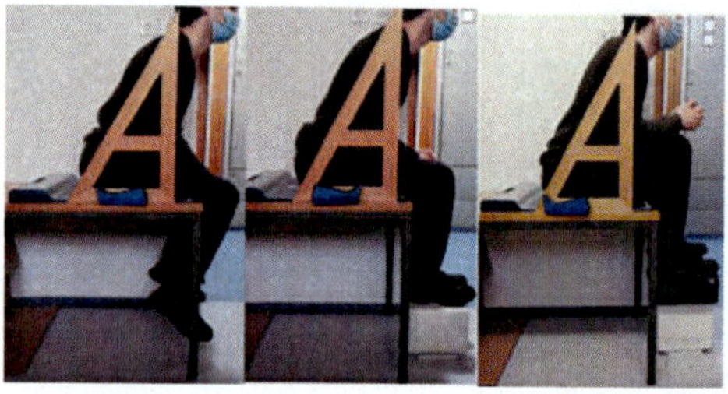

Fig. 6 Sitting with a 60° forward trunk tilt

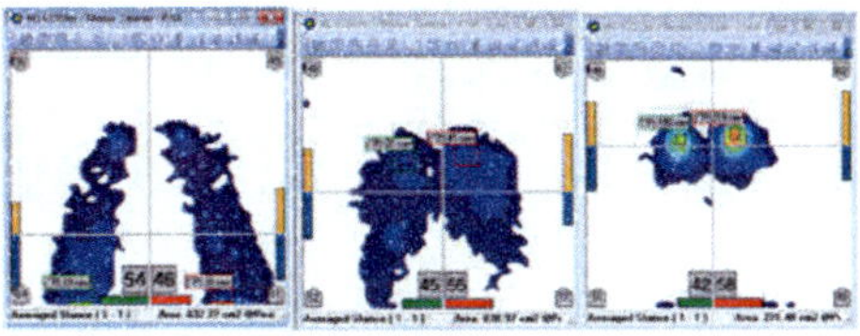

Fig. 7. 2D representation of the contact surfaces with pressure distribution (trunk at an angle of 60°)

Fig. 8. 3D representation of the pressure distribution (trunk at an angle of 60°)

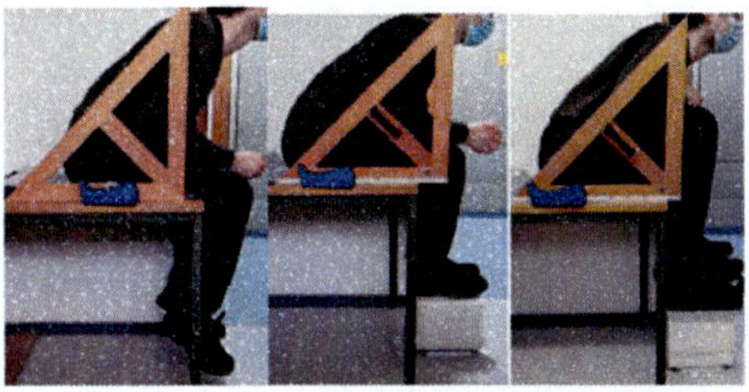

Fig. 9 Sitting with a 45° forward trunk tilt

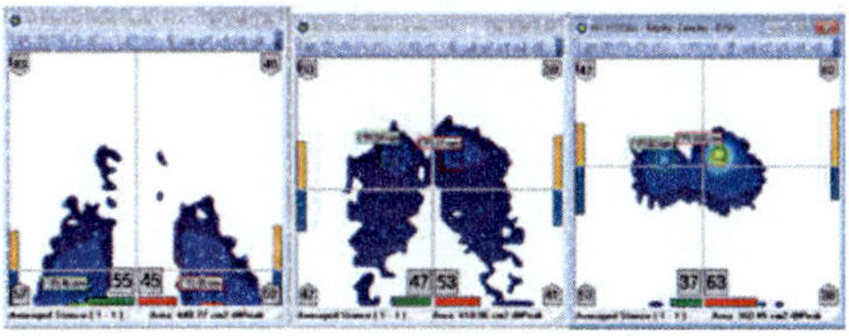

Fig. 10. 2D representation of the contact surfaces with pressure distribution (trunk at an angle of 45°)

Fig. 11. 3D representation of the pressure distribution (trunk at an angle of 45°)

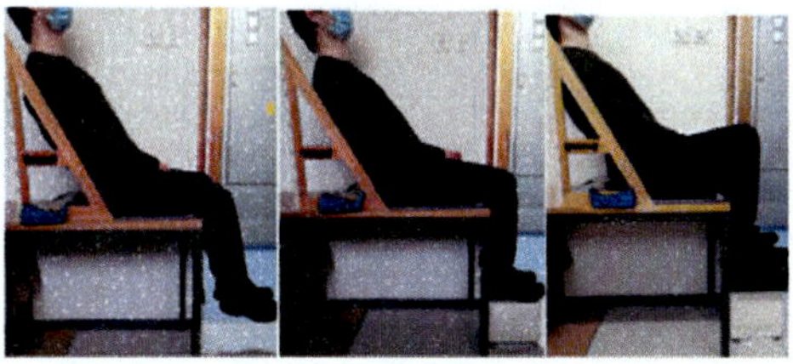

Fig. 12 Sitting with a trunk tilt of 120°

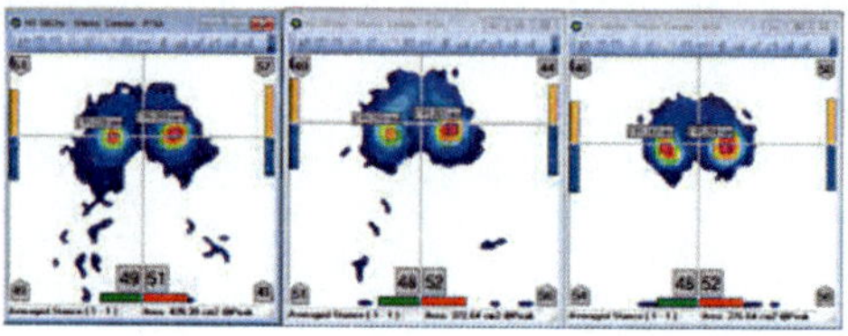

Fig. 13. 2D representation of the contact surfaces with pressure distribution (trunk at an angle of 120°)

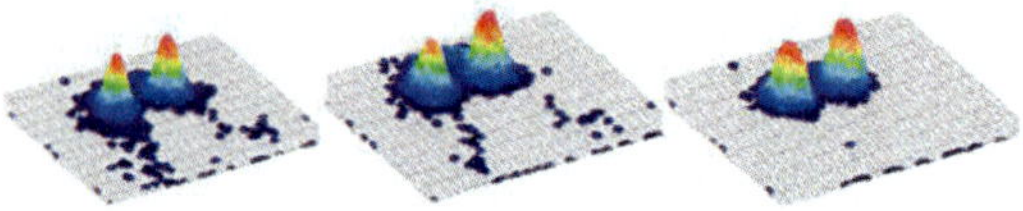

Fig. 14. 3D representation of the pressure distribution (trunk at an angle of 120°)

3 Results

In order to be able to compare the results, Table 2 shows the results of all measurements as percentage values compared to the area and pressure values when sitting at the normal height of the chair at a trunk tilt of 90° (they amount to 100%) because in the literature [5] states that this is the optimal way of sitting (Figs. 15, 16, 17 and 18).

Table 2 Percentage values of the results of measurements

		Contact area (%)	Maximum pressure – left (%)	Maximum pressure – right (%)
Trunk tilt angle 90°	low chair	119,14	92,8	93,8
	normal chair	100	100	100
	high chair	45,3	113,1	108
Trunk tilt angle 60°	low chair	130,4	13,5	17,4
	normal chair	131,7	18,5	13,8
	high chair	56,0	96,4	67,0
Trunk tilt angle 45°	low chair	90,9	15,8	20,5
	normal chair	125,7	12,2	12,9
	high chair	62,3	74,8	36,6
Trunk tilt angle 120°	low chair	88,3	114,1	102,2
	normal chair	76,8	114,0	93,8
	high chair	57,0	114,4	108,5
Sitting cross-legged	over thigh	105,7	74,8	72,3
	over knee	76,8	114,0	84,4
	coccyx	97,2	84,2	100 (coccyx!)

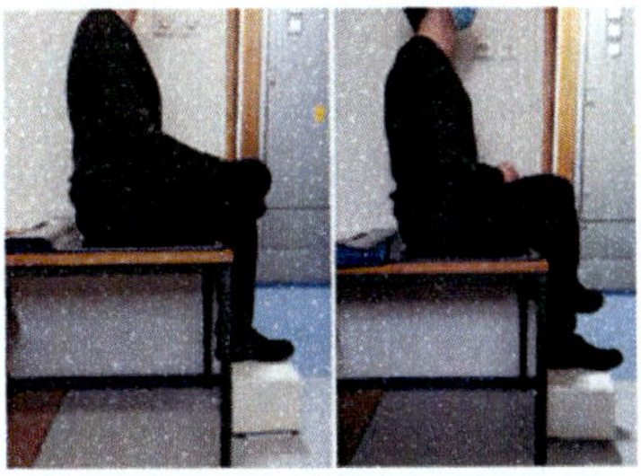

Fig. 15 Sitting with crossed legs

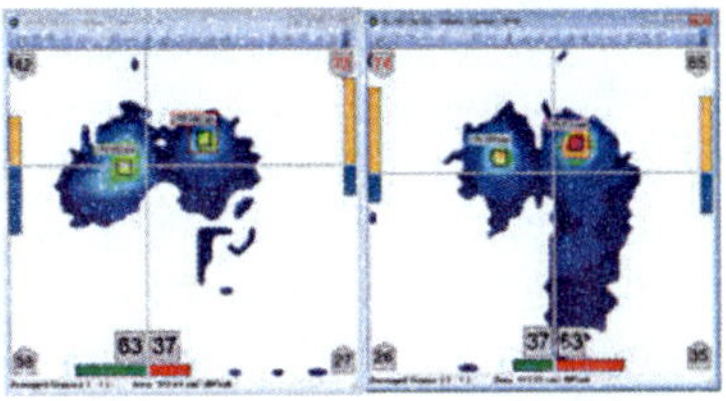

Fig. 16. 2D contact surfaces with marked zones of highest pressure when sitting with crossed legs

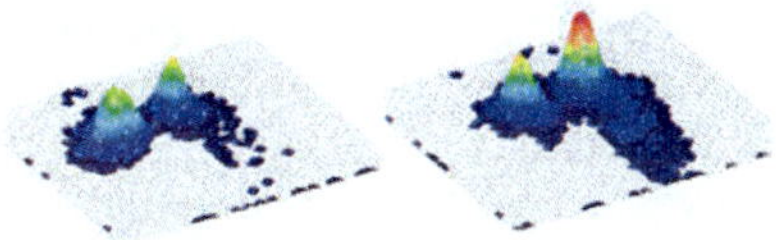

Fig. 17. 3D representation of pressure distribution when sitting with crossed legs

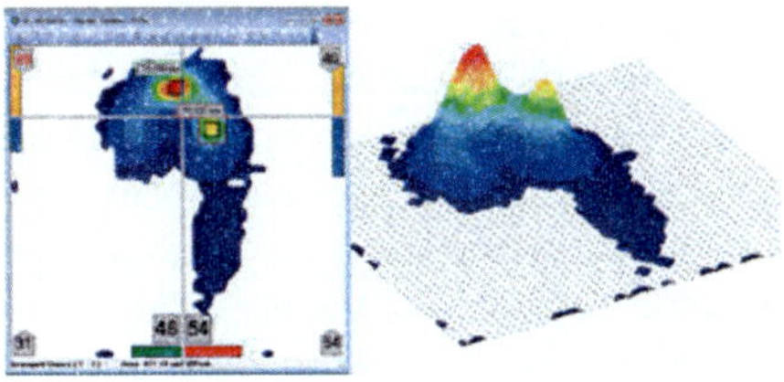

Fig. 18 Representation of an extreme situation when the body leans back; distribution of pressure under the coccyx

4 Conclusion

The results of this study, which was conducted for only one subject, show the influence of the way of sitting on the distribution of pressure on the contact surface of the seat. According to the literature [6], the optimal way to sit is with a normal chair height, and the upper body in an upright position, but the results of this research did not show the least pressure in that case. When moving the trunk forward, the contact pressure is mainly reduced due to the forward movement of the center of gravity and the increase of the contact surface, but there is also an additional load below the knees and an unnatural position of the spine. An exception is sitting on a low chair, where the contact surface decreases and the contact pressure increases. Leaning the trunk back shows an increase in the contact pressure under the sitting bones, while the maximum value of the contact pressure decreases with a normal chair height.

When sitting on high surfaces, the thighs bear a greater load than the buttocks, while the opposite is true for low and normal seat heights. Depending on whether the body is in an upright position or is tilted forward or backward, the greatest pressure occurs when sitting in an upright position. An exception is sitting on a high chair with the trunk tilted back. Also, when presenting the results, an asymmetry in the distribution of pressure between the left and right sides was noticed, which is directly related to the poor posture of the subjects and the deformation of the spine.

Sitting with crossed legs significantly affects the movement of the center of gravity towards the left or right side of the contact surface. When sitting with legs crossed over the upper leg, the contact surface increases. Also, the projection of the center of gravity moves towards the crossed leg and there is a reduction of pressure under the sitting bones, which is why this type of sitting is mostly pleasant. In the case of sitting cross-legged with the leg crossed over the knee, the center of gravity moves towards the leg on which it rests and there is a significant difference in the pressures under the left and right sides of the contact surface, i.e. the sitting bones. As a result of leaning back while sitting with the legs crossed when the leg is crossed over the knee, the pressure under the coccyx increased, which confirms that the coccyx also has an influence in the distribution of pressure.

A limitation of this study is that the measurement was performed for only one subject. Future research should be conducted on a statistically significant sample of male and female respondents. Due to the difference in the structure of the male and female pelvis, it is necessary to determine the differences in the measurement results that depend not only on gender, but also on other factors such as the weight or structure of the person. The size of the contact surface depends on the soft tissue and the distance between the seated bones; a larger contact area results in a lower maximum pressure, for example people with a higher BMI have more mass, and thus a larger contact area, so the maximum pressure that occurs in such people should be lower compared to people with a lower BMI. Although it is believed that of all the known indexes, the body mass index is most closely related to the amount of fat tissue in the human body [7], this index is not a perfect measure because it is possible that taller and/or more muscular people generally have a higher body mass index, even though they the proportion of fat tissue is relatively low.

References

1. Staarink, H.A.M.: Sitting posture, Comfort and Pressure: Assessing the Quality of Wheelchair cushions, Doctoral thesis, Delft University Press, (1995),
2. Horvath, P.G., Antal, R.M., Domljan, D., Denes, L.: Body pressure distribution maps used for sitting comfort visualization. Sigurnost **59**(2), 123–132 (2017)
3. Naddeo, A., Califrano, R., Vink, P.: The effect of posture and load distribution on (dis)comfort perceived by students seated on school chairs. Int. J. Interact. Des. Manuf. **12**(4), 1179–1188 (2018)

4. Tekscan Homepage. https://www.tekscan.com/products-solutions/systems/mobilemat. Accessed 21 Nov 2022
5. Li, W., Mo, R., Yu, S., Chu, J., Hu, Y., Wang, L.: The effects of the seat cushion contour and the sitting posture on surface pressure distribution and comfort during seated work. Int. J. Occup. Med. Environ. Health **33**(5), 675–689 (2020),
6. Verver, M.M., van Hoof, J., Oomens, C.W.J., Wismans, J.S.H.M., Baaijens, F.P.T.: A finite element model of the human buttocks for prediction of seat pressure distributions. Comput. Methods Biomech. Biomed. Eng. **7**(4), 193–203 (2004)
7. Mahadevan, S., Ali, I.: Is body mass index a good indicator of obesity? Int. J. Diab. Dev. Countries **36**, 140–142 (2016)

Ergonomic Risk Assessment for Concrete Formwork Activities in Construction Using Digital Human Modeling

Vigneshkumar Chellappa, Jitesh Singh Chauhan, and Zuzana Strukova

Abstract Construction workers are at high risk of musculoskeletal disorders (MSDs) during formwork operations. This study adopted a mixed-method approach, including site observations and evaluation of worker postures using digital human modelling (DHM) to assess the MSDs risk involved in formwork operations. The CATIA V5 software was used to conduct Rapid Upper Limb Assessment (RULA), as the workers associated with formwork operations mostly use their upper body. A video camera was used to record the working cycle postures of workers during their routine tasks. Three significant postures were selected that could develop a high risk of MSDs during formwork operations. The findings indicated a high level of discomfort to the arms, neck, trunk, and wrist, which could result in long-lasting harm to the workers. The recommendations are proposed to avoid and reduce the signs of MSD risk caused by the original operations. This study is essential since an ergonomically well-designed environment increases workers' comfort and productivity.

Keywords Concrete formwork · Ergonomics · MSDs · DHM

1 Introduction

One of construction workers' most prevalent occupational health issues is work-related musculoskeletal disorders (MSDs; [1]). Construction workers are frequently involved in multiple ergonomic risk factors, despite the enormous physical work demand, long working hours, whole-body vibration, prolonged awkward static/

V. Chellappa (✉)
Amity School of Design, Noida 201303, India
e-mail: cvkumar@amity.edu

J. S. Chauhan
Center for Ergonomics: User-Centered Design and Occupational Wellbeing, Indian Institute of Technology Guwahati, Guwahati 781039, Assam, India

Z. Strukova
Faculty of Civil Engineering, Technical University of Košice, 042 00 Košice, Slovakia

I. Salopek Čubrić et al. (eds.), *Proceedings of the 9th International Ergonomics Conference*, Lecture Notes in Networks and Systems 701,
https://doi.org/10.1007/978-3-031-33986-8_7

repetitive postures, and unfavorable working environments [2]. Therefore, the main contributing factor to non-fatal injuries in the construction industry is MSDs [3]. MSDs accounted for approximately 33% of the US construction industry's total absenteeism in 2018 [4]. In addition to causing work absenteeism, compensation claims, and schedule delays, the high prevalence of MSDs also raises the recruitment/training costs of the construction industry [1]. According to Valero et al. [5], workers in the construction industry are most prone to develop MSDs if they are regularly required to perform awkward postures and activities such as twisting, bending, lifting, bending, and twisting. Formwork operations have the highest risk of MSDs among all construction activities [2].

The primary contributing element to MSDs is the physical demands [6]. Formwork activities require a lot of physical demand and have a high prevalence of ergonomic injuries [7]. Carpenters are often in charge of assembling formwork panels and must bend their knees, squat, and carry large weights, which puts them at significant risk of MSDs [8]. Injuries to the wrist, hand, shoulder, hip, knee, low back, foot, leg, and ankle are most frequently sustained by the carpenters, and they also have a high prevalence of "sprains and strains" [9].

Numerous studies on the causes, mitigation, and control of MSDs in construction operations have been undertaken throughout the years. However, limited studies [2, 10] discussed the risks of MSDs in formwork operations, particularly traditional formwork [2]. The current study focused on MSD risks for workers associated with vertical formwork operations (i.e., column formwork) during steel formwork systems in the Indian construction industry. Since steel formwork systems are increasingly being used in the industry, it is essential to investigate the risks of MSD for workers interacting with steel formwork to prevent risky exposures and ensure their safety.

2 Digital Human Modeling

An efficient technique for designing, analyzing, and predicting ergonomics is digital human modelling (DHM; [11]). When combined with computer-aided engineering (CAE) software, DHM becomes more effective [12]. For instance, the Dassault Systems Company combined the standalone SAFEWORK with CAE as a DHM software [13]. The creation of 2D and 3D human models from anthropometric data of targeted users or populations for ergonomic assessment of virtual human ft to virtual workstation components is carried out using DHM software [14]. The integration makes it possible for the human and product models to coexist in the same environment. As a result, designing and simulating various human postures or tasks is made simpler, more precise, realistic, and time-efficient [15]. Rapid Upper Limp Assessment (RULA), which is based on the work of McAtamne and Corlett [16], is integrated with the DHM module of the CATIA V5 software. The upper part is the human body area that is assessed for comfort in this way.

The ergonomic assessment of products and workplaces in various sectors, such as aerospace-aviation, defence, transportation, healthcare, and manufacturing, has drawn attention to DHM software [17]. DHM has been used in the workplace to reduce the MSDs of fishermen by designing small fishing vessels [18], reduce awkward postures by redesigning work accessories in Indian shop floor workstations [19], and so on. A few examples where DHM has been used successfully include the modified cycle rickshaws [20], shoe racks [21], adjustable walking canes [22], etc. This study evaluated the worker postures during vertical formwork operations using DHM to assess the MSDs.

3 Methodology

3.1 Onsite Observation

The study was performed on two construction projects in northeast India. It was evident from the literature that the worker's tasks mostly use his upper body and only minimally involve his legs. Consequently, it was determined that RULA was a suitable method for ergonomics study under the scenario depicted. Workers were chosen because of the repetitive nature of the work and the awkward postures. The workers' task was to build formwork panels weighing 108 kg. The assembled panels were then manually lifted to the site for erection. Around the rebar, they erect the panels there and lock them with bolts and nuts. A video camera captured the working cycle postures while performing his routine task. The postures considered significant for the current study are depicted in Figs. 1 and 2.

3.2 Postural Analysis

Following the observation, DHM Software was used to model the selected postures. The CATIA V5 software was used to carry out the RULA. In CATIA V5, a module allows inserting a Manikin that represents a human model. In the beginning, the

Fig. 1 Panel lifting

Fig. 2 Panel erection

Fig. 3 DHM model for postures 1–3

manikin was standing. The model's entire body can be changed to reflect any real posture. After replicating the actual posture, the assessment was carried out as an integrated feature. The three significant postures were modeled (see Fig. 3), and the Human Model in CATIA V5 does not display which body parts are less comfortable.

4 Results and Discussion

Each posture underwent RULA analysis using the CATIA V5 software. Table 1 summarizes the findings for the three postures. With an overall score of 7 for all the postures, the study showed that the postures are awkward and the worker were at high risk. The RULA analysis for the postures is shown in Figs. 4–6. For the outcomes that were achieved, immediate investigation and modifications are required. The wrist and arm are revealed to be the areas with the most risk in all the postures.

It was clear from the analysis that the workers were at high risk of developing MSDs when they were repeatedly involved in panel erection and lifting tasks. Formwork system manufacturers could develop ergonomic interventions to prevent MSDs from workers during formwork operations. Based on the study findings and resources

Table 1 RULA score

Posture	Score	Risk level	Action to be taken
1, 2 and 3	7	High	Investigate and change immediately

Fig. 4 RULA score for posture 1

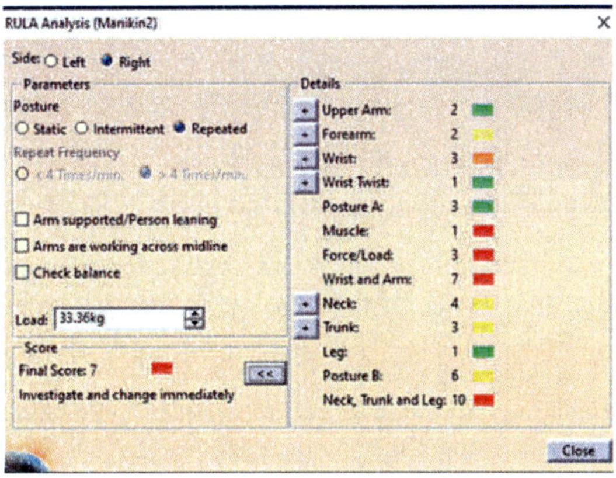

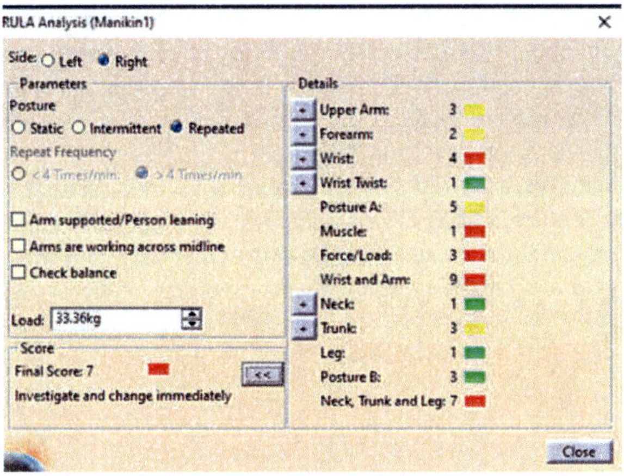

Fig. 5 RULA score for posture 2

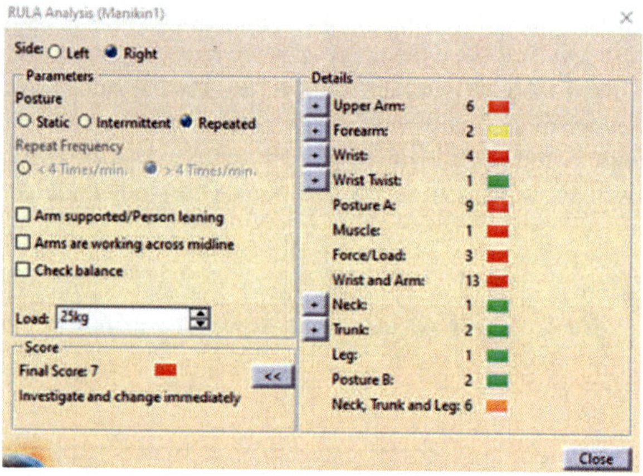

Fig. 6 RULA score for posture 3

provided by NIOSH [23] and Gambatese and Jin [2], the recommendations below may aid in preventing MSDs that occur during formwork operations.

- Construction organizations should try to minimize lifting and overreaching requirements for their formwork workers by providing lifting devices such as cranes, hoists, forklifts, and hand tools with extension clamps that minimize bending and twisting.
- The weight of the form components should be marked or labelled to inform their workers about the components' weight.
- To reduce wrist-related disorders in form workers, formwork system manufacturers should incorporate handles or grips into formwork components without unduly increasing the weight.
- Manufacturers should design ergonomic tools to eliminate the need for form workers to outreach and bend. For instance, extension clamps could be included in tools to allow users to do these activities while standing up, without elevating their shoulders, and with minimal bending.
- To avoid repetitive activities for an extended period, rotate workers among various tasks during a shift. The research team observed on the job site that employers frequently allocate form workers one or more specific form components at the beginning of the workday; rotating workers between tasks is extremely uncommon.

5 Conclusion

The study assessed the MSDs risks of using steel formwork in the Indian construction industry. The results indicated that the formwork operations develop a high risk of MSDs in workers in the arms, neck, trunk, shoulders, and wrists. Recommendations are given to manufacturers and construction contractors based on the research findings. The recommendations include numerous chances, resources, and methods to prevent MSDs in form workers. These practices adhere to the global norm for a good workplace. These recommendations are expected to minimize MSDs if they are implemented. A small number of projects and limited locations are the main drawbacks of this study. These limitations hamper the generalization of the study's conclusions. Future research is required to estimate the risk of MSDs, including studies involving different form systems such as traditional, aluminum, and modular construction sites and various ergonomic assessment approaches such as REBA (Rapid Entire Body Assessment).

References

1. Inyang, N., Al-Hussein, M., El-Rich, M., Al-Jibouri, S.: Ergonomic analysis and the need for its integration for planning and assessing construction tasks. J. Constr. Eng. Manag. **138**(12), 1370–1376 (2012)
2. Gambatese, A., Jin, Z.: Identification and Assessment of Musculoskeletal Disorders (MSDs) Risk for Concrete Formwork Systems. Technical Report. CPWR-The Center for Construction Research and Training, USA (2021)
3. Wang, D., Dai, F., Ning, X.: Risk assessment of work-related musculoskeletal disorders in construction: state-of-the-art review. J. Constr. Eng. Manag. **141**(6), 04015008 (2015)
4. BLS (2020). Fact Sheet: Occupational injuries and illnesses resulting in musculoskeletal disorders (MSDS). https://www.bls.gov/iif/oshwc/case/msds.htm. Accessed 18 Jan 2022
5. Valero, E., Sivanathan, A., Bosché, F., Abdel-Wahab, M.: Musculoskeletal disorders in construction: a review and a novel system for activity tracking with body area network. Appl. Ergon. **54**, 120–130 (2016)
6. Sobeih, T., Salem, O., Genaidy, A., Abdelhamid, T., Shell, R.: Psychosocial factors and musculoskeletal disorders in the construction industry. J. Constr. Eng. Manag. **135**(4), 267–277 (2009)
7. Schneider, S., Susi, P.: Ergonomics and construction: A review of potential hazards in new construction. Am. Ind. Hyg. Assoc. J. **55**(7), 635–649 (1994)
8. Hajaghazadeh, M., Marvi-milan, H., Khalkhali, H., Mohebbi, I.: Assessing the ergonomic exposure for construction workers during construction of residential buildings. Work **62**(3), 411–419 (2019)
9. Welch, L.S., Hunting, K.L., Anderson, J.: Concrete form work, injury and musculoskeletal disorders. In Proceedings of the Human Factors and Ergonomics Society Annual Meeting, pp. 5–647–5–650. SAGE Publications, Los Angeles (2000)
10. Ahankoob, A., Charehzehi, A.: Mitigating ergonomic injuries in construction industry. IOSR J. Mech. Civil Eng. **6**(2), 36–42 (2013)
11. Vigneshkumar, C., Salve, U.R.: A prototype of safety knowledge management for the construction companies. In: Ginzburg, A., Galina, K. (eds.) Building Life-cycle Management. Lecture Notes in Civil Engineering, vol. 231, pp. 277–284. Springer, Cham (2022). https://doi.org/10.1007/978-3-030-96206-7_29

12. Demirel, H.O., Duffy, V.G.: Applications of digital human modeling in industry. In: Duffy, V.G. (ed.) Digital Human Modeling. Lecture Notes in Computer Science, vol. 4561, pp. 824–832. Springer, Berlin (2007). https://doi.org/10.1007/978-3-540-73321-8_93
13. Kim, S.H., Lee, K.: Development of discomfort evaluation method for car ingress motion. Int. J. Automot. Technol. **10**(5), 619–627 (2009)
14. Chafn, D.B.: Improving digital human modeling for proactive ergonomics in design. Ergonomics **48**, 478–491 (2005)
15. Fritzsche, L.: Ergonomics risk assessment with digital human models in car assembly: simulation versus real life. Human Factors Ergon. Manuf. Serv. Ind. **20**, 287–299 (2010)
16. Mcatamney, L., Corlett, E.N.: RULA: a survey method for the investigation of work-related upper limb disorders. Appl. Ergon. **24**(2), 91–99 (1993)
17. Maurya, C.M., Karmakar, S., Das, A.K.: Digital human modeling (DHM) for improving work environment for specially-abled and elderly. SN Appl. Sci. **1**(11), 1326 (2019)
18. Álvarez-Casado, E., Zhang, B., Sandoval, S.T., Pedro, M.: Using ergonomic digital human modeling in evaluation of workplace design and prevention of work-related musculoskeletal disorders aboard small fishing vessels. Human Factors and Ergon. Manuf. Serv. Industr. **26**, 463–472 (2016)
19. Sanjog, J., Baruah, R.L., Patel, T., Karmakar, S.: Redesign of work-accessories towards minimizing awkward posture and reduction of work cycle elements in an Indian shop-foor workstation. In: Rebelo, F., Soares, M. (eds.) Advances in ergonomics in design. Advances in intelligent systems and computing, vol. 485, pp. 219–232. Springer, Cham (2016). https://doi.org/10.1007/978-3-319-41983-1_20
20. Mohapatra, S.: Application of digital human modeling and work posture analysis of human powered cycle rickshaw pullers on Odisha, India - a case study. In: Proceedings of 13th International Conference on Humanizing Work and Work Environment, Mumbai (2015)
21. Sanjog, J., Karmakar, S., Agarwal, H., Patil, C.D.: Designing and ergonomic evaluation of a shoe-rack in CAD environment. Int. J. Comput. Appl. **49**, 38–41 (2012)
22. Sahoo, B.B., Yein, N., Pal. S.: Design concept of adjustable cane-chair for elderly in virtual environment. In: Proceedings of 13th International Conference on Humanizing Work and Work Environment, Mumbai (2015)
23. NIOSH. Ergonomics and Musculoskeletal Disorders (2018). https://www.cdc.gov/niosh/topics/ergonomics/default.html. Accessed 03 Aug 2022

Evaluating the Tactile Comfort of Knitted Sportswear Depending on the Gender of the Participants

Goran Čubrić, Ivana Salopek Čubrić, and Andrea Bojko

Abstract Given that customers make sportswear purchases based on their visual and sensory perceptions, research that examines consumers' perceptions of tactile comfort is crucial in the field of sportswear. As a result, more effort must be put into understanding sportswear characteristics in relation to the human senses. This paper involved 20 assessors, aged 20 to 25, including 10 males and 10 women that evaluated the tactile comfort of knitted sportswear. Four significant bipolar properties are assessed: stiffness-softness, roughness-smoothness, non-stretchable-stretchable, and thick-thin. The study indicated that female assessors gave on average higher grades than male assessors. In addition, both genders correctly indicated the material with the highest proportion of elastane as the most stretchable.

Keywords yarn · material · tactile comfort · subjective method · sportswear · knitted fabric

1 Introduction

Advances in technology and the development of high-performance materials are developing sportswear with improved properties. The investigations covering the perception of tactile comfort are nowadays an important topic in the sportswear research domain, due to the fact that consumer purchases of sportswear garments are driven by visual and sensory perception [1–3]. Therefore, additional efforts have to be done towards the interpretation of sportswear attributes with respect to human senses

G. Čubrić (✉) · I. Salopek Čubrić · A. Bojko
Faculty of Textile Technology, University of Zagreb, Zagreb, Croatia
e-mail: goran.cubric@ttf.unizg.hr

I. Salopek Čubrić
e-mail: ivana.salopek@ttf.unizg.hr

A. Bojko
e-mail: andrea.bojko@ttf.unizg.hr

I. Salopek Čubrić et al. (eds.), *Proceedings of the 9th International Ergonomics Conference*, Lecture Notes in Networks and Systems 701,
https://doi.org/10.1007/978-3-031-33986-8_8

[4, 5]. The subjective evaluation of the clothing comfort involve complex processes consisting of a large number of different stimuli from the clothing that communicate with the brain through different channels of sensory responses [6]. To investigate fabric tactile comfort, a series of fabrics with varying constructional (weave type, blend, finish, etc.) and physical (mass per unit area, thickness, thread density, etc.) parameters may be studied [7–9]. Researchers point out that fabric tactile comfort is a nuclear factor related to the comfort of textile and clothing [10]. Therefore, strong efforts have been done towards the interpretation of sensory attributes with respect to human senses [11, 12]. For example, in a study conducted by Hue et al., fifteen subjects participated and evaluated all the specimens. And five comfort related sensory sensations, including softness, smoothness, prickliness, warmth and dampness, were evaluated in this study [13]. Furthermore, Xue et al. proposed a mathematical algorithm that shows efficiency in exploring the interactive mechanism between fabric tactile properties and the visual method [14]. Table 1 presents an overview of the studies conducted focusing at a profile of evaluators participating in the studies covering the sensory evaluation of fabric attributes. As can be seen, the total number of evaluators in the studies varies from 7 to 42 and mostly includes both female and male volunteers. In a number of studies, the age range of participating evaluators is quite wide (20–54). As can be seen from the Table, the other specifics important for the definition of the profile, such as origin or profession are not taken into consideration in all the studies.

This study aims to introduce an approach on conducting evaluation of sportswear with rather material through a subjective method with participation of young evaluators with a rather narrow age range. For this purpose, four fabric sensory attributes of 9 sportswear fabrics were assessed.

2 Experiment

2.1 Materials

For the investigation is carefully selected a set consisting of nine representative fabrics used for the production of sportswear. The fabrics consist of polyamide (PA) and elastane yarn in different compositions. The fabric ID, as well as surface mass and thickness, are given in Table 2.

2.2 Attributes Assessment

Total of 20 assessors (10 men and 10 women) took part in the assessment of selected 9 materials. Unlike in other studies, intention here was to get insight into the perception of young people that are active in sports and leisure activities and therefore, assessors

Table 1 The profile of evaluators in conducted studies

Number of evaluators	Age	Gender (female: male)	Origin	Other specifics	Main researcher
9	20–50	–	–	–	Bensaid [15]
12	–	10:12	–	–	Bacci [16]
12	–	5:7	–	–	Bertaux [17]
35	Less than 30-more than 60	24:11	28 of 35 Germans	5 blind or with visual disorder	Wiskott [18]
12	–	12:0	–	–	Niedermann, [19]
21	–	–	–	Sight disorders	Grinevičiute [20]
10	23–54	–	Same nationality	–	Tadesse [8]
40	23–60	24:16	–	–	Ozcelik [21]
30	–	20:10	–		Hui [22]
15		10:5		Employees at the same company	Cardello [23]
42	23–55	30:12	–	–	Xue [14]
15	24–45	7:8	–	–	Naima [24]
11	–	–	–	–	Philippe [25]
7	–	7:0	–	–	Wang [26]
29	–	–	Japanese	–	Kanai [27]
23	21–45	15:8	–	–	Xue [28]
28	23–56	14:14	From Asia, Africa and Europe	Stayed in Europe for at least 1 month	Musa [1]
23	20–30	12:11	–	–	Xiao [29]
20	25–41	11:9	–	Untrained	Xue [12]
31	19–38	16:15	–	–	Asad [30]
13	Around 20	–	–	–	Park [31]
8	24–35	–	–	–	Wang [32]
10	23–54	–	–	Same university	Tadesse [8]

20–25 years old with an average of 23 years were recruited. In order to evaluate the material attributes, the tested materials were cut to a size of 100 × 100 mm. The evaluators were informed before the test about the properties they would evaluate. The evaluators evaluated four bi-polar material attributes, one by one:

- Stiffness - softness
- Roughness - smoothness
- Non-stretchable-Stretchable and

Table 2 Overview of selected materials

Fabric ID	Surface mass, g/m^2	Fabric thickness, mm	Composition
S1	187.81	0.655	80% PA + 20% Elastane
S2	171.82	0.665	80% PA + 20% Elastane
S3	198.10	0.561	78% PA + 22% Elastane
S4	209.60	0.527	78% PA + 22% Elastane
S5	226.53	0.495	59% PA + 41% Elastane
S6	113.12	0.358	73% PA + 27% Elastane
S7	216.51	0.502	80% PA + 20% Elastane
S8	161.02	0.404	72% PA + 28% Elastane
S9	184.12	0.450	71% PA + 29% Elastane

- Thick-thin.

A seven-point scale marked with numbers from 1 to 7 was used to evaluate each attribute. For each attribute that is assessed, there were precise different instructions for testing and manipulation during the assessment. The overview is presented in Table 3.

3 Results and Discussion

The results of the subjective assessment are presented in Figs. 1–4.

Observing the results from all the graphs, it can be seen that female assessors gave higher grades on average than male assessors. This is especially evident by the roughness-smoothness and non-stretchable-stretchable attributes, where materials S4 and S7 were given a 1.0 or higher rating for both attributes.

It can be seen from Fig. 3 that both genders indicated that the S5 material was the most stretchable, which was to be expected. This material has the highest proportion of elastane at 41%. After that, materials S6, S8 and S9 (27–29% of elastane) follow in terms of elastane content, but assessors recognized only material S6 as elastic. Materials S1 and S3, which have 20% elastane, were characterized by assessors as more elastic than S8 and S9.

Table 3 Bipolar attributes with graphical instructions for manipulation

Bipolar Attributes	Instruction for manipulation during assessment
Stiffness-Softness	 
Roughness-Smoothness	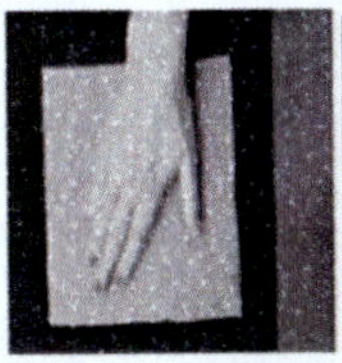 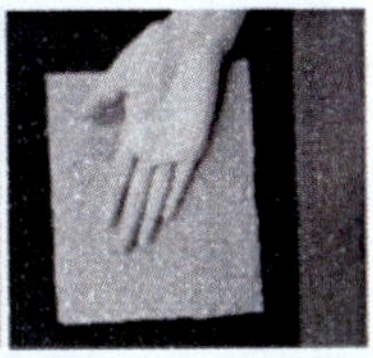
Non-stretchable-Stretchable	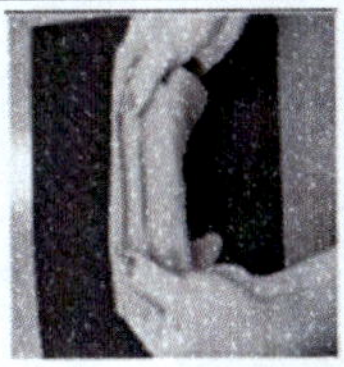 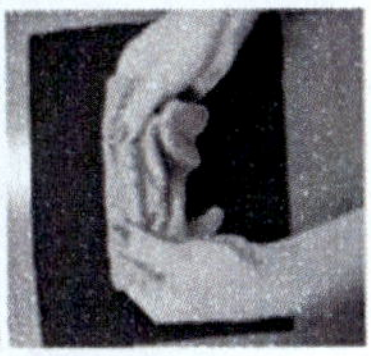
Thick-Thin	

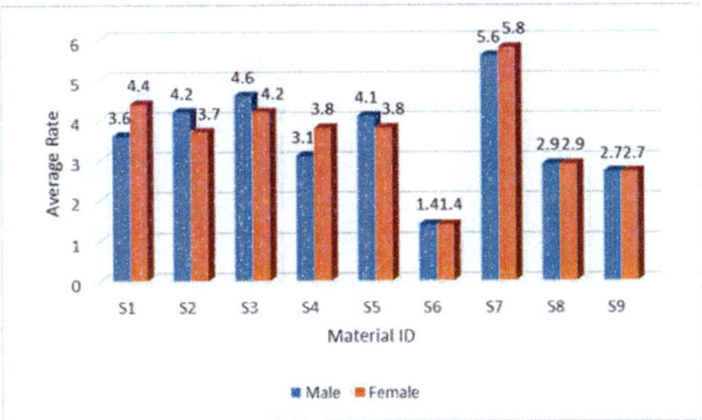

Fig. 1 Assessment for bi-polar attributes stiffness-softness

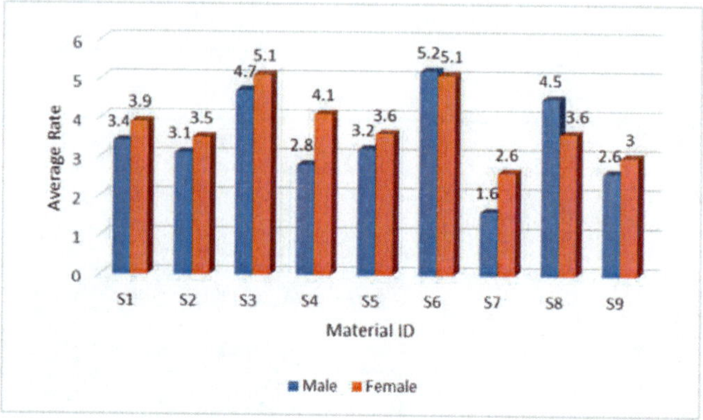

Fig. 2 Assessment for bi-polar attributes roughness-smoothness

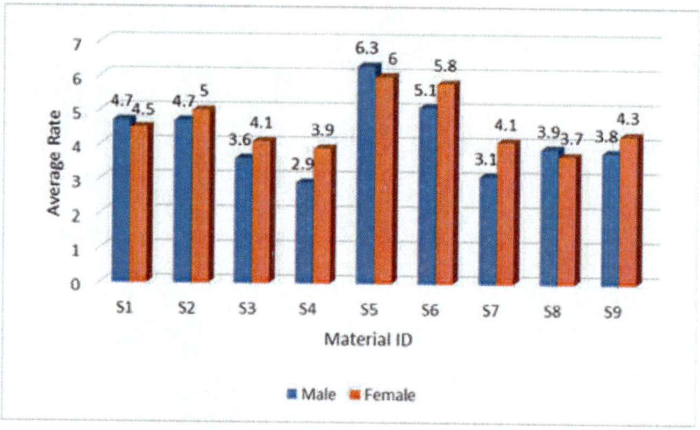

Fig. 3 Assessment for bi-polar attributes non-stretchable-stretchable

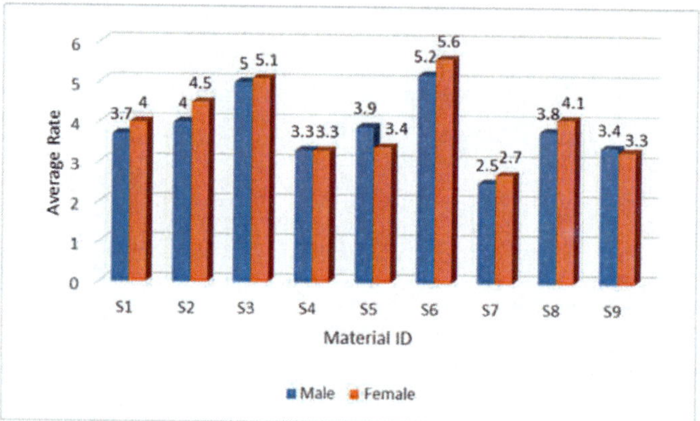

Fig. 4 Assessment for bi-polar attributes thick-thin

The attribute thick-thin assessment is very similar and we can say that this attribute is gender neutral (Fig. 4). It is interesting to see that both gender rated the material S6 as the thinnest, which corresponds to the fact that this material is really the thinnest of all the tested materials (Table 2). The thickest materials are S1 and S2, but the respondents did not notice this, but rated material S7 as the thickest.

4 Conclusion

The evaluation of the tactile comfort of comparable materials used to make sportswear clothing items is the main topic of this study. Therefore, a panel of assessors consisting of male and female evaluators with a rather narrow age range analyzed knitted fabrics made from diverse yarn compositions. The study indicated the following:

- female assessors gave on average higher grades than male assessors,
- both genders correctly indicated the material with the highest proportion of elastane as the most stretchable,
- they also managed to perceive the thinnest material among the presented materials as such.

The presented study contributes in terms of utilization the outcomes in the creation of sportswear clothing could. This is expected to be important for manufacturers planning to create the final goods while taking into account the opinions of young consumers. The findings of this study are anticipated to be valuable for market segmentation and the creation of sportswear for specific group of consumers. In further work, we will concentrate to other age group as well as level of education to see what are the difference and similarity between groups.

Acknowledgements HrZZ Croatian Science Foundation This work has been fully supported by Croatian Science Foundation under the project IP-2020-02-5041 „Textile Materials for Enhanced Comfort in Sports" – TEMPO.

References

1. Musa, A.B.H., Malengier, B., Vasile, S., Van Langenhove, L.: A comprehensive approach for human hand evaluation of split or large set of fabrics. Text. Res. J. **89**(19–20), 4239–4252 (2019)
2. Salopek Čubrić, I., Skenderi, Z.: Evaluating thermophysiological comfort using the principles of sensory analysis. Coll. Antropol. **37**(1), 57–64 (2013)
3. Salopek Čubrić, I., Čubrić, G., Majumdar, A.: Sensory attributes of knitted fabrics intended for next-to-skin clothing. J. Text. Inst., First Online (2022)

4. Milun, K., Čubrić, G., Salopek Čubrić, I.: Evaluating the tactile comfort of sportswear fabrics. In: Ćorić, D., Šolić, S., Ivušić, F. (eds.) Proceedings MATRIB 2021, pp. 376–384. HDTM, Zagreb (2021)
5. Geršak, J., Grujić, D.: Vpliv oblačila na toplotno fiziološko udobje človeka pri različnih obremenitvah in klimatskih pogojih, Tekstilec **46**(1-2), 183–190 (2003)
6. Li, Y., Wong, A.S.W.: Clothing Biosensory Engineering. Woodhead Publishing, Washington (2006)
7. Salopek Čubrić, I., Čubrić, G., Perry, P.: Assessment of knitted fabric smoothness and softness based on paired comparison. Fibers Polym. **20**(3), 656–667 (2019)
8. Tadesse, M.G., Harpa, R., Chen, Y., Wang, L., Nierstrasz, V., Loghin, C.: Assessing the comfort of functional fabrics for smart clothing using subjective evaluation. J. Ind. Text. **48**(8), 1310–1326 (2019)
9. Zeng, X., Ruan, D., Koehl, L.: Intelligent sensory evaluation: concepts, implementations, and applications. Math. Comput. Simul. **77**(5–6), 443–452 (2008)
10. Liao, X., Li, Y.: A simultaneous measurement method to characterize touch properties of textile materials. Fibers Polym. **5**(7), 1548–1559 (2014)
11. Pensé-Lhéritier, A.M., Guilabert, C., Bueno, M.A., Sahnoun, M., Renner, M.: Sensory evaluation of the touch of a great number of fabrics. Food Qual. Prefer. **17**(6), 482–488 (2006)
12. Xue, Z., Zeng, X., Koehl, L., Shen, L.: Consistency and reliability of untrained consumers' perceptions of fabric hand of men's suiting. Text. Res. J. **86**(13), 1425–1442 (2016)
13. Hu, J.Y., Hes, L., Li, Y., Yeung, K.W., Yao, B.G.: Fabric Touch Tester: Integrated evaluation of thermal–mechanical sensory properties of polymeric materials. Polym. Test. **25**(8), 1081–1090 (2006)
14. Xue, Z., Zeng, X., Koehl, L., Chen, Y.: Extracting fabric hand information from visual representations of flared skirts. Text. Res. J. **84**(3), 246–266 (2014)
15. Bensaid, S., Osselin, J.F., Schacher, L., Adolphe, D.: The effect of pattern construction on the tactile feeling evaluated through sensory analysis. J. Text. Inst. **97**(2), 137–145 (2006)
16. Bacci, L., Camilli, F., Drago, S., Magli, M., Vagnoni, E., Mauro, A., Predi, S.: Sensory evaluation and instrumental measurements to determine tactile properties of wool fabrics. Text. Res. J. **82**(14), 1430–1441 (2012)
17. Bertaux, E., Lewandowski, M., Derler, S.: Relationship between friction and tactile properties for woven and knitted fabrics. Text. Res. J. **77**(6), 387–396 (2007)
18. Wiskott, S., Weber, M.O., Heimlich, F., Kyosev, Y.: Effect of pattern elements of weft knitting on haptic preferences regarding winter garments. Text. Res. J. **88**(15), 1689–1709 (2018)
19. Niedermann, R., Rossi, R.M.: Objective and subjective evaluation of the human thermal sensation of wet fabrics. Text. Res. J. **82**(4), 374–384 (2012)
20. Grinevičiute, D., Daukantiene, V., Gutauskas, M.: Textile hand: comparison of two evaluation methods. Mater. Sci. (Medžiagotyra) **11**(1), 57–63 (2005)
21. Ozcelik, G., Supuren, G., Gulumser, T., Tarakcioglu, I.: A study on subjective and objective evaluation of the handle properties of shirt fabric. Fibres Text. Eastern Europe **68**(3), 56–62 (2008)
22. Hui, C.L., Lau, T.W., Ng, S.F., Chan, K.C.C.: Neural network prediction of human psychological perceptions of fabric hand. Text. Res. J. **74**(5), 375–383 (2004)
23. Cardello, A.V., Winterhalter, C., Schutz, H.G.: Predicting the handle and comfort of military clothing fabrics from sensory and instrumental data: development and application of new psychophysical methods. Text. Res. J. **73**(3), 221–237 (2003)
24. Naima, H.A., Mehdi, S., Morched, C.H.: Training and control of performance of a tactile sensory panel. J. Appl. Sci. **13**(3), 366–376 (2013)
25. Philippe, F., Schacher, L., Adolphe, D.C., Dacremont, C.: Tactile feeling: sensory analysis applied to textile goods. Text. Res. J. **74**(12), 1066–1072 (2004)
26. Wang, Y., Wang, Y., Zhao, M., Yu, M., Li, J.: Experimental study of clothing tactile comfort based on electro-neurophysiology. Fibres Text. Eastern Europe **22**(4), 102–106 (2014)
27. Kanai, H., Kimura, H., Morishima, M., Shouji, U., Nishimatsu, T., Shibata, K., Yamamoto, T.: Microscopic photometry and its parameterization for objective evaluation of aesthetics of woven fabrics. Text. Res. J. **82**(19), 1982–1995 (2012)

28. Xue, Z., Zeng, X., Koehl, L.: Development of a method based on fuzzy comprehensive evaluation and genetic algorithm to study relations between tactile properties and total preference of textile products. J. Text. Inst. **108**(7), 1085–1094 (2017)
29. Xiao, L., Li, Y., Hu, J., Wu, X., Li, Q.: A simultaneous measurement method to characterize touch properties of textile materials. Fibers Polym. **15**(7), 1548–1559 (2014)
30. Asad, R.A., Yu, W., Siddiqui, Q., Vincent, M., Wang, Q.: Subjective evaluations of fabric-evoked prickle using the unidimensional rating scale from different body areas. Text. Res. J. **86**(4), 350–364 (2016)
31. Park, J., Yoo, H.-S., Hong, K.H., Kim, E.: Knitted fabric properties influencing coolness to the touch and the relationship between subjective and objective coolness measurements. Text. Res. J. **88**(17), 1931–1942 (2018)
32. Wang, Q., et al.: Representations of fabric hand attributes in the cerebral cortices based on the Automated Anatomical Labelling atlas. Text. Res. J. **89**(18), 3768–3778 (2019)

The Usefulness of Eye-Tracking Glasses in the Technological Upgrade of the Manual Workplace – An Ergonomic Aspect

Tina Cvahte Ojsteršek and Brigita Gajšek

Abstract Ergonomics of workplaces in manufacturing are coming to the forefront of research in Industry 4.0 and automatization processes that involve human factors. The paper presents a case study based research approach for analyzing a workplace which has potential to be partially automated. Manual processing of complex metal surfaces was chosen as an example of an ergonomically less human-friendly workplace demanding to automate. Ergonomic work assessments are rarely used in such work environments, although all legal requirements for safe work are met. The pilot technological upgrade of manual processing of design-demanding surfaces includes hand over of a part of the processing to a collaborative robot. The workplace was assessed using the REBA ergonomic risk assessment method before the upgrade which found several issues such as a non-ergonomic body position and repetitive movements. After the upgrade to a shared workspace with a collaborative robot, eye-tracking glasses were used to explore their potential for assessing work ergonomics. It was found that a large portion of the time when setting up the robot is spent looking at the contact of the robot hand attachment to the material surface, and when working alongside the robot, the robot presents an important distraction to the work process of the worker. We conclude that eye tracking is a very useful tool in assessing workplaces and work settings where human factors and especially their visual attention play a major role in the analyzed process.

Keywords eye-tracking · workplace · worker · collaborative robot · ergonomics

T. Cvahte Ojsteršek · B. Gajšek (✉)
Faculty of Logistics, University of Maribor, Celje, Slovenia
e-mail: brigita.gajsek@um.si

T. Cvahte Ojsteršek
e-mail: tina.cvahte@um.si

I. Salopek Čubrić et al. (eds.), *Proceedings of the 9th International Ergonomics Conference*, Lecture Notes in Networks and Systems 701,
https://doi.org/10.1007/978-3-031-33986-8_9

1 Introduction

Inspired by the idea of Industry 4.0, we expect a technological boom in all industries. Almost a decade after establishing the definition of Industry 4.0, technological upgrading is carried out at different speeds in different industries. In the electronic, electrical, and automotive industries, it is happening much faster than, for example, in toolmaking and mechanical engineering. The paper focuses on a specific workplace in a toolmaking company. The European tooling industry mainly comprises SMEs, creating around 12 billion Euro annual turnover by more than 8,100 companies [1]. The mentioned industry sees one of the biggest challenges by 2030 in advanced automation and production cells, namely in developments in collaborative robotics and advanced control systems, and new sensor technologies. The production process is mainly automatized but still holds a considerable proportion of manual, repetitive work, mostly in the assembly area [2]. An example of a demanding manual job is hand polishing the tool's functional surface. The achieved surface quality depends on the skills of a toolmaker. As a result of manual work, this operation is a potential place for errors and requires repetitive and hard physical work. As such, it is recognised as a potential for ergonomic assessment and presents an opportunity for improvement. In the article, we first present an ergonomic evaluation of the mentioned workplace to justify further efforts to improve the work at the workplace by polishing the functional surfaces of the tool.

The workplace polishing of the tool's functional surfaces represents the potential for transformation into a hybrid workplace, where a collaborative robot takes over part of the repetitive, labor-intensive polishing operations. The realization of the opportunity opens up a new question of the impact of the coexistence of a robot with a human on a person's physical and psychological burden.

Human–robot interaction is one of the most discussed topics in modern robotic solutions [3]. In the coexistence of a human and a collaborative robot, attention is paid to three types of interaction between them. The first type of interaction refers only to their coexistence in the same space. In doing so, they do not need to synchronize their activities but only be aware of each other and pay attention to the other's actions. The second type of interaction refers to the collaborative performance of tasks. This is the most demanding form of interaction because it is necessary to adjust the robot's movements in real time with human movement [3]. The third type of interaction is robot learning, during which the robot and the human do not perform the work for which the workplace is designed.

The essential starting point for establishing workplaces where humans and robots are collocated is that safety is guaranteed. Safety issues in human–robot interaction are regulated by standards [4–6]. Thus, there are no safety barriers to the widespread use of collaborative robots in industrial settings. General safety standards handle four collaborative directions, which cover the main operation approaches in human–robot interaction during their collaboration [7, 8], namely "Safety-rated monitored stop," "Hand guiding," "Speed and separation monitoring," and "Power and force limiting." These four perspectives are meant to all be present in each working scenario, not as individual perspectives of collaboration between humans and collaborative robots [3].

Among researchers, research into the effectiveness of human–robot interaction is widespread. In general, human activity is efficient when the work is done effectively (in the shortest amount of time) and economically (costs are minimal). Replacing human work with the work of a collaborative robot is justifiable in the case that the robot takes over physically demanding and dangerous work from a human and, at the same time, does not affect or improve the profitability of the work. Of course, all this can be achieved, but the question we are addressing is whether it would be reasonable to also study the mental effort of the worker due to the need to change their eye focus between his work and the work of the collaborative robot and to direct some of their cognitive attention to the robot.

Eye tracking as a tool for researching human interaction with their environment has evolved from the beginning stages of intricate and large machines to today's wearable unobtrusive technologies. Its main goal is to track eye movements and fixations in relation to the coordinate system of a wearer's field of vision, which gives important insight into the visual and cognitive attention of the wearer. Visual attention, meaning focusing on specific elements in a person's visual field, can be overt (direct) or covert (in the peripheral region of the visual field) [9]. Eye tracking focuses on overt visual attention, a good indicator of general attention to an area or task in an industry setting [10]. Its use in Industry 4.0 engineering and manufacturing concepts mostly focuses on product design and validation [11]. On the other hand, its use in measuring and analysing the interaction between workers and collaborative robots is still rare [12]. With this article, we want to contribute to this field based on the understanding that the interaction processes between humans and machines/robots in a manufacturing environment is fundamental for ensuring a good working environment and improving factory performance [13].

As a first step towards studying human mental load while working alongside a collaborative robot, in the article, we present a case study of the suitability of eye-tracking technology for studying a worker's overt attention during the polishing of the tool's functional surface. Before utilising eye tracking, a general ergonomic assessment of the workplace using the Rapid Entire Body Assessment (REBA) will be performed. With a combination of methods, a final assessment of the case study workplace and an estimate of the usability of eye tracking in combination with a more traditional ergonomic assessment tool will be presented.

2 Methodology

The research is based on a manual workplace for polishing metal 'tools' functional surfaces. This workplace (see Fig. 1) represents the potential for transformation into a hybrid workplace, where a collaborative robot could take over part of the repetitive, labor-intensive polishing operations. The realization of the opportunity opens up a

Fig. 1 Existing manual workplace

new question of the impact of the coexistence of a robot with a human on a worker's physical and psychological burden.

The research questions posed are:

RQ1: Is the worker at the workplace exposed to ergonomic risks while polishing the functional surfaces of the tool?

RQ2: Could eye tracking technology be used to understand the changing focus of a worker's gaze during work?

The practical part of the research was divided into two parts. The first deals with the general ergonomic assessment of work at the workplace, while the second deals with a detailed assessment of the worker's overt visual attention. The latter was performed to study potential additional cognitive load while simultaneously testing the usefulness of eye-tracking technology in the workplace that has been technologically upgraded with a collaborative robot.

2.1 REBA Ergonomic Assessment

The Rapid Entire Body Assessment (REBA) ergonomic method was used for the ergonomic evaluation. REBA tool uses a systematic process to evaluate both upper and lower parts of the musculoskeletal system for biomechanical and musculoskeletal disorders (MSD) risks associated with the evaluated job task [14]. The output of REBA tool usage is the REBA Score. This single REBA Score represents the level of MSD risk for the evaluated task. The minimum REBA Score is 1 (negligible risk, no action is required), and the maximum REBA Score is 15 (very high risk, change must be implemented) [15].

2.2 Eye Tracking Analysis

Tobii Pro Glasses 3 were used for eye tracking analysis. The glasses capture a video of the field of vision of the subject wearing them as well as coordinates of their gaze points in relation to the field of vision. This enables researchers to gain exact insight into gaze directions of the subject in any given moment in relation to their field of vision. Results from recording were imported into Tobii Pro Lab, a dedicated software for analysis of results [16]. Two intervals of interest were selected from the recording of the work process: an interval where the worker placed the material slab and trained the robot hand on processing the material, and an interval where the worker worked on their own material slab alongside the robot hand, and a situational snapshot was prepared for each scenario. For further analysis, the main areas of interest [AOI] were defined in the situational snapshot, as shown in Fig. 2 for robot setting and training, and in Fig. 3 for working alongside the robot.

Gaze points from the raw recordings were manually mapped onto the static snapshots for the two selected intervals. Information about fixations and area of interest visits were obtained for each interval of interest, and gaze heatmaps were generated using mapped data. A fixation is defined as a time when both eyes and their gaze are static and focusing on a single point [17], and a visit is defined as a connected set of fixations inside an area of interest that starts with the first fixation inside the AOI and ends with the last fixation inside it before the gaze moves to a point outside of the AOI [18]. Information about the number of visits and fixations, average and total duration for each AOI in each time of interest were obtained with the analysis. Heatmaps that show the concentration of fixations by their relative number and relative duration compared to the whole interval of interest were also exported.

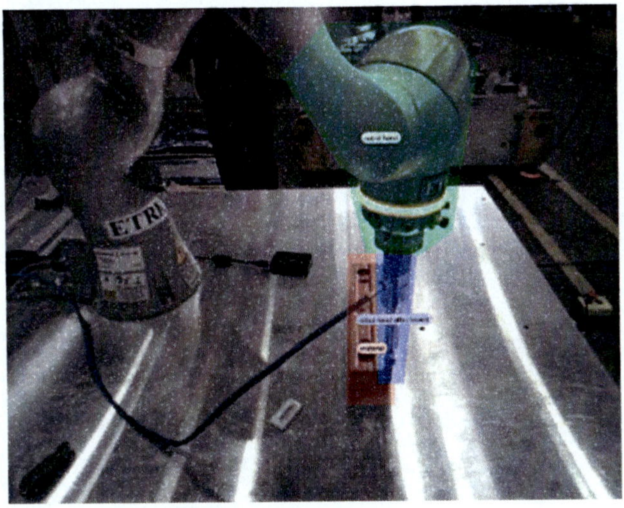

Fig. 2 Snapshot and defined areas of interest for the training interval

Fig. 3 Snapshot and defined areas of interest for the collaborative working interval

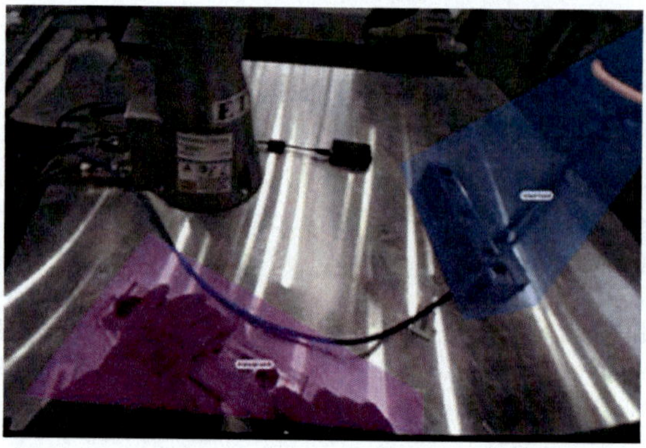

3 Results

3.1 Ergonomic Risks at the Workplace Polishing Functional Surfaces of Tools

All tools' functional surfaces must be polished after the final machine mechanical processing. Due to the complex shapes of the engravings, this work is entirely manual and requires long periods of pressing the sanding belt or whetstone by hand against the surface of the tool and gliding back and forth (see Fig. 1).

The REBA methodology divides body parts into group A (arm and wrist) and group B (neck, trunk, and legs). Scores B were considerably higher than scores A because of high neck strain (see Fig. 4). Standing posture also negatively affects the feet. When analysing REBA scores for different body areas, the most loaded part of a worker's body is the back, on account of the standing posture.

REBA assessment revealed that the worker's working height is not adjusted to his height. Consequently, the worker does not adopt an ergonomic position. His back is bent forward, and his neck is flexed (see Fig. 1), which creates muscular tension.

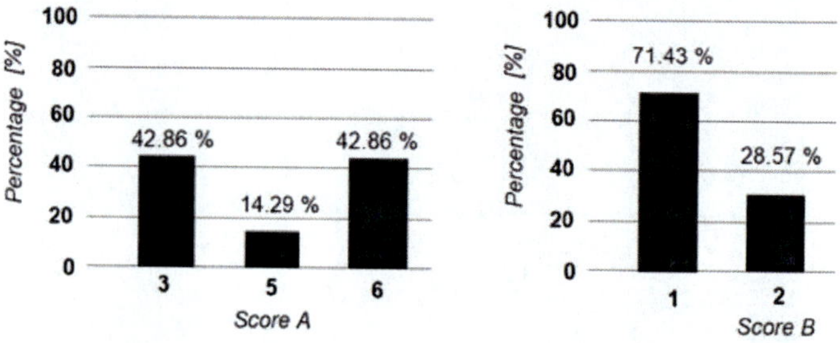

Fig. 4 REBA assessment results

Uncomfortable posture and repetitive movements scored the REBA assessment as medium ergonomic risk in 71% of the analysed postures.

Although the worker does not execute an arduous task, he is exposed to significant ergonomic risks generated by standing posture, static work, and repetitive hand and arm movements.

Identified problems in the analysed workplace were the non-adjustable working surface height, repetitive movements because of gliding back and forth, working in a bent posture, repeating movements, working in a painful position, and wrist twisting.

Changing the non-adjustable height of the working surface to an adjustable one would help a lot. The worker's back would not be bent forward anymore, and his neck would not be flexed, lowering muscular tension. Adding job rotation could help reduce repetitions of movements. Of course, the latter is difficult to do because all toolmaker workplaces involve manual polishing. Consequently, the introduction of a collaborative robot makes sense in this setting since it could take over part of the polishing in more or less flat places without demanding engravings and consequently reduce the most repetitive motions for the worker. In the next step, the previously assessed workplace upgraded with a collaborative robot will be analysed from the perspective of the worker from the viewpoint of his visual attention (which in turn also points to their cognitive attention).

3.2 *Visual Behaviour in Setting up and Collaborating with the Collaborative Robot*

The first analysed stage regarding the visual attention of the worker was the preparation stage, in which the worker "teaches" the collaborative robot task execution. The interval of preparing the working area and training the robot to perform the selected task took 75 s. During this time, the visual attention of the worker as measured by the duration of fixations was targeted to the material to be worked on by the robot in a total duration of 4 s, to the robot hand attachment for 39 s and to the robot hand itself for 19 s. The average duration of fixations and visits was the longest when fixating on the robot hand attachment. Detailed metrics are given in .

Table 1.

The heatmap of the relative number of fixations while preparing and "training" the robot is shown in Fig. 5. Concentrations of fixations occur on the point of contact between the robot hand attachment and the material slab, and in the area of the robot hand where the setting buttons and indicators are located.

The heatmap of the relative duration of fixations while preparing and "training" the robot is shown in Fig. 6. The results show that even though the number of fixations to the robot hand attachment was almost double to the number of fixations onto the robot hand itself, the duration of fixations to the latter was notably shorter, indicating only short fixations to that area, while fixations to the robot hand were

Table 1 Visual attention metrics for the robot training time interval

Area of interest	total duration of fixations [ms]	average duration of fixations [ms]	number of fixations	total duration of visit [ms]	average duration of visit [ms]	number of visits
material	3967	331	12	4107	587	7
robot hand attachment	39,291	2619	15	39,612	4951	8
robot hand	18,614	665	28	19,235	2748	7

Source: Own

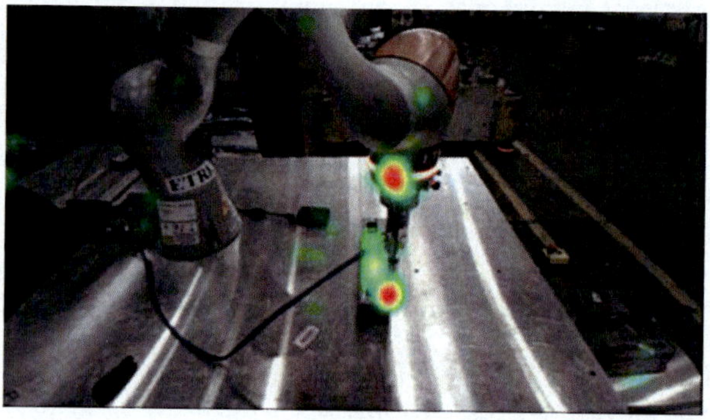

Fig. 5 Heatmap of the relative number of fixations while preparing and "training" the robot

made with stronger intent and concentration and consequently lasted much longer, as also evidenced by the metrics shown in Table 1.

The second stage of the collaboration was the stage where a worker manually processed a material slab while the collaborative robot did the same task alongside. For assessing visual attention during this stage, a 67 s interval was selected. During this time, fixations of the worker were targeted to his area of manual labour for 51 s and to the robot hand itself for 11 s. Detailed metrics of the interval are presented in Table 2.

The heatmap of the relative number of fixations while working alongside the robot is shown in Fig. 7.

The heatmap of the relative duration of fixations while working alongside the robot is shown in Fig. 8. The difference between the distribution of concentration areas between the number and the duration of fixations is not as notable as it was in the training period due to the smaller difference in average fixation duration among the two AOIs.

A pattern of fixations was found in this phase where the worker focused on their work and on the material slab for approximately 10 s (which is also indicated by the average visit duration to the manual work AOI of 8.8 s) and then performed a quick

Fig. 6 Heatmap of the relative duration of fixations while preparing and "training" the robot

Table 2 Visual attention metrics for the collaborative work time interval

Area of interest	total duration of fixations (ms)	average duration of fixations (ms)	number of fixations	total duration of visit (ms)	average duration of visit (ms)	number of visits
robot hand	10,679	821	13	11,140	2785	4
manual work	50,736	1103	46	53,041	8840	6

Source: Own

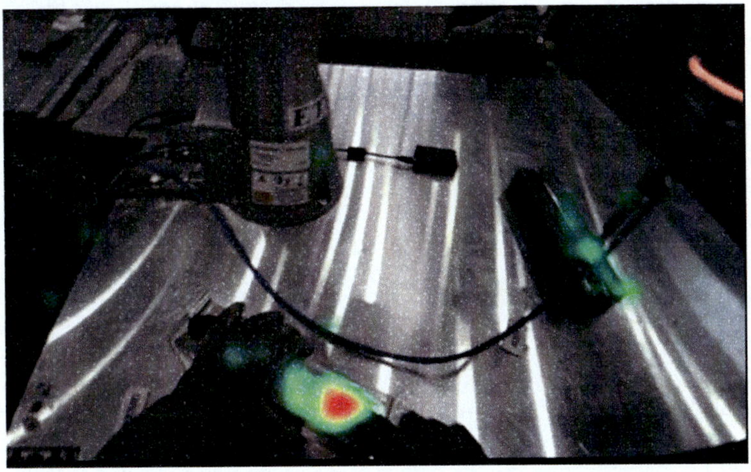

Fig. 7 Heatmap of the relative number of fixations while working alongside the robot

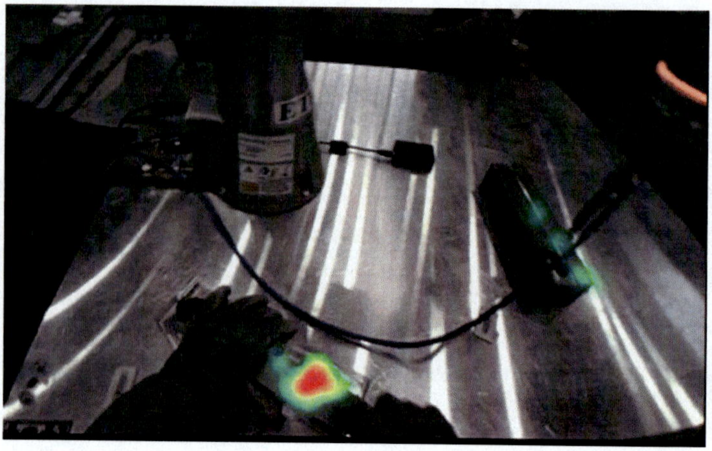

Fig. 8 Heatmap of the relative duration of fixations while working alongside the robot

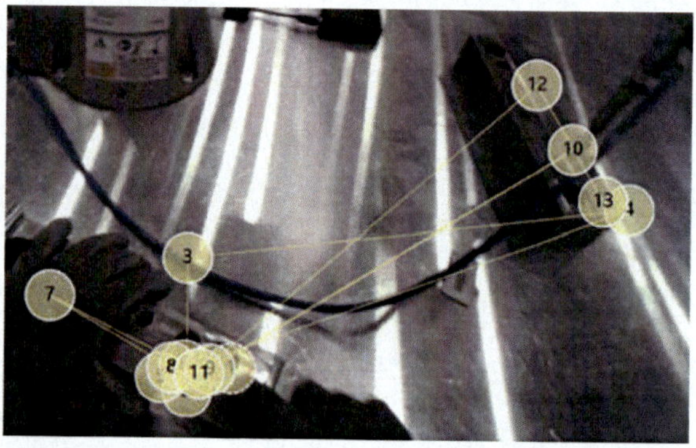

Fig. 9 Gaze plot of fixations in a sample of time while working alongside the robot hand

glance towards the robot hand and its operating area, then quickly returned to fixate on their manual work. A gaze pattern for a selected 20 s timeframe of this pattern is shown in Fig. 9, where the numbers in the circles mark the order of performed fixations.

4 Discussion

The focal company envisioned a technological upgrade of the workplace in the form of a collaborative robot, which would bring an increase in productivity due to the robot processing surfaces faster than a human, even though more complex shapes can still be processed only by the human hand. If work is to be handed over to a collaborative robot, a human must first hand over the task to the robot by showing the robot the machining area's boundaries and setting the processing type.

REBA assessment revealed problems because of the non-adjustable working surface height, repetitive movements because of gliding back and forth, working in a bent posture, working in a painful position, and wrist twisting. Some of these risks could be eliminated even without the technological upgrade of the workplace, but they would not eliminate the most critical risk, repetitive hand movements. The work requires a certain number of movement repetitions, which cannot be changed.

A technological upgrade with a collaborative robot makes sense because there are 30% of areas that do not have design-challenging engravings and are appropriate for using the robot. This would in turn reduce the duration of repetitive movements performed by workers in this workplace. We can hypothesize that the increase in mental effort due to frequent control of the collaborative robot may compensate for the increase in physical fatigue. For proof it is necessary to know what a person actually pays attention during sharing space and tasks with a collaborative robot. Through the presented case study, the results show that eye-tracking technology could be used to understand the changing focus of a worker's gaze during work. Monitoring the worker's eye gaze direction gave us insight into worker's overt attention during individual tasks. During the training of the collaborative robot, worker placed the greatest emphasis on monitoring the robot's status through light messages on the robot's arm in the form of longer gaze fixations. Worker's gazes were directed towards the point of contact between the tool and the material for the longest period, in the form of short but frequent gaze fixations, to make sure that he is showing the robot the appropriate places on the material and that the tool really touches the required points.

When the worker and the robot then begin to perform work in parallel, the worker directs his gaze to his work surface for 76% of the time and pays attention to his work. However, 12% of the time, the worker's attention was directed to the collaborative robot's work surface. At this time, he did not stop working but only shifted his gaze for 2.7 s on average away from his working area. Could this reduce a person's safety at work? A worker is not threatened by a collaborative robot, as safety standards guarantee this, but he can be threatened by the separation of attention between his work and the work of the collaborative robot.

As this was only a case study based research, one of the major limitations is a smaller dataset. The information was gathered during a worker's first collaborative session with the robot, so we currently cannot deduct what worker's attention focus would be on prolonged or multiple sessions of collaborative working. This will need to be studied in depth with multiple cases and multiple work settings.

5 Conclusion

The research recognised eye tracking as a tool for efficiently and precisely tracking the direction of the visual gaze of a worker, as the technology provider also states. We can detect worker's interaction with collaborative robot placed in a shared workplace. In using collaborative robots, humans remain a major part of the work but also work alongside robots. By studying visual attention in these settings, we can understand human behaviour in these settings better and design more efficient and ergonomically sounder workplaces and work processes.

The research concludes that there is potential to investigate the impact of adding a collaborative robot to a human workplace on a human's mental and physical workload and the actual change in productivity with the help of eye-tracking glasses.

Further research should extend the presented research in several main directions. As mentioned previously, this was a case study based research which should be repeated using the same two step methodology in different settings and different collaboration scenarios to increase in validity. Another aspect is to research the effect of experience of previous work with a collaborative robot on the overt attention of workers working alongside one, as well as the effects of different demographic parameters such as age or work experience. In an ergonomics aspect, an in-depth analysis should be made into more ergonomics metrics to compare the situations where a worker is working alone and when they are working alongside a robot.

Funding This research was partially funded by the Republic of Slovenia and the European Union from the European Regional Development Fund, ROBOTOOL-1 project, grant number OP20.03540.

References

1. Wollbrink, M.: Tooling 2030 - Strategic Roadmap for Industry. https://toolingplatform.manufuturenet.eu/images/artigos/presentation_roadmap.pdf. Accessed 11 Sep 2022
2. Radić, I., Rupnik, B., Šinko, S., Kramberger, T., Gajšek, B.: Redesign of the workplace for toolmakers towards industry 4.0. In: Karabegović, I. (ed.) Handbook of research on integrating Industry 4.0 in business and manufacturing. IGI Global, Hershey, Pennsylvania, USA, pp. 492–511 (2020)
3. Galin, R.R., Meshcheryakov, R.V.: Human-robot interaction efficiency and human-robot collaboration. In: Kravets, A.G. (ed.) Robotics: Industry 4.0 Issues & New Intelligent Control Paradigms SSDC, vol. 272, pp. 55–63. Springer, Cham (2020). https://doi.org/10.1007/978-3-030-37841-7_5
4. International Standardization Organization: ISO/TS 15066:2016–02 (e) Robots and robotic devices—Collaborative robots. International Standardization Organization, Geneva, Switzerland (2016)
5. International Standardization Organization: ISO 10218–2:2011 Robots and robotic devices—Safety requirements for industrial robots—Part 1, 2: Robot systems and integration. International Standardization Organization, Geneva, Switzerland (2011)

6. Villani, V., Pini, F., Leali, F., Secchi, C.: Survey on human–robot collaboration in industrial settings: safety, intuitive interfaces and applications. Mechatronics **55**, 248–266 (2018)
7. Khalid, A., Kirisci, P., Ghrairi, Z., Thoben, K-D., Pannek, J.: Towards implementing safety and security concepts for human-robot collaboration in the context of Industry 4.0. In: 39th International MATADOR Conference on Advanced Manufacturing, pp. 0–7. University of Manchester, Manchester, UK (2017)
8. Lazarte, M.: Robots and humans can work together with new ISO guidance, https://www.iso.org/news/2016/03/Ref2057.html. Accessed 13 Apr 2022
9. Bisley, J.W.: The neural basis of visual attention. J. Physiol. **589**(1), 49–57 (2011)
10. Shotton, T., Kim, J.H.: Assessing differences on eye fixations by attention levels in an assembly environment. In: Ayaz, H., Asgher, U. (eds.) Advances in Neuroergonomics and Cognitive Engineering, pp. 417–423. Springer International Publishing, Cham (2021). https://doi.org/10.1007/978-3-030-51041-1_55
11. Borgianni, Y., Rauch, E., Maccioni, L., Mark, B.G.: User experience analysis in industry 4.0 - the use of biometric devices in engineering design and manufacturing. In: IEEE International Conference on Industrial Engineering and Engineering Management, pp. 192–196. IEEE, Bankok, Thailand (2018)
12. Weiss, A., Huber, A., Minichberger, J., Ikeda, M.: First application of robot teaching in an existing industry 4.0 environment: does it really work? Societies **6**(3), 20 (2016). https://doi.org/10.3390/soc6030020
13. Peruzzini, M., Grandi, F., Pellicciari, M.: Benchmarking of tools for user experience analysis in industry 4.0. Proc. Manuf. **11**, 806–813 (2017). https://doi.org/10.1016/j.promfg.2017.07.182
14. ErgoPlus, A Step-by-Step Guide to the REBA Assessment Tool. https://ergo-plus.com/reba-assessment-tool-guide/. Accessed 9 Nov 2022
15. Šinko, S., Draghici, A., Boatca, M.E., Đukić, G., Gajšek, B.: Raising workplace sustainability with ergonomics assessment methods. Acta technica napocensis (in press)
16. Tobii Pro AB, Tobii Pro Lab (version 1.181.37603 (x64)). Tobii Pro AB, Danderyd, Sweden (2021).
17. Rayner, K.: Eye movements and attention in reading, scene perception, and visual search. Q. J. Exp. Psychol. **62**(8), 1457–1506 (2009)
18. Sears, M., Alruwaythi, O., Goodrum, P.: Evidence of inconsistent results using current eye tracking glance and visit analysis standards. Autom. Constr. **132**, 103951 (2021). https://doi.org/10.1016/j.autcon.2021.103951

Ergonomics in Small Home-Based Industries: Batik Making in Malaysia

D. D. I. Daruis, D. Mohamad, and N. K. Khamis

Abstract This paper aims to review the ergonomics study related to the batik industry, especially in Malaysia. The cross-sectional study is carried out using a 10 years' filter and keywords such as batik AND ergonomic AND Malaysia. The searches were carried out mainly on Google scholar website but also cross-referenced with searches from Scopus. There were 300 plus papers when the keyword Malaysia was included, but 1,200 papers when that keyword was omitted. More than 90% of batik studies are from the Indonesian industry. Focus of the papers were on physical ergonomics, mostly associated batik workers with at least one of musculoskeletal disorders such as shoulder or back pain and CTD (although not as popular as the former). Two site visits were also mentioned to give a snapshot of ergonomics hazards that are common in small and medium batik workplace. In conclusion, more in-depth study to use objective measurements are needed to validate/verify these claims from subjective perception studies and pen & paper analyses.

Keywords hand-drawn batik · painting · musculoskeletal disorders · WRMSD · occupational health · ergonomic case studies

1 Introduction

Experts and connoisseurs in the textile industry would know that batik is not the product name but rather a technique of fabric coloring or rather the process of fabrics or textile decoration which uses wax as a resist containing various colors and patterns

D. D. I. Daruis (✉)
National Defense University of Malaysia, Kuala Lumpur, Malaysia
e-mail: dian@upnm.edu.my

D. Mohamad
Universiti Malaysia Kelantan, Kota Bharu, Malaysia

N. K. Khamis
Universiti Kebangsaan Malaysia, Bangi, Malaysia

I. Salopek Čubrić et al. (eds.), *Proceedings of the 9th International Ergonomics Conference*, Lecture Notes in Networks and Systems 701,
https://doi.org/10.1007/978-3-031-33986-8_10

[1, 2]. However, it is widely accepted as noun as well where a batik is actually an object or product (usually cloth or clothes) made using the batik technique. Batik patterns can be produced using specially carved block of stamps, hand drawings or a more recent technique, digital prints. Historically, there were evidences of batik that can be traced back to 2000 years ago, in places in the Middle East, Central Asia and India [3].

Batik handicrafts are mostly synonym with Indonesia and Malaysia as the craftsmanship is currently produced as national enterprise in both countries. The batik industries are strongly supported by the NGO and their respective government bodies. Locally, batik is the most popular handicraft product among individual buyers and the third most popular after pewter and wood crafts among the corporate buyers [1]. Albeit the international reach of the batik commercialization, batik industries in both countries are very much small-medium scaled, home-based and mostly traditional. Generally, batik craftsman draws or stamps the patterns with hot wax and then colors or dyes the fabric, which is usually carried out multiple times, air dries the fabrics (usually in 2 m or 4 m' piece of cloth), soaks the fabric in medium that seals the colors, rinses the fabric and then removes the wax in boiling water, immediately soaks it again in cold water then later fabric softener before finally air dries it again. Turning the plain cloth to a batik art through drawing or stamping requires the most effort and time from the craftsman. The traditional nature of producing artwork using the batik technique resulted in health and ergonomic issues on the workers. There have been many research interests on the ergonomics from various aspects of the batik production [4]. Therefore, this paper aims to revisit literatures published between the year 2012 to 2022 focusing on ergonomics study related to batik, and try to relate the findings from two visits to batik artisan in Kelantan, one of the states in east Malaysia where batik is prominent.

2 Methodology

In the beginning of the literature search, Google Scholar was used to search for two types of keywords. Although this crawler-based is not the most accurate and functions differently than the other bibliographic database like Scopus [5], however it is the easiest and fastest to access and respond. The keywords were Batik AND Ergonomic and Batik AND Ergonomic AND Malaysia, both were performed using advanced search, limited to between the years 2012 to 2022. The Boolean use of AND is important to filter out non-batik studies. Then the search was continued on Scopus database, using the same keywords as before. For further analysis from Google Scholar results, more keywords were then included to filter down the findings among others like musculoskeletal, back pain, shoulder and leg.

Other than secondary data analyses, a visit to two home-based batik industries in Kelantan (a state east in Malaysia) allowed the researcher to witness first-hand the work method that the craftsmen practiced. Observations were focused on the workers or the batik craftsmen posture and also their workspace design that might have been the risk factors that cause any musculoskeletal prevalence and other ergonomic issues.

3 Results

Through Google scholar search, about 1,200 papers (as of 30th May 2022) from 2012 to 2022 were listed when Batik AND Ergonomic keywords were used. However, it is important to highlight that Boolean usage on Google Scholar does not always work appropriately and that this platform can be used as secondary resources but not worth as primary resources [5]. Nevertheless, manual screening was carried out and the few of the main findings representing different kind of study categories are discussed in the subsections. From Scopus database, out of the 1,200 findings by Google Scholar, only fourteen are listed, the papers are also featured in Google Scholar search result.

Next section, the visit to the two batik artisans resulted in few videos and photographs that allowed ergonomics analysis like RULA and REBA to be performed. Also, the layout of the workstations and the tools that they use at the two different places showed that it actually affects the posture of the batik craftsmen when performing their work. However, the ergonomic analyses results are for discussion in a future paper.

3.1 Batik and Ergonomic

First few pages of the 1,200 papers results showed that the studies found under these two keywords can be categorized into studies of; posture assessment or prevalence of Musculoskeletal Disorders (MSD) and the ergonomic risk factors (almost 39%), tools design (25%), participatory ergonomics or ergonomics intervention (11%), environmental ergonomics and workspace design (7%), material handling/lifting (5%), productivity (9%) and occupational ergonomics (4%). The Scopus database listed the fourteen papers in almost similar categories breakdown. The highest number of papers are from MSD studies (five papers) followed by tools design (four papers) and the rest two and one each from productivity, material handling/ lifting and environmental ergonomics and workspace design categories.

For the highest number of papers category, the most observed musculoskeletal disorders were found to be mainly at the back; shoulders and lumbar area [6–8] and these are mostly from Indonesian batik workplaces where the craftsmen do their work seated either on small stool or with both legs at one side on the floor (*bersimpuh*). The Malaysian batik musculoskeletal disorders observations showed the highest prevalence is at the shoulders, lower back and followed by the ankles/leg/ sole since the Malaysian batik craftsmen work posture are standing [9–11]. A recent paper highlighted that there are complaints of Carpal Tunnel Syndrome (CTS) among Sumenep (Indonesia) batik craftsmen due to "*canting*" (holding special batik tool made of metal to draw batik patterns using wax) where they use the canting and work for long hours [12]. Another paper related to the keywords reported on postural evaluation and prevalence of musculoskeletal disorders of the hand [13]. Interestingly, the approaches used are distinctive than other studies, the LUBA method; an assessment technique for postural loading on the upper body based on joint motion discomfort and maximum holding time, Job Strain Index (JSI) and ACGIH HAL (Hand Activity Level). Few selected papers of different categories were explored too. One paper that the authors categorized under the work tools design, Anjani et al. [14] suggested an optimal stool design that they named as *dingklik*. The ergonomic stool, designed with measurement of the users' anthropometry, aimed to help the batik crafter to do their work while in sitting position comfortably in long hours, as that is the preferred posture among most Indonesian batik maker. Another paper studied the production process improvement using ergonomic checkpoints for stamping station work facilities [15]. The study named the hand tools and machine safety as the primary aspects associated with the work table; the cooker for the wax and the heavy metal stamp block that affects the craftsmen work posture. Finally, a paper by Muslimah et al. [16] which studied on green ergonomic and eco-efficiency in Solo batik industry, trying to introduce tools and work procedures to improve the Eco-Efficiency Index (EEI) of the batik workshop whilst providing comfort to the workers. The focus was particularly on the issue of wax removal workstation and replacing the furnace that uses firewood.

3.2 Batik and Ergonomic and Malaysia

When the keyword Malaysia is included, Scopus database listed five documents, however when manually scrutinized further, only two out of the five studies that are from Malaysian Batik scene; Darliana et al. [9] and Shaharuddin et al. [17]. Whereas, Google Scholar gave 300 plus papers, although when looked at closely, many studies are still from Indonesia. An engineering intervention was to help reduce lifting loads among stamping batik craftsmen [9]. The researchers studied the problem of the workshop, evaluated their problems ergonomically and provided a tool design solution. Whilst the fundamental of thermal and structural characteristics of the wax was investigated in the aim to design better *canting* tool for batik making enhancement [17].

3.3 Two Batik Artisan Case Study

The site visit was carried out in mid-September 2022. Both batik companies are situated at Kota Bharu, the capital city of Kelantan state. It was observed that as experienced and passionate craftsmen, the more senior craftsman performed the drawing and painting tasks while standing straight, relaxed and not much awkward

Fig. 1 **a** Case study 1 shows a junior batik artist drawing the pattern with *canting* **b** Case study 2 shows a senior batik artist standing upright while coloring the pattern (photos were taken during site visit September, 2022)

postures. However, the younger craftsman tends to bend so much (between 45° to 90°) at the drawing and coloring stage. Figure 1(a) showed one of the junior artist on the left versus Fig. 1(b) the experienced artist on the right, both working on their batik drawings.

Another main contributor to the awkward standing posture is that the workstation height. For batik *canting* or drawn/painted batik, the fabric is usually pinned or held to batik traditional frame (wood frame only on the sides, like the concept of ram for embroidery). The batik artisan on the left (Fig. 2 (a), although the place's size is bigger and airy (no walls)), the workstations have fixed height. On the contrary, the smaller place has a different system where the frame can be adjusted to different heights by moving the rope attached to the ram vertically along a nearby column to higher or lower height according the comfort and preference of the craftsmen (circled in Fig. 1 (b)). The column has nails or hooks at different levels to secure and stop the rope from sliding.

Both places require more lighting as it was very dim inside the workshops (less than 500 lx). The smaller workshop (Figs. 1 and 2 (b)) is really stuffy during hot season. The distances between one hanged fabric to another workstation are very close to each other, and it is a potential safety hazard.

Fig. 2 **a** Case 1 shows the workstation with fixed height **b** Case 2 shows the smaller and closed work area but with height-adjustable workstations (photos were taken during site visit)

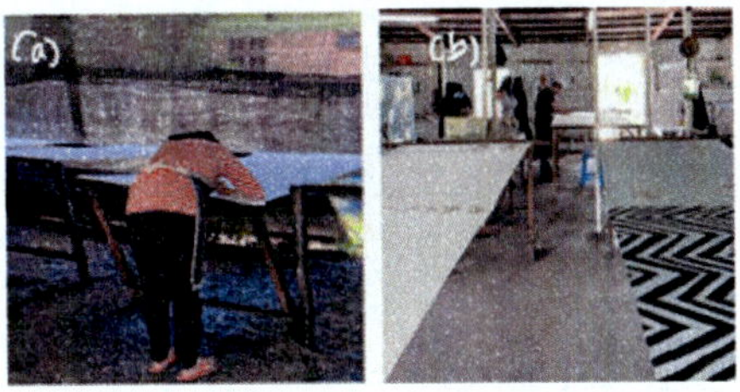

4 Conclusion

There are quite a number of ergonomics research on batik, however the majority of the studies were in Indonesia, which has different work practice from Malaysian batik industry. More investigations are needed on ergonomic at the workplace and design of the tools & equipment to close the gaps between batik industry of the two countries. Two ergonomic walkthrough observations exposed many initial problems. Further improvement approaches could be explored through ergonomic risk assessments as well as design and engineering solutions such as ergonomic batik work height/table/ seat, that ultimately will benefit the whole batik industry survival.

Acknowledgements The research was funded by the FPE Impact Grant from Foundation for Professional Ergonomics (FPE): www.ergofoundation.org.

References

1. Nordin, R., Abu Bakar, S.S.: Malaysian batik industry: protecting local batik design by copyright and industrial design laws. Int. J. Bus. Soc. **13**(2), 117–132 (2012)
2. Hitchcock, M., Nuryanti, W.: Building on Batik: The Globalization of a Craft Community. Routledge (2016). https://doi.org/10.4324/9781315261164
3. Gaffney, D.: The History of Batik. The Batik Guild. http://batikguild.org.uk/batik/history. Accessed 15 Aug 2022
4. Shaharuddin, S.I.S., Shamsuddin, M.S., Drahman, M.H., Hasan, Z., Mohd Asri, N.A., et al.: A Review on the Malaysian & Indonesian Batik Production, Challenges & Innovations in the 21st Century, pp. 1–19. SAGE Open July-September (2021)
5. Gusenbauer, M., Haddaway, N.R.: Which academic search systems are suitable for systematic reviews or meta-analyses? Evaluating retrieval qualities of Google Scholar, PubMed, and 26 other resources. Res. Synth. Methods **11**(2), 181–217 (2020). https://doi.org/10.1002/jrsm.1378
6. Arifin, A.N., Rahayu, S.R., Rustiana, E.R.: Analysis of postures towards musculoskeletal disorders experienced by batik artisans of Sukamaju Giriloyo Yogyakarta. Public Health Perspect. J. **5**(1) (2020)
7. Prabarukmi, G.S., Widajati, N.: The correlation of ergonomic risk factor with musculoskeletal complaints in Batik Workers Hubungan Faktor Risiko Ergonomi dengan Keluhan Muskuloskeletal pada Pembatik. Indonesian J. Occup. Saf. Health **9**(3), 269–278 (2020)
8. Santosa, A., Ariska, D.K.: Faktor-faktor yang berhubungan dengan kejadian musculoskeletal disorders pada pekerja batik di kecamatan Sokaraja Banyumas. Medisains **16**(1), 42–46 (2018)
9. Darliana, M., Hanisa, H., AzmulFadhli, K., Nurulahda, S., DianDarinaIndah, D.: Ergonomic engineering intervention of batik stamping work to reduce lifting load. Malaysian J. Public Health Med. **20**(Special1), 124–127 (2020). https://doi.org/10.37268/mjphm/vol.20/no.Special1/art.680
10. Yusof, N., Yusof, R., AhmatBasri, M.F., Soin, N.: Ergnomics evaluation of postural assessment among "Canting" Batik workers. Adv Eng. Forum **10**, 226–230 (2013)
11. Musa, R., Kyi, W., Rampal, K.G.: Work-related musculoskeletal symptoms among batik workers in Kelantan. Malaysian J. Med. Sci. **7**(2), 13–17 (2000)
12. Khomairah, N., Widajati, N.: The relationship of age, working periode and work attitude with complaints of carpal tunnel syndrome on workers in the Sumenep batik industry Indonesia. Indian J. Forensic Med. Toxicol. **14**(2), 1377–1382 (2020)

13. Pratiwi, I., Brillyanto, V., Fitriadi, R., Anis, M., Abdol Rahman, M.N.: Postural evaluation ad hand activity level at Batik cap process using LUBA and ACGIH HAL methods. Int. J. Recent Technol. Eng. **8**(3), 2552–2560 (2019)
14. Anjani, S., Hidayati, R., Adlan, Y.A., Suzianti, A., Hapsari, R.T.V.: Design of ergonomic stool (Dingklik) for Batik Crafters. Int. J. of Technol. **4**(3), 299–305 (2013)
15. Suhardi, B. Zoratoshi, N., Laksono, P.W., Custodio, B.: Ergonomic Checkpoints as the Base of Stamping Station Work Facilities Improvement. In: 2nd Int'l Conf. of Industrial, Mechanical, Electrical and Chemical Eng., ICIME. ISBN 978-146738504-6 (2016)
16. Muslimah, E., Abdullah, I., Suparman, S., Yanuwiadi, B., Riniwati, H.: Pendekatan Green Ergonomic di UKM Batik X. IENACO (Industrial Eng National Conference) 7, (2019)
17. Shaharuddin, S.I.S., Nurul, A.T.B.S.T., Khairusshima, M.K.N.: Thermal Modelling and analysis of batik canting design thermal modeling and analysis of batik canting design. Proc. Eng. 184 (2017)

Study of the Impact of Quantitative Lighting Parameters as a Physical Agent to the Health of Workers During the Manufacturing Process

Darina Duplakova, Jan Duplak, Ljubica Janjetovic, and Dejan Kojić

Abstract Lighting conditions in an indoor working environment are affected by two basic attributes. The first attribute is the exterior lighting conditions, which are most often provided in the workplaces by upper or side lighting openings, and the second attribute is the constructional and architectural solution of the workplace. The basis for the correct assessment of lighting conditions in the workplace is the implementation of quantitative and qualitative analyzes of lighting components. The presented article is devoted to the assessment of lighting conditions within a specific production process in the working environment of the production hall. The introductory part of the article provides a theoretical overview of the issues with the determination of regulations and mathematical expressions that are necessary for the analysis and processing of the obtained data. The main part of the article is devoted to the description of input conditions for the implementation of in situ measurements, presentation of the obtained data, their analysis and evaluation. In the performed experiment, 820 values were collected from measurements in two series, while sampling was determined in the range of 5 s. Data analysis was focused on assessing light intensity, uniformity, direction error of luxmeter, and standard deviation. For the determination of the resulting values and the subsequent determination of specific rationalization measures, the average value was determined due to the possibility of comparison with regulatory regulations and standards. The conclusion of the article summarizes the results of the research and provides an overview of specific rationalization measures.

Keywords Working conditions · Lighting analysis · Measurements · Production process

D. Duplakova (✉) · J. Duplak
Technical University of Kosice, Faculty of Manufacturing Technologies with a Seat in Presov, Bayerova 1, 080 01 Presov, Slovakia
e-mail: darina.duplakova@tuke.sk

L. Janjetovic · D. Kojić
PIM University Banja Luka, Technical Faculty, Despota Stefana Lazarevića bb, 78 000 Banja Luka, Bosnia and Herzegovina

I. Salopek Čubrić et al. (eds.), *Proceedings of the 9th International Ergonomics Conference*, Lecture Notes in Networks and Systems 701,
https://doi.org/10.1007/978-3-031-33986-8_11

1 Introduction

Lighting is one of the basic factors that affect a person's creative activity. We define lighting as the use of light to achieve the visibility of a scene, objects or their surroundings. Adequate and appropriate lighting must be provided for people to perform visual tasks efficiently and accurately. The purpose of this article is analysis of lighting conditions within a specific production process with a focus on assessing the quality parameters of lighting on the health of workers with a comprehensive summary of rationalization measures. Lighting can be provided by daylight, artificial light or a combination of both. As additional types of lighting, it is possible to identify general lighting, graduated lighting, and local lighting [1, 2].

When assessing and designing a lighting system, priority should be given to creating optimal conditions for visual performance in a particular work environment to ensure maximum visual well-being, which is defined as a psychological condition in which the visual system performs its function optimally. [3] by ensuring optimal lighting, the resolution increases, which, last but not least, affects the time of performance of the work with a significant reduction in visual fatigue. [4].

Due to basic principles of safety and health management, working conditions can be divided into material, which represent climatic conditions, equipping the workplace with machines, equipment, tools, furniture and at the same time their spatial and aesthetic solution; and social conditions, which include interpersonal relationships within the team and relationships. [5] optimal working conditions during the assessment of workplace lighting are solved by analyzes, studies, and subsequent modifications of luminaires and lighting openings, which form the main part of the illuminated spaces. [6] illuminated spaces can be divided according to functions and types of activities, which are implemented in them as follows [7]:

Workspaces: This is a general term for industrial premises where most of the main work activities associated with specific visual requirements are determined - by the type of work performed or the purpose of the premises. Areas that can be included in this classification:

- production and assembly halls,
- factory and commercial spaces,
- warehouses and halls,
- offices and administrative spaces.

Social spaces: They are used for rest, recreation, entertainment, or distraction. These are spaces such as living rooms, restaurants and the like. In these rooms, the lighting requirements are lower than for industrial premises. Lights with lower light output and warm light colour are used here. Various sports grounds, stadiums, and areas for sports events can also be considered as social spaces. Social spaces are divided into space categories IV and V, where non-shielded, partially shaded and shaded lighting is used according to the specific type of these spaces.

Other spaces: special-purpose rooms or rooms intended for minor auxiliary tasks. These are, for example, corridors, staircases, shop windows, areas for handling, etc.

2 Material and Methods

To ensure effective analysis of light parameters during the production process with regard to working conditions, the basic mathematical relationships for the subsequent implementation of the necessary calculations for the collected data were determined in the first place. The most frequently assessed lighting parameter in practice is illuminance, which is determined by a mathematical formula [8]:

$$E = \frac{CP \times \omega}{A} \quad [\text{lX}] \tag{1}$$

where

CP – candle power [lumen/steradian].

ω – solid angle [steradian].

A- area [m^2].

The average value of illuminance, from the obtained data, can be calculated using the following mathematical expression:

$$\overline{E} = \frac{\sum_{i=1}^{n} E_i}{n} \tag{2}$$

where:

$\overline{E}$ - illuminance (average value) [lx].

E_i – illuminance for i-measurement point.

n– number of a measurement point.

For the complexity of the analysis, it is necessary to determine the uniformity of the lighting space, which is determined by three basic formulas. The first formula is represented as a min–max ratio by the following equation [9]:

$$U_1 = \frac{E_{min}}{E_{max}} \tag{3}$$

The second formula is written as a min-avg ratio by the following equation [9]:

$$U_2 = \frac{E_{min}}{n \sum_{i=1}^{n} E_i} \tag{4}$$

The third interpretation is based on the context reflected in the standard error based on the implementation of local point measurements. The third interpretation of the uniformity parameter can be expressed mathematically as follows [9]:

$$U_3 = \frac{\overline{E}}{\overline{E} + \sigma_E} \tag{5}$$

where

σ_E - illuminance standard deviation.
$\overline{E}$ - illuminance (average value) [lx].

In the first part of the research, the collected data were analyzed using special software supplied directly with the instrument, which interprets the measured values in basic numerical and graphical form.

Due to the basic possibilities of the collected data interpretation by the software solution, for the needs of the obtained data complex processing, except for the graphic output, all data were exported to a spreadsheet, where they were subsequently supplemented with the necessary analyzes.

3 Analysis of Lighting Parameters During the Manufacturing Process

At the workplace of machining non-standardized moulds, the lighting conditions when working on CNC equipment were analysed. The measuring point (Fig. 1) was placed directly on the CNC panel of the machine, i.e. in the place where the worker operates for the longest time.

A luxmeter with a measurement range of up to 200,000 lx, linearity to 2%, and with a resolution of 0.01 was used for data collection. It was set the temporal measuring mode and the time interval for sampling was 5 s due to standard setting of practical sampling measurements and according to capabilities of device.

Due to the unevenness of working conditions in terms of lighting, two sets of measurements were performed at the above measuring point - the first series in the morning (during the morning work shift), the second series in the afternoon (during the afternoon work shift). The first set of measurements was specific to the lighting installed exclusively by the ceiling lights, while the second set of measurements carried out in-situ in the afternoon had, in addition to the ceiling lights, additional lighting directly on the machine. Data were collected in both sets at a time interval of 5 s. A total of 820 measurements were performed, 410 in each set.

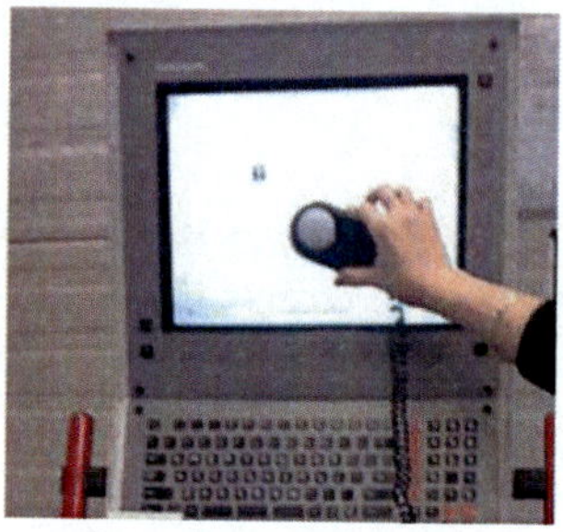

Fig. 1 Measurement point – lighting intensity in-situ measurement

The uniformity values of the workspace lighting were also determined using a specialized software solution, which was verified by mathematical calculations. The obtained data are presented in Table 1.

In the analysis of the uniformity, in comparison with the required values, an insufficient value was found in the first series of measurements within the second interpretation of the uniformity of the assessed illuminated space.

Each device that is used for measurement has standard as well as specific errors. Errors entering the measurement result were determined for the correct application output, specifically, the directional error of the luxmeter was analyzed - an error caused by inaccurate evaluation of the effects of light incident on the photometric head from a direction other than perpendicular and standard deviation. The directional error can be determined mathematically based on the mathematical formula [10]:

$$f = \frac{1}{4}\Sigma_{j=0}^{3} f\left(\varphi = j\frac{\pi}{2}\right) \tag{6}$$

where after mathematical modifications of functional integration relations the expression is formed [10]:

$$f(\varepsilon, \varphi) = \frac{Y(\varepsilon, \varphi)}{Y(0, \varphi)} - 1 \tag{7}$$

where

ε - the angle between the normal to the sensor surface and the direction of impact,

φ - the azimuthal angle,

Y - a luxmeter signal.

f - the directional error of the luxmeter.

The standard deviation evaluation was represented in the sampling interval every 12 measurements, i.e. every 60 s, concerning the number of samples. The assessment of lighting intensity and the analysis of attributes related to direct in-situ measurements were recorded in Table 2 with subsequent graphical extended interpretation (Fig. 2).

By identifying the mean values, the value of the mean of the lighting intensity averages was subsequently determined, on the basis of which it was possible to carry out the verification with the currently valid standards and regulations. For a specific production area, the condition set out in regulations and norms is to ensure a

Table 1 Uniformity determination of evaluated workspace

	The first set of measurements			The second set of measurements		
	Calculated value	Required value	Difference	Calculated value	Required value	Difference
U_1	0.87	0.7	0.17	0.92	0.7	0.22
U_2	0.69		**−0.06**	0.85		0.15
U_3	0.99		0.29	1.00		0.30

Table 2 Average values of lighting intensity during the morning work shift

Measurement range		$\overline{E}$ [lx]	σ_E	Measurement range		$\overline{E}$ [lx]	σ_E
11:01:02	11:01:57	206.12	0.82	11:18:02	11:18:57	220.03	1.29
11:02:02	11:02:57	205.10	0.82	11:19:02	11:19:57	223.13	1.04
11:03:02	11:03:57	206.18	0.46	11:20:02	11:20:57	225.40	0.87
11:04:02	11:04:57	204.01	1.44	11:21:02	11:21:57	228.31	1.13
11:05:02	11:05:57	204.43	0.74	11:22:02	11:22:57	233.36	0.99
11:06:02	11:06:57	205.89	0.44	11:23:02	11:23:57	236.78	1.22
11:07:02	11:07:57	205.88	1.03	11:24:02	11:24:57	239.27	0.91
11:08:02	11:08:57	206.15	1.24	11:25:02	11:25:57	243.65	2.02
11:09:02	11:09:57	208.39	0.30	11:26:02	11:26:57	249.95	1.45
11:10:02	11:10:57	208.67	0.48	11:27:02	11:27:57	256.05	1.97
11:11:02	11:11:57	209.89	0.96	11:28:02	11:28:57	261.26	1.52
11:12:02	11:12:57	212.53	0.35	11:29:02	11:29:57	268.62	3.16
11:13:02	11:13:57	213.70	0.66	11:30:02	11:30:57	276.91	2.28
11:14:02	11:14:57	213.62	0.43	11:31:02	11:31:57	284.62	2.08
11:15:02	11:15:57	216.24	0.81	11:32:02	11:32:57	292.45	3.10
11:16:02	11:16:57	217.12	0.38	11:33:02	11:33:57	303.46	2.56
11:17:02	11:17:57	217.18	0.64	11:34:02	11:34:57	311.77	2.96

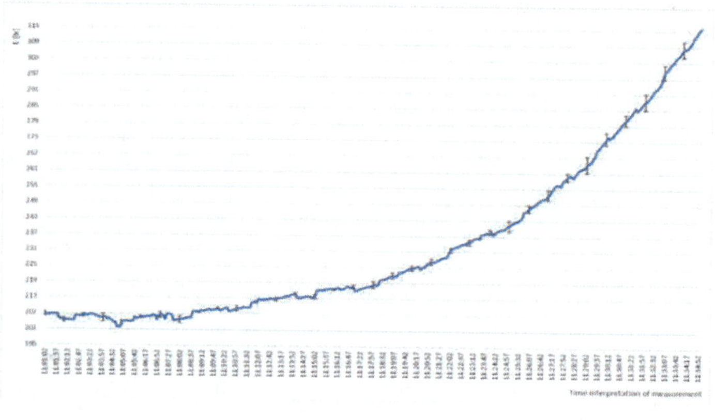

Fig. 2 Line chart of lighting intensity from in-situ measurement – morning work shift (x axe - time interpretation of measurement; y - axe lighting intensity)

minimum value of lighting intensity of 500 lx, while in the first set of measurements a mean value of 232.7 lx was determined, which does not represent half the value required to meet the conditions set out in the standards.

When assessing the second set of measurements (Table 3), the mean of the lighting intensity averages of 335.36 lx was determined in a similar way, which also represents a value significantly lower than the value normatively conditioned. Compared to the morning work shift, a slight increase in values can be observed, but this increase is due to the additional lighting equipment used by the afternoon shift.

In assessing the values for the afternoon work shift, the standard deviation and the directional error of the luxmeter were similarly identified. The standard error of the luxmeter was determined to be 1.3%, which is within the permissible error specified by the manufacturer with a value of 6%.

On the above graphical interpretation (Fig. 3) of the obtained values of in-situ measurements of lighting intensity for the afternoon work shift, it is possible to observe a bounce - a decrease in lighting intensity, which could have been caused by a stroboscopic effect due to the existence of an older type of ballast. Based on the evaluated data, conclusions were drawn with the identification of possible corrective measures, which are listed in the final part of the article.

Table 3 Average values of lighting intensity during the afternoon work shift

Measurement range		$\overline{E}$ [lx]	σ_E	Measurement range		$\overline{E}$ [lx]	σ_E
14:31:04	14:31:59	305.16	1.27	14:48:04	14:48:59	337.63	0.31
14:32:04	14:32:59	308.29	0.98	14:49:04	14:49:59	338.46	0.33
14:33:04	14:33:59	311.34	0.83	14:50:04	14:50:59	339.76	0.56
14:34:04	14:34:59	314.58	1.49	14:51:04	14:51:59	341.17	0.45
14:35:04	14:35:59	317.91	0.73	14:52:04	14:52:59	342.24	0.20
14:36:04	14:36:59	322.49	1.04	14:53:04	14:53:59	342.91	0.18
14:37:04	14:37:59	325.93	1.06	14:54:04	14:54:59	343.97	0.37
14:38:04	14:38:59	329.93	1.10	14:55:04	14:55:59	345.03	0.23
14:39:04	14:39:59	334.12	0.81	14:56:04	14:56:59	345.88	0.39
14:40:04	14:40:59	337.02	1.26	14:57:04	14:57:59	347.38	0.34
14:41:04	14:41:59	331.98	3.23	14:58:04	14:58:59	348.21	0.23
14:42:04	14:42:59	331.43	0.17	14:59:04	14:59:59	349.37	0.40
14:43:04	14:43:59	332.13	0.28	15:00:04	15:00:59	350.44	0.44
14:44:04	14:44:59	308.29	0.98	15:01:04	15:01:59	351.58	0.17
14:45:04	14:45:59	334.54	0.38	15:02:04	15:02:59	352.33	0.31
14:46:04	14:46:59	335.62	0.40	15:03:04	15:03:59	353.61	0.40
14:47:04	14:47:59	336.77	0.25	15:04:04	15:04:59	354.67	0.26

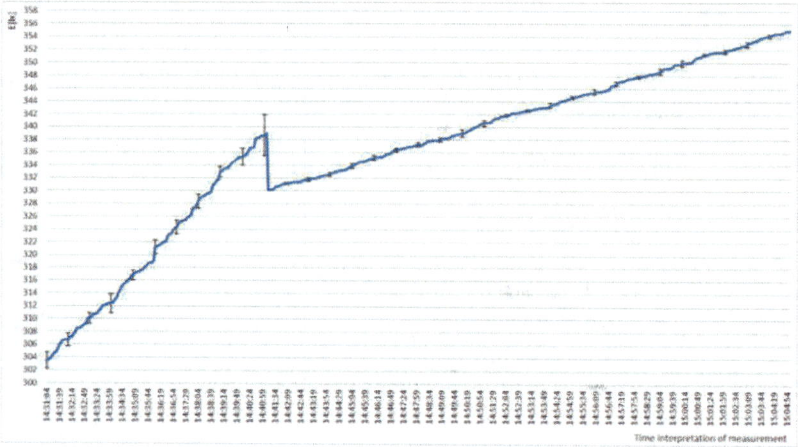

Fig. 3 Line chart of lighting intensity from in-situ measurement – afternoon work shift (x axe - time interpretation of measurement; y - axe lighting intensity)

4 Conclusion

The field of lighting technology is currently expanding in the context of addressing cost-effective industrial measures related to digitization. Outdated lighting systems adversely affect the health of employees, which also has a negative impact on production. The analysis of the conditions of the machining process of non-standardized moulds showed that:

- the values of lighting intensity analysed in both series of measurements do not comply with the required values stated in the standards,
- the identification of a stroboscopic phenomenon of artificial lighting, which has a negative impact on the health of workers,
- low maintenance of installed lighting system, which affects the intensity of lighting,
- the use of outdated lighting system technology, which has a significant impact on energy and economic attributes.

To eliminate the above negative effects, it is necessary to ensure:

- supplementation of additional lighting sources to ensure an increase in the values of lighting intensity and at the same time to ensure compliance with the required values specified in the standards,
- replacement of current lighting sources to eliminate the stroboscopic phenomenon,
- implementation of new SMART technological solutions for lighting systems to increase energy and economic efficiency,
- creation of a maintenance system for the lighting system,

- creation of a new suitable layout solution for a new rationalized lighting source.

When renovating lighting systems, it is generally necessary to proceed in accordance with the established norms, standards and methodology, to subsequently avoid further rationalization needs in the given issue. This analysis has the potential to further explore the possibility of creating a methodology for assessing primarily SMART technological solutions for lighting systems with the provision of sustainable, energy and economically efficient production.

Acknowledgements *This article was supported by research grants VEGA 1/0431/21 and KEGA 018TUKE-4/2021.*

References

1. Budak, V.P., Aizenberg, J.B.: The light field and the scope of light science. Light Eng. **29**(1), 4–10 (2021)
2. Zhang, R., Yang, Y., Fang, Q., Liu, Y., Zhu, X., Wang, M., Su, L.: Effect of indoors artificial lighting conditions on computer-based learning performance. Int. J. Environ. Res. Public Health **17**(7), 1–11 (2020)
3. Kocisko, M., Teliskova, M., Baron, P., Zajac, J.: An integrated working environment using advanced augmented reality techniques. In Proceedings of the 4th International Conference on Industrial Engineering and Applications (ICIEA), Nagoya, Japan, pp. 279–283 (2017)
4. Tureková, I., Lukáčová, D., Bánesz, G.: Quality assessment of the university classroom lighting-a case study. TEM J. **7**(4), 829–836 (2018)
5. Zulkarnaen, A.H.: Implementation of safety and health management system in a fire factory in Kosambi Tangerang. SAR J.-Sci. Res. **1**(4), 128–134 (2018)
6. Purba, M., Purba, A.P., Purba, A.M.: Employer perceptions of job skills in manufacturing industry. SAR J.-Sci. Res. **3**(4), 178–182 (2020)
7. Burak, S., Veljovic, F.: Ergonomic analysis and redesign of workspace in order to minimize workers' workload and optimize their nutrition. TEM J. **8**(2), 572–576 (2019)
8. Sitamarhi Insitute of Technology: Fundamentals of Illumination. https://www.sitsitamarhi.ac.in/wp-content/uploads/2020/04/file_5e8ef00b06190.pdf
9. Hu, Y., Luo, M.R., Yang, Y.: A study on lighting uniformity for LED smart lighting system. In: 2015 12th China International Forum on Solid State Lighting (SSLCHINA). IEEE, pp. 127–130 (2015)
10. ISO/CIE 19476:2014 Characterization of the performance of illuminance meters and luminance meters

Qualitative Assessment of the Occupational Health and Safety Knowledge Management Practices of Hungarian Companies

Ferenc Faragó and Gyula Szabó

Abstract In light of the accident indicators, the necessary and appropriate knowledge is questionable, as is the efficiency of knowledge sharing and knowledge management related to occupational health and safety. This research aims to examine the adequacy of occupational health and safety knowledge and the factors influencing the effectiveness of occupational health and safety education. We explore the typical problems of knowledge sharing and the gaps in the knowledge management of companies related to occupational health and safety. Based on in-depth expert interviews, we examined occupational safety knowledge management of medium-sized and large companies at a micro level among Hungarian occupational safety specialists and managers. We used the grounded theory method to analyze the interviews. The research pointed out that knowledge as such is a determining but not sufficient factor in companies' occupational health and safety performance. However, applying knowledge determines success, and the corporate culture encourages it. Based on the results of our study, we created a theoretical model of effective occupational health and safety knowledge management. The limitations of our study are the small sample size and the fact that the research only covered Hungarian companies. We propose further research to validate the model and to find a possibility to measure employees' occupational safety knowledge.

Keywords Occupational Health and Safety · Knowledge Management · Value Creation · Business Success Factors · OHS Performance

1 Introduction

The process began around the turn of the millennium, during which the book value of the company's assets and the non-financial items that was not reflected in the accounting but significantly modified the company's value and, thus primarily the

F. Faragó (✉) · G. Szabó
Óbuda University, Budapest, Hungary
e-mail: farago.ferenc@uni-obuda.hu

I. Salopek Čubrić et al. (eds.), *Proceedings of the 9th International Ergonomics Conference*, Lecture Notes in Networks and Systems 701,
https://doi.org/10.1007/978-3-031-33986-8_12

company's intellectual capital, were increasingly separated in the field of management research [1]. Among the macro factors that determine competitiveness, cultural factors have become essential, in which Chikán and Czakó have already emphasized the importance of knowledge and the role of knowledge [2]. Recognizing the value of intellectual capital and knowledge and understanding its competitive advantage led to the development of knowledge management techniques. Research in this area has identified three main components of intellectual capital: human capital, which represents the competence of employees; organizational capital, which is created as a result of intellectual work and appears in the form of a product or service but also includes, for example, corporate culture, and customer capital, which includes the company's customer base.

In this situation, the value of knowledge, the recognition of knowledge, and the importance of managing knowledge have increased tremendously [1]. Researchers created the basic concepts related to knowledge (data, information, knowledge), and knowledge management was defined. According to Szeleczki, knowledge management is a management approach that manages different forms of knowledge to provide a competitive advantage/business value for a given organization [1].

As antecedents to the development of knowledge management, Klimkó distinguishes three main areas: the resource-based approach to corporate competitiveness, organizational learning, and the measurement of intellectual capital [3]. The recognition and interpretation of knowledge and its management as a corporate resource created the necessity of knowledge management [4, 5]. The goal-oriented, learning-oriented, process-oriented and technicist approaches were distinguished as the main trends of knowledge management [3].

Organizational knowledge gained a significant role, which companies viewed as strategic capital and one of the determining factors of business success. Success requires conscious and efficient creation, collection, accumulation, sharing and application of organizational knowledge [6]. The development and application of effective knowledge management practice and its measurement and evaluation required the development of methodologically sound procedures. Knowing the processes of knowledge acquisition and sharing has greatly helped companies design and implement knowledge management consciously. Anklam distinguished three generations of the development of knowledge management [7].

Information is widely available with the development of information technology, primarily with the rapid expansion of the World Wide Web; as Szeghegyi writes, the challenge today is not to access information but the effective mapping, use and sharing of available data, information and knowledge [8].

Similarly, occupational health and safety contribute to effectiveness; however, knowledge sharing could be more emphasized. There is an aspiration for a high level of occupational health and safety performance [9], but the accident statistics are depressing, affecting 2.4% of the employed in the EU [10]. In Hungary, employers reported 9,776 accidents in the first half of 2022, which is 3.5% higher than in the same period of the previous year [11]. Work-related accidents and health problems resulting from occupational diseases have a detrimental effect on work efficiency. They significantly contribute to absenteeism and therefore represent a severe problem for employees and employers [12] [13]. Inadequate workplace safety and health protection practices result in higher costs resulting from sick leave, lower work quality and lost production, i.e. lower efficiency.

The regulations related to occupational health and safety determine employers' information and education obligations. For example, the Occupational Safety and Health Act [14] mentions the employers' education obligation in at least eight paragraphs. Accordingly, the employer is obliged to educate the employees:

- the conditions of safe and non-threatening working conditions,
- the dangers and how to protect against them,
- the employer's occupational health and safety rules,
- the rules for the use of personal protective equipment,
- all the knowledge necessary to perform the work safely.

The OSH training must include the results of the risk assessments and accident investigations, the modification of the risk assessment, and the corrective measures [15].

According to Hungarian regulations [14], more than one-time education is needed, and intelligent knowledge management can ensure the maintenance of essential knowledge during the entire period of employment.

The Occupational Safety and Health Act requires occupational safety specialists to participate in education [14]. In contrast to the risk assessment mentioned in the same place of the law, training doesn't require particular OSH competency since the sole responsibility of the occupational safety specialist is the composition of the training content.

While operating the occupational health and safety management system (OSHAS), the company also creates an educational plan, which already puts occupational health and safety knowledge sharing into the system. However, this doesn't mean assessing the effectiveness and efficiency of education, so its added value to performance remains unclear. The number of training, as a leading indicator in occupational health and safety performance measurement, only rarely appears [9]; this may indicate that companies need to take advantage of the added value of knowledge to company performance.

In occupational health and safety management, researchers often use quantitative research to examine the numerical information in key performance indicators designed for occupational health and safety education. Still, these do not provide an opportunity for a deeper understanding of problems, processes, and organizational and employee (personal) behaviour [16–18]. For a complex interpretation of the problem, we found it necessary to investigate using a qualitative research method to learn about the organizational processes and human phenomena typical of corporate occupational health and safety education to collect and analyze information on attitudes and effectiveness. According to Lampek and Horváthné [19], qualitative studies provide an opportunity for a thorough, in-depth assessment of the researched area and, usually, a small sample used, in which representativeness is not the goal.

One of the essential tasks of qualitative research is highlighting individual characteristics [16]. According to Sántha, qualitative analysis is instrumental if it can provide additional information generated by the unexpressed language level [20].

Kovács described in interview methods and techniques [21] the atmosphere of such an in-depth interview conversation is open, supportive, and free of manipulation [16]. As Kelemen points out, according to Seidman [22], the purpose of interviewing is to understand other people's experiences and how these experiences are interpreted [16].

The objective of this research was to examine the factors determining the effectiveness of occupational health and safety knowledge management, the expectations of companies regarding occupational health and safety knowledge, the impact of knowledge on company performance, and the adequacy of the training material at all levels of the organization. We consider knowledge is an interpreted, subjective form of information, which includes the ability to understand, apply and use it [1].

This paper wants to answer the question of how the necessary knowledge should be adapted to each employee group's level so that they can adequately understand and utilize the information provided during their work.

2 Methodology

During the research, we used expert interviews of people who know and apply occupational safety processes and are responsible for the data, typically the occupational safety specialists of the companies or the managers of the given area. Accordingly, we conducted the research with the heads of the specialized fields (occupational safety specialists and EHS managers) in the first semester of 2022.

The criteria for selecting the interviewees was that, based on their education and work, they had basic professional knowledge, professional skills, theoretical and practical knowledge, and personal experience concerning domestic, occupational health and safety practice. We contacted the interviewees through personal acquaintances and professional relationships. We made ten inquiries in April and May 2022, completing seven interviews. The interviews varied between 55–85 min, but 75 per cent were at least one hour long. The interviews were conducted in person or online, anonymously, taking data protection regulations into account. We recorded the sessions only for this research.

We interviewed a total of seven people, four men, aged between 34 and 56 years old, who had a secondary or higher occupational safety education and worked as an occupational safety specialist, EHS manager or occupational safety inspector.

During the interviews, we asked the experts the following questions:

Q1: In your opinion, how important is occupational health and safety knowledge in companies?

Q2: Does and to what extent does occupational safety knowledge influence the organization's occupational safety performance?

Q3: Do the employees at your company have adequate knowledge?
Q4: Do you consider the domestic practice of occupational health and safety education adequate?

To analyze the interviews, we used scientific analytical methods, the content analysis of the reports and the grounded theory method, with combined categorization, corresponding to a mixed methodology [20, 23]. We used report condensation to transform and categorize the main points of the interviews into short forms. After coding, the long texts of the interviews were structured into a diagram using word cloud creation software. The applied qualitative content analysis procedures, deductive and inductive category formation and coding ensure the scientific nature of the research [16, 17, 20].

We used Wordclouds.com free online word cloud generator, to perform a word cloud method to process the interviews and content analysis.

3 Results

Overall, based on the interviews, occupational safety experts unanimously consider knowledge related to occupational safety as fundamental. However, according to experts, knowledge is an essential but insufficient condition for effective company occupational health and safety. The decisive success factor is knowledge application, which requires adequate knowledge to use during work at all levels of the company, as well as a corporate culture that supports and encourages the application of knowledge.

In connection with occupational health and safety knowledge transfer and knowledge application, our experts highlighted the qualities and attitudes of the employee brought and acquired. They emphasized in the interviews that education can expand the acquired qualities. Still, it is imperative to influence the attitude because the individual's behaviour has multiple impacts on the result. Transcripts and extracts of the interviews are available from the authors upon request.

According to the unanimous opinion of the interviewed experts, the employees do not have the necessary knowledge for safe work. Even though training occurs at all companies, occupational safety knowledge transfer efficiency is low. The main reason is that employers need a well-thought-out training strategy for sharing occupational health and safety knowledge; there is no knowledge map, and they need to assess the need for knowledge too. This way, adequate, differentiated knowledge does not develop within the organization. Individual attitude on the employee side is typically not safety-oriented, making knowledge transfer difficult. Late safety-conscious education hinders the assimilation of knowledge. Occupational safety specialists must have adequate pedagogical and educational competencies and should create differentiated training.

According to our experts, the necessary occupational health and safety knowledge is absent; accordingly, knowledge application does not take place to the desired extent, which also determines the companies' pursuit of occupational health and safety.

Based on the research results, we created a possible theoretical model of effective knowledge management, illustrated in Fig. 1. Elements of the occupational health and safety knowledge management model:

- a corporate culture that actively supports the pursuit of safety
- risk assessment providing adequate information and risk management procedures aimed at reducing risks
- knowledge map, knowledge management aiming at the transfer of practical, differentiated knowledge and knowledge application
- control and feedback in the direction of risk management for the continuous development
- an early education system that develops employees' attitudes
- vocational training ensuring the pedagogical and educational technical competencies of occupational safety specialists

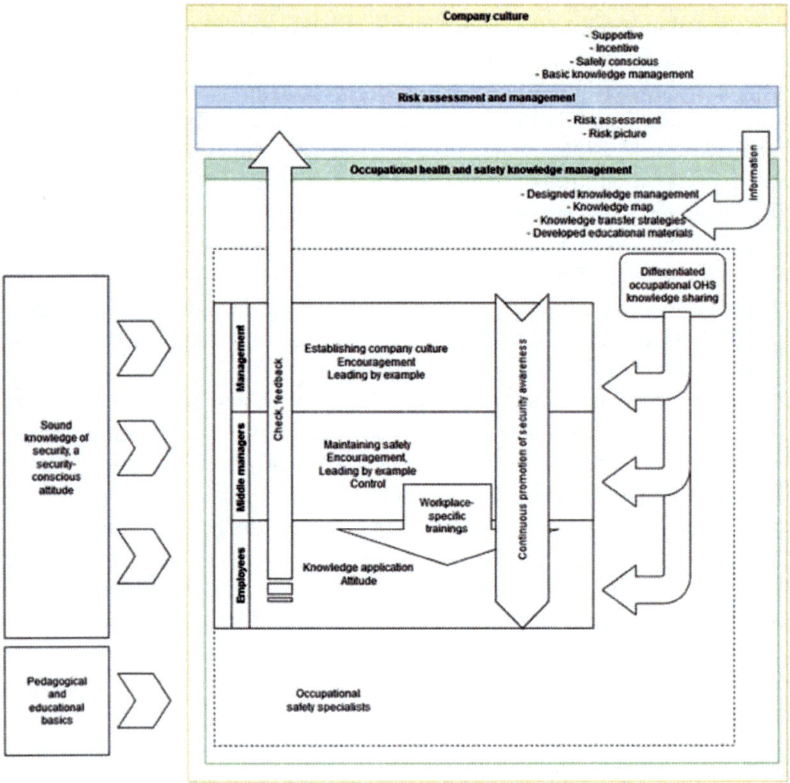

Fig. 1 The occupational health and safety knowledge management model

The limitations of our research are the small sample size and the geographical boundaries of the study.

We propose further research to validate the theoretical model of occupational health and safety knowledge management, to learn about the expectations and strategic thinking of managers representing the employer concerning occupational health and safety, and to determine the attitude of employees toward safety. It is necessary to examine the company's practice of occupational safety knowledge sharing, as well as the possibilities of measuring knowledge.

4 Conclusion

This research aimed to assess Hungarian companies' occupational health and safety knowledge management practices, examine the adequacy of occupational health and safety knowledge, and determine the factors influencing training effectiveness. Several factors influence the success of companies' work safety value-creating activities. In the present research, we studied the adequacy of occupational safety training and the impact of individual and organizational knowledge related to occupational safety on occupational safety performance.

Healthy and safe workplaces include adequate occupational safety and health capabilities for the given situation at all levels and in all company parts. The source of knowledge is primarily risk assessment, while knowledge management reflects culturally determined solutions that already work at the company.

Following safety culture and behaviour-based OSH approaches, we can assign some basic OSH activities and skills to different company levels.

Our research partially confirmed our assumption that occupational health and safety knowledge is a determining element of effective corporate occupational health and safety. We propose identifying organizational factors determining the effectiveness of occupational health and safety knowledge management.

Our study suggests that the knowledge of employees regarding occupational health and safety is of fundamental importance from the point of view of the companies' occupational health and safety efforts. However, more than acquiring OSH knowledge, success requires applying a knowledge management system. The results of the qualitative research pointed out the typical problems of knowledge sharing, the solution of which can make the knowledge management of companies related to labour protection more effective.

Based on the results of the investigations, we formulated the theoretical model of effective occupational health and safety knowledge management, defined the further possible research directions and proposed the expansion of the research.

References

1. Szeleczki, Z.: A tudásmenedzsment koncepciója és háttere. Vezetéstudomány, **XXX**(12), 22–30 (1999)
2. Chikán, A., Czakó, E.: Kis nemzetgazdaságok versenyképessége a globális gazdaságban - a magyar példa. In: Chikán, A., Czakó, E., Zoltayné, P.Z. (eds.) Vállalati versenyképesség a globlizálódó magyar gazdaságban, pp. 27–39. Akadémia Kiadó, Budapest (2002)
3. Klimkó, G.: A tudásmenedzsment megközelítési módjai. Vezetéstudomány **XXXII**(4), 14–20 (2001)
4. Winter, S.G.: Knowledge and competence as strategic assets. In: Teece, D.J. (ed.) The Competitive Challenge: Strategies for Industrial Innovation and Renewal, 1st edn., pp. 159–184. Ballinger, New York (1987)
5. Liebeskind, J.P.: Knowledge strategy and theory of the firm. Strategy Manage. J. **17**, 93–107 (1996)
6. Gaál, Z., Szabó, L., Obermayer-Kovács, N.: Tudásmenedzsment-profil érettségi modell. Vezetéstudomány **XI**(6), 2–15 (2009)
7. Anklam, P.: The Social Network Toolkit: Building Organisational Performance through Collaborative Communities. Ark Group, London (2005)
8. Szeghegyi, Á.: A tudásmenedzsment stratégiai szerepe a vállalatoknál. Vállalkozásfejlesztés a XXI. században, 53–68 (2011)
9. Faragó, F.: Munkavédelmi teljesítménymérés és szervezeti kultúra kvalitatív felmérése közép- és nagyvállalati környezetben. Biztonságtudományi Szemle. **4**(1), 131–141 (2022)
10. Eurostat Homegpage, 2.4% of EU people reported at least one work accident in 2020. https://ec.europa.eu/eurostat/en/web/products-eurostat-news/-/ddn-20211012-2. Accessed 10 Sept 2022
11. Innovánciós és Technológiai Minisztérium Munkavédelmi Irányítási Főosztály, "Tájékoztató a munkabalesetek alakulásáról a feldolgozott munkabaleseti jegyzőkönyvek alapján," Budapest (2021). http://www.ommf.gov.hu/index.php?akt_menu=223. Accessed 04 Sept 2022
12. European Agency for Safety and Health at Work: The value of occupational safety and health and the societal costs of work-related injuries and diseases. Report (2017)
13. Dorman, P.: The Economics of Safety, Health, and Well-Being at Work: An Overview. InFocus Program on SafeWork, International Labour Organisation (2000)
14. Jogtár Homepage, "1993. évi XCIII. törvény a munkavédelemről." https://net.jogtar.hu/jogszabaly?docid=99300093.tv. Accessed 25 Sept 2021
15. Szabó, G.: The characteristics of industrial safety risk management. In: Arezes, P., Boring, R. (eds.) Advances in Safety Management and Human Performance, AHFE 2020. Advances in Intelligent Systems and Computing, vol. 1204, pp. 47–52. Springer, Cham (2020). https://doi.org/10.1007/978-3-030-50946-0_7
16. Kelemen-Erdős, A., Ariel, M.Z.: Tematikus szolgáltatásélmény art- és romkocsma környezetben. Turisztikai és Vidékfejlesztési Tanulmányok **2**(3), 58–73 (2017)
17. Kelemen-Erdős, A., Molnár, A.: Cooperation or conflict? the nature of the collaboration of marketing and sales organizational units. Econ. Cult. **16**(1), 58–69 (2019)
18. Kelemen-Erdős, A.: Dead-end development or real progress? paradigm shift initiatives in marketing theory. In: International May Conference on Strategic Management, vol. XV, no. 1, pp. 26–38 (2019)
19. Boncz, I.: Kutatásmódszertani alapismeretek. Pécsi Tudományegyetem Egészségtudományi Kar, Pécs (2015)
20. Sántha, K.: Kvalitatív tartalomelemzés.Eötvös József Könyvkiadó, Budapest (2022)
21. Szabadbölcsészet Homegpage, Feischmidt, M.: Kvalitatív módszerek az empirikus társadalom és kultúrakutatásban. (2006). http://mmi.elte.hu/szabadbolcseszet/mmi.elte.hu/szabadbolcseszet/index72c4.html?option=com_tanelem&id_tanelem=835&tip=0. Accessed 10 Mar 2022
22. Seidman, I.: Az interjú mint kvalitatív kutatási módszer. Műszkai könyvkiadó, Budapest (2002)
23. Neulinger, Á.: Több-módszertanú és vegyes módszertanú kutatások. Vezetéstudomány, **XLVII**, 63–66 (2016)

Powered Smart Textile-Based Exoskeleton for Human Support Movement

César Ferreira, Aureliano Fertuzinhos, Ricardo Silva, Miguel Ramalho, Bruno Vale, João Silva, Luani Costa, Cristina Oliveira, Ana Ramôa, Fabiana Aguiar, André Pilastri, Arthur Matta, Paula Dias, Rosane Sampaio, Dário Machado, Marta Costa, Ana Roças, Pedro Madureira, Juliana Moreira, João Rui Pereira, Carla Pereira, and Fernando Bruno Pereira

Abstract Industrial environments require people's physical efforts related to excessive force, poor posture, and repetitive tasks that increase the risk of developing musculoskeletal diseases. Due to the high prevalence of postural complaints and consequent global health burdens, a significant demand has been conducted to build and implement powered lower and upper limb devices for human enhancement, considering the working layout/position, anthropometric and ergonomic data to generate valuable information for posture correction guidance.

STVgoDigital aims to support repetitive seamstress movements by developing a textile-based exoskeleton that reduces physical efforts, unnecessary movements, fatigue symptomatology, and pain. The main goal of STVgoDigital is to support the

C. Ferreira · A. Fertuzinhos (✉) · R. Silva · M. Ramalho · B. Vale · J. Silva
CeNTI – Centre for Nanotechnology and Smart Materials, V. N. Famalicão, Portugal
e-mail: afertuzinhos@centi.pt

L. Costa · C. Oliveira · A. Ramôa
CITEVE – Centro Tecnológico das Indústrias Têxtil e Vestuário de Portugal, V. N. Famalicão, Portugal

F. Aguiar · A. Pilastri · A. Matta · P. Dias · R. Sampaio · D. Machado
Associação C.C.G /ZGDV Centro de Computação Gráfica, Guimarães, Portugal

M. Costa · A. Roças
LMA – Leandro Manuel Araújo, S.A., Santo Tirso, Portugal

P. Madureira · J. Moreira
Fraunhofer Portugal AICOS, Porto, Portugal

J. Moreira
Center for Rehabilitation Research – Human Movement System (Re)habilitation Area, Department of Physiotherapy, School of Health, Polytechnic of Porto, Porto, Portugal

J. R. Pereira · C. Pereira · F. B. Pereira
PAFIL, V. N. Famalicão, Portugal

I. Salopek Čubrić et al. (eds.), *Proceedings of the 9th International Ergonomics Conference*, Lecture Notes in Networks and Systems 701,
https://doi.org/10.1007/978-3-031-33986-8_13

transition to Worker 4.0 generation through the development of an exoskeleton that integrates a sensing component that monitors workers' movements in real-time, identifies their ergonomic posture, and feeds the actuation component. This exoskeleton will support the execution of movements inherent to some professional activities.

The approach identified specific worker needs (physical and cognitive ergonomic, thermic, and sensorial comfort), which were later used to design the best structure for the exoskeleton. The active systems will be connected to a global communication device responsible for providing the respective outputs. This solution will be tested on-site, measuring the response of electromyography sensors to overcome physical discomfort and alignment issues with the human body anatomy, which are typical tradeoffs of these innovative products. STVgoDigital aims to adapt to flexible, dynamic, and data-enriched environments, contributing to greater integration of an IoT ecosystem in a garment.

Keywords Exoskeleton · Sensing · Actuation · Muscles · Textile garments

1 Introduction

The European Textile Industry includes 171.000 enterprises employing 1.7 million people [1]. STVgoDigital Worker 4.0 proposes an innovative solution aiming to improve safety, comfort, and to support the repetitive movement of textile operators in their workspace with a flexible and comfortable device. Many efforts have been conducted to develop and optimize processes for increasing productivity and product quality, reducing production costs, and supporting human working tasks. Nowadays, this is possible with the rise of IoT technologies (Internet of Things) applied to exoskeletons for the trunk and upper limbs.

One of the most advanced technologies in this field is a shoulder and elbow positioning device that includes sensing, actuation, and control electronics, developed by NASA, which allows elbow flexion/extension with and without robotic-assisted shoulder stabilization [2]. Another example is the Ekso by Ekso Bionics [3] designed to help the shoulder and arm in the rehabilitation of patients suffering from muscle weakness. More recently, a new device Myoshirt [4], which includes pneumatic actuation and shape memory alloys (SMAs), in a flexible and wearable exoskeleton, has been constructed at ETH Zurich to assist shoulder abduction and horizontal flexion/extension movements. This exoskeleton was implemented with a Deep Learning model (LSTM) that has shown promising results in anticipating risk movements [5].

As for passive-acting solutions for the lumbar region, there are several alternatives in terms of rehabilitation, postural correction, sports, and well-being. Back Braces stand out commercially, including postural correction vests, lumbar, and sacroiliac belts. Even so, there is still the opportunity for the development of innovative solutions capable of responding properly in short periods.

Textile Industry is one of the most affected sectors due to the repetitiveness and precision work performed by seamstresses in their daily activities. Long work turn duration, sustained low-level work, and precise handwork are the main risk factors for neck-shoulder muscular complaints. During sustained and/or repetitive muscle contractions, muscle fatigue can be perceived not only by applying a surface electromyogram but also by the qualitative feedback of the workers that are much more sensitive to proprioception changes. In a previous study, a total of 18 healthy seamstresses pointed out that sewing operations are related to high sustained static load on neck-shoulder muscles and several risk factors can be observed after 200 min of working operation [6]. In this research, an ergonomic and biomechanical study in a factory environment was performed to determine the positioning of the actuation system in the shoulder and lumbar regions in flatlock, lockstitch, sewing, and tapping activities to assist seamstress movements correctly.

2 Methodology

2.1 Ergonomic Study

A pilot study was carried out to observe employees performing their activities in a daily work rhythm. Body mass index (BMI) was also calculated during data collection. The electromyography activity of the main muscles was captured using the Bitalino MuscleBIT BT equipment, with Bluetooth communication. The software measured the muscular load during a certain task by applying the Amplitude Probability Density Function method. Employees were observed anteriorly, laterally, and posteriorly, with video and photographic records. Critical segments from an ergonomic point of view were identified by applying ISO 11226 and REBA standards.

2.2 Biomechanical Study

The maximum force intensity of the shoulder muscles during seamstress activities was calculated from 9 eligible participants. These values correspond to the required forces to hold the arm in the extended position near the horizontal. This analysis includes gravity effects, reaction forces at the glenohumeral joint, and the forces at the left and right deltoid muscles.

2.3 *Prototyping*

2.3.1 Actuation Axes

The artificial muscles would provide proprioceptive information to the superficial muscles of the body. As such, they must support the supraspinatus and infraspinatus muscles. The first pneumatic actuator would be at the glenohumeral joint, which will act on the anteroposterior axis, supporting abduction in the frontal plane; the second one will act along the scapula, in the longitudinal axis, supporting rotation in the transverse plane. In total, 4 pneumatic actuators will be placed in a Y-shaped polycarbonate structure that will support the entire exoskeleton.

2.3.2 Actuation Elements

The electrovalves and cables were evenly distributed throughout the structure. An orientation path was created for each of the independent actuators of the shoulder to increase muscular proprioception, ensuring greater compliance with the user's movement, as well as an elbow pad to fix the pneumatics distally on the arm.

The electronic modules were positioned on the back of the chair where the seamstress will be, freeing the central and lumbar areas from excessive mechanical efforts. In addition, a lumbar support structure was designed to adjust the flexion/extension and rotation of the lumbar segments.

2.3.3 Exoskeleton Structure and Materials

Based on the requirements of the sensing and actuation components, which must have very precise locations about the body's operator, a wearable exoskeleton was developed to ensure the physiological comfort of the operator, using concepts of body mapping and made with technical knitted and woven textiles with properties such as soft touch, lightness, elasticity, moisture management, and easy drying properties, breathability, dampening of rigid elements, at the same time presenting resistance to the functioning of the active elements.

3 Architecture

The proposed exoskeleton system architecture, presented in Fig. 1, is divided into three core activities, one focused on real-time sensing of human body posture, and the others are the data processing platform with risk evaluation algorithms that activate the supporting musculoskeletal actuation system.

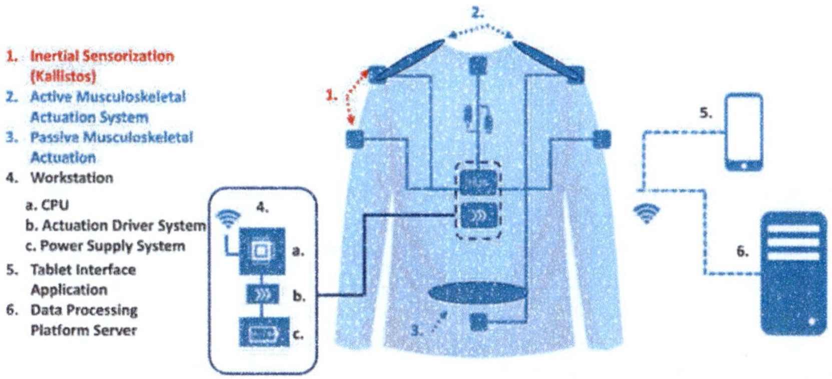

Fig. 1 The exoskeleton system architecture

Real-time human body posture sensing is based on six inertial sensors, integrated into KALLISTO©'s (developed by Fraunhofer AICOS) modules. Data regarding the movement and position of workers is collected and communicated to the data processing platform via magnetic USB.

The data processing platform is composed of three parts: a local mini-CPU, a data processing platform server, and a monitoring interface application. The local mini-CPU is responsible for collecting data from the sensing system and locally processing it to generate outputs (based on dedicated intelligent algorithms) to activate the actuation system according to current worker needs. The second stage of the data processing platform is the creation of specific risk evaluation algorithms based on the sensing data (about ergonomic posture) collected during the worker process.

The actuation system unit will have two main functions to develop activities in the longitudinal and transversal planes, according to orders of action occurring from the processing of data algorithms. This enables individualized control of each exoskeleton actuator muscle.

Pneumatic artificial muscles (PAMs) were developed to relieve the operator's muscles. PAMs applied in this project are called McKibben's muscles. These consist of an inflatable bladder and an inelastic braid with constricted endings. These constrictions imply that the bladder will only expand radially, and the braid will transduce this movement in axial movement. The applied muscles used a braid of 24° with a nominal diameter of 25 mm and the working pressure set to 0,8 bar.

4 Results and Remarks

At this point the flexible and wearable exoskeleton is composed of 4 layers: (1) a textile structure in direct contact with the seamstress; (2) a comfort layer, which is a 3D textile; (3) a semi-rigid structure, in polycarbonate, to support the actuation systems; (4) a textile to adjust to the body. With the HUB system fully encapsulated, it will be possible to detach the entire Y-structure together with the actuators to sanitize the textile layers.

The angles in Table 1 refers to those of the reaction force at the joint in relation to horizontal axis. Using mean values of the mass and the radius of rotation as a percentage of the length of each segment, the muscular strength in the shoulder was calculated [7]. The deltoid muscle force ranged between 154 and 376 N, as shown in Table 1. Decreasing the shoulder elevation angle about the trunk, there is also a reduction in the force needed to support the arm, ensuring that the actuators support a percentage of the effort promoted by the employee.

According to ISO 11226 and REBA standards, sewing and tapping activities were the most prominent to develop muscular complaints, especially in the arm and wrist (REBA score 9). Workers from these operations also gave some feedback on shoulders pain at a certain point, elbow constraints, and general muscle tingling. Abduction and elevation movements must be highlighted. The ergonomic risk is increased due to numerous repetitions of the same movement, which at the end of the day (8 h shift) may cause fatigue. To better understand the actuator's behavior used in this work, a setup (Fig. 2) was prepared to apply different known loads to the muscles.

The pressure was regulated and stabilized using a manual pressure regulator with a scaled barometer. Another experiment consisted of actuating the muscle with a 6 kg load at different pressures. The results of these experiments are presented in Fig. 3.

The experiments have shown that the muscle's maximum displacement is 10 cm and its behavior is much more dependent on the weight than on the applied pressure. For the actuating pressure of 0,8 bar, the actuator's behavior seems to not be affected.

Another important factor to evaluate is the lifetime of the actuators to ensure that they have the required resistance to withhold the number of cycles. Thus, each

Table 1 Forces at the deltoid muscles and reaction forces at the glenohumeral joints

Participants	Estimated Arm and Forearm Mass (kg)	Deltoid Muscular Force (N)		Glenohumeral Joint Reaction (N)		Angle (°)	
		LEFT	RIGHT	LEFT	RIGHT	LEFT	RIGHT
P1	3,0	225,7	239,5	222,0	235,7	1,000	1,000
P2	2,3	164,6	156,4	161,8	153,8	0,053	0,046
P3	4,0	228,5	264,6	224,4	260,1	0,021	0,045
P4	5,1	357,1	376,1	351,1	370,0	0,053	0,060
P5	2,5	187,8	163,5	184,7	160,7	0,061	0,041
P6	3,3	208,1	268,3	204,5	264,1	0,036	0,072
P7	2,9	226,0	213,5	222,4	210,0	0,066	0,059
P8	2,8	190,4	183,6	187,2	180,5	0,047	0,041
P9	2,3	154,1	174,4	151,5	171,6	0,046	0,063

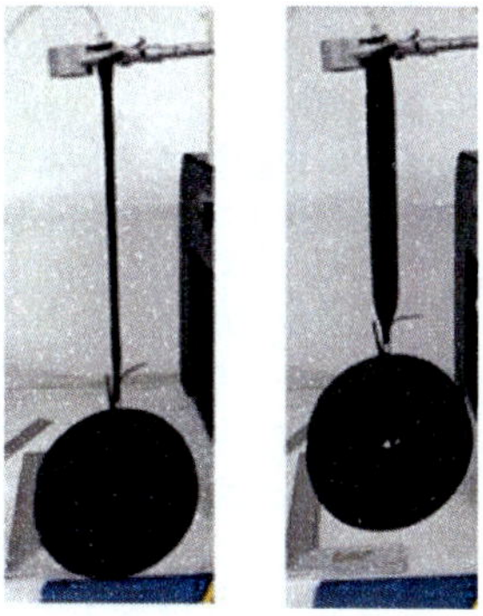

Fig. 2 Setup to evaluate PAM performance

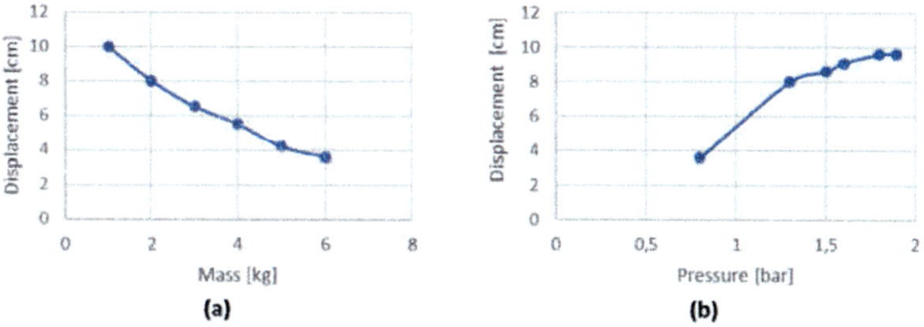

Fig. 3 Displacement measured according to applied load at constant pressure of 0,7 bar (a) and to applied pressure with constant load of 6 kg (b)

actuator was submitted to continuous lifting and releasing of a 2 kg free load. Table 2 shows the results of the actuator's resistance up to 370 cycles.

Comparing this solution with commercial ones that use electric motors and mechanical force transfer systems, despite having a slower response, less versatility to displacement adjustments, and the need to use compressed air, this solution stands out for its lower weight, volume, and greater conformity with the natural movement of the arms. This textile exoskeleton is disruptive compared to current solutions because of its modularity which allows cleaning and sanitizing after use. Furthermore, the modular components will permit some adjustment to the distance

Table 2 Results of the fatigue experiment and final state of the PAMs

Observation	Elapsed time	Cycles	State
P0	–	0	OK
P1	1 h	60	OK
P2	2 h	120	OK
P3	4 h	240	OK
P4	6 h	370	Air leakage

between shoulders so that PAMs work in the same relative position, and the first textile layer will have an adjustment mechanism above the chest and along the waist guaranteeing their use on people with different anthropometrics and body weights.

Another innovative feature lies in the fact that most of the used components were produced by 3D printing technologies, in a more flexible material and directly in a 3D mesh surface, which allows better adjustment of the structure to the body. In addition, it stimulates the proprioception of the muscles because the actuators are exactly positioned in the required location for support, including the same axis and anatomical plane.

Despite all the actuation and other rigid and semi-rigid components of the exoskeleton, the combination of the shape and the four layers of technical textile materials selected for its construction allowed the development of a wearable exoskeleton that looks like a normal cloth, making it much more user-friendly and not disturbing the user's movements during their work.

5 Conclusions and Future Work

The results obtained so far using testing setups highlighted the increased stimulation of the muscles in abduction and elevation movements using PAMs. However, additional work must be done to determine the final dimension of each actuator, so the user feels comfortable doing other natural movements without constraints. Regarding the electromyography tests, results have shown promising outputs in measuring the response of the muscles either at rest or in movement, capturing the muscular contractile response of different muscles. The future work relies on the integration of all information of monitoring and control, with established hardware and firmware communication, and a mobile app to resume the user's output.

Acknowledgements This article is a result of the project STVgoDigital - Digitalization of the Textile & Clothing sector (POCI-01-0247-FEDER-046086), supported by COMPETE 2020, under the PORTUGAL 2020 Partnership Agreement, through the European Regional Development Fund (ERDF).

References

1. www.eea.europa.eu. (n.d.). Textiles in Europe's circular economy — European Environment Agency. https://www.eea.europa.eu/publications/textiles-in-europes-circular-economy. Accessed 29 Aug 2022
2. Kadivar, Z., Beck, C.E., Rovekamp, R.N., O'Malley, M.K.: Single limb cable driven wearable robotic device for upper extremity movement support after traumatic brain injury. J. Rehab. Assistive Technol. Eng. **8**, 20556683211002450 (2021). https://doi.org/10.1177/20556683211002448

3. https://eksobionics.com/eksohealth/. The next step in NeuroRehabilitation. Accessed 23 Aug 2022
4. https://futurism.com/the-byte/wearable-muscles-strap-stronger. Accessed 30 Aug 2022
5. Fernandes, C., et al.: A deep learning approach to prevent problematic movements of industrial workers based on inertial sensors. In: 2022 International Joint Conference on Neural Networks (IJCNN), pp. 01–08. IEEE (2022)
6. Zhang, F.-R., He, L.-H., Shan-shan, W., Li, J.-Y., Ye, K.-P., Wang, S.: Quantify work load and muscle functional activation patterns in neck-shoulder muscles of female sewing machine operators using surface electromyogram. Chin. Med. J. **124**(22), 3731–3737 (2011)
7. António, C., Fonseca, F.: Fundamentos de biomecânica músculo-esquelética e ortopédica. Publindústria, Porto (2011)

Influence of Train Route Setting Functions Automation on Railway Traffic Efficiency

H. Haramina, I. Toš, and F. Batrla

Abstract The paper analyses the impact of train route setting individual functions automation on the efficiency of railway traffic. The research was conducted on the railway infrastructure equipped with the Croatian national train control system and different versions of station interlocking systems regarding their degree of automation. The results of the research prove that automation of certain process functions due to reduce of human errors and significant acceleration of train and traffic control process can improve railway infrastructure capacity, quality of timetable realization and railway traffic safety.

Keywords railway automation · human factors · railway efficiency · traffic safety

1 Introduction

On most of railway networks for train control purposes there are applied several different generations of interlocking systems. Namely, results of recent research show that, although dwindling, legacy interlocking systems still play an important part in the current European railway network [1]. There are in use interlocking devices of various generations, which comply different methods of train control where some of them have a bad impact on capacity of railway infrastructure and traffic safety [2, 3]. In this paper, data collected from three railway stations with different methods of train control are processed to perform the comparative analysis of train control methods in railway stations of Popovača, Sunja and Hrvatska Dubica in order to draw conclusions related to the influence of the automation of train route setting functions on the efficiency of railway traffic for the purpose of the possible improvement of traffic safety and capacity of railway infrastructure.

H. Haramina (✉) · I. Toš · F. Batrla
Faculty of Transport and Traffic Sciences, University of Zagreb, Zagreb, Croatia
e-mail: hharamina@fpz.hr

I. Salopek Čubrić et al. (eds.), *Proceedings of the 9th International Ergonomics Conference*, Lecture Notes in Networks and Systems 701,
https://doi.org/10.1007/978-3-031-33986-8_14

2 An Analysis of the Train Control Method in the Popovača Railway Station

In the Popovača railway station is equipped with an electro-relay interlocking device. This electro-relay device enables central control of main signals, switches, level crossing devices and devices of automatic block system of adjacent interstation line sections, and it performs all actions related to train route setting. The main benefit of this kind of interlocking system is possibility of application of automatic supervision of tracks occupancy and its relationship with other route elements setting (e.g., switches, signals etc.) to avoid human errors related to traffic safety.

In the analysis of the train control method used in this station, a case study will be applied, which includes the meeting of five trains in the station. It is about the meeting of the following trains: freight train number 47598 and passenger train number 2018 from the direction of adjacent station Moslavačka Gračenica with trains, freight number 81953, passenger number 2017 and fast number 745 from the opposite direction from adjacent Ludina station. Figure 1 shows the secured entry route and its overlap allocated for passenger train 2018 on the second track, the occupied third track by passenger train 2017, the occupied fourth track by freight train 47,598 and the departure of express train 745 from Ludina station. At 4:00 p.m., freight train 47,598 departs from the Moslavačka Gračenica station, where the train operator of this station sends a phonogram to the train operator of the Popovača station. To secure the entry of freight train 47,598 on the fourth track, simultaneous pressing of the input key "U" and the track number key on the interlocking interface desk sets the input signal "A" for the permitted run and blocks the entry route including its overlap. At 4:08 p.m., train 47,598 arrives at Popovača station. After stopping the train, by simultaneously pressing the button marked as "PV" and the number of the track occupied by the train 47,598, the overlap is unblocked so that switches which were included in it are released and ready to be set for the next train, then by pressing the button marked as "TP" it can be requested for permission to change the direction from Popovača station to the station Moslavina Gračenica. At 4:03 p.m., freight train 81,953 departs from Ludina station, and the Ludina train operator gives a phonogram to the train operator in Popovača station. A passage for that train, should be ensured through the Popovača station because it is going to cross with passenger train 2018 in the station Moslavačka Gračenica. At 4:09 p.m., by simultaneously pressing the exit button "I" and the track number button, the exit route for train 81,953 is set from the exit signal "C2". In order to allow the train to enter, the level crossing in the station must be activated by simultaneously pressing the group button marked with "GP" and the button for activation/deactivation of the level crossing placed in the station "B1". Then by simultaneously pressing the home route button "U" and the track number button, the home route from signal "B" for train 81,953 is set. At 4:12 p.m., train 81,953 passes through the Popovača station and with its passage release the used train route. After this, by simultaneously pressing the group button "GP" and the level crossing activation/deactivation button in the station "B1", the level

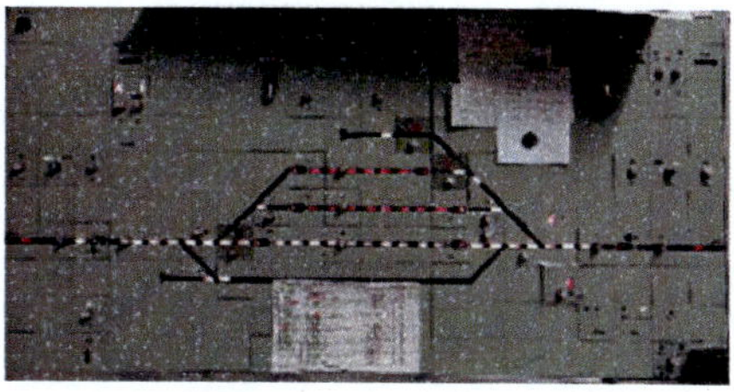

Fig. 1 Interlocking interface in Popovača railway station

crossing is deactivated, and then the train operator of the Popovača station reports the departure of train 81,953 to the train dispatcher in Moslavačka Gračenica station.

3 An Analysis of the Train Control Method in the Sunja Railway Station

The Sunja railway station is secured mechanically, whereas the railway station is not secured with a block device. There is no home signal "A" in the direction from the adjacent Novska station, and the home signals from adjacent Majur station "B" and Sisak Caprag station "C" are not valid and the signal marks "Signal not valid" are placed on them.

The Sunja railway station (see Fig. 2) is protected by the signal signs "Slow" and "Stop", signalled by yellow and red signal batons located on the pre-signal and entrance signal poles. The switches are installed and locked manually, and they are secured with "Robel" lock keys. In the station, switches are settled manually on the spot because there is no device for securing the station, but the security is carried out by evidence communication between the train operator and the switch operator via telephone lines and this conversation is recorded. The record of all orders of the train operators to the switch operators and their actions are recorded in the phonogram book of the train routes. The meeting of three passenger trains from three different directions was analysed. The first train number 5117 which is coming from the direction of the Sisak Caprag station, second one was train number 5202 from the direction of the Hrvatska Dubica station and third one train number 5122 from the direction of the Majur station. At 6:50 p.m., the station dispatcher in Sisak Caprag requests permission for train 5117 from the station dispatcher in Sunja. The station dispatcher in Sunja gives permission to receive train 5117. Train 5117 leaves the Sisak Caprag station at 18:54, which the station dispatcher in Sisak Caprag announces by phonogram: "Train number 5117 leaves at 18:54. (surname of the train driver). The running time of the passenger train between Sisak Caprag station and Sunja station is 23 min [3]. In order for the switch operators to be notified in time for the setting of switches and securing the train route at 7:04 p.m., the station dispatcher calls the switch operators of block 1 and 2 on the switch line, and their conversation is recorded with a voice recorder. The switch operator calls and the station dispatcher

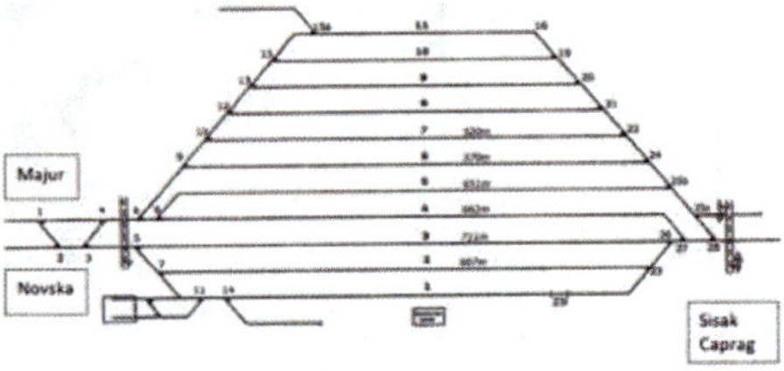

Fig. 2 Track layout in Sunja railway station

Fig. 3 Switch keys of block 2 in Sunja railway station

simultaneously gives them an order to set the home route for an upcoming train with a phonogram: "Entry for train number 5117 on the third track". After receiving the order, switch operators take the necessary switch keys and go to turn switches in the right position for the intended train run. In this case, since train 5117 enters the direction of the third track, all the switches in the train route remain in their regular position for the strait train run, but regardless switch operators must check visually on the ground and through the keys located in the block in such a way as to check whether all the keys of those switch 28, 27 and 26 with (+) mark and rectangular end are in their prescribed place (see Fig. 3).

At 7:06 p.m., the switch operators call the station dispatcher and the switch operator of the block 2 on the entrance side of the station reports that the route for train 5117 is secured and gives the control number, while the switch operator of the block 1 says that he is present, what he confirms with the control number, because the switches in the overlap should remain in their regular position, so he only checked this. At 7:12 p.m., the station dispatcher gives the order to the swich operator of block 2 to give the movement authority to train 5117 with the manual signal "Forward" at the home signal "C". At 7:17 p.m., the train 5117 arrives at Sunja station and at 7:18 p.m. the station dispatcher gives information about it to the to the station dispatcher in the Sisak Caprag railway station. At 6:53 p.m., a station dispatcher i Hrvatska Dubica station requests permission for train 5202 from the station dispatcher in Sunja who gives his permission to receive this train. The train than departs from Hrvatska Dubica station at 18:57. The running time of this passenger train between Hrvatska Dubica station and Sunja railway stations is 35 min [3]. After the arrival of train 5117 and the return of the switch operator of block 2 to his block house, the train dispatcher calls the switch operators of blocks 1 and 2 to set the entrance route for the train 5202 which is coming from the direction of Hrvatska Dubica railway station. At 7:22 p.m., the station dispatcher gives an order to the switch operators of blocks 1 and 2 to set a route for train 5202. After receiving the order, the switchers take the necessary switch keys and go to set the switches in the correct position for the intended train run. Since train 5202 is entering the second track, only switch number five must be set from its regular position in the direction to the correct position for the intended train run. After receiving the command from the station dispatcher, the switch operator of block 1 i.e. on the entrance side for train 5202 takes the (+) key of the switch number 5 and unlocks the switch from the position for driving in the straight direction, sets it in the right position and locks it with the (−) key that locks the switch in the position for driving on the diverging track. This (−) key was in the "Robel" lock while the switch was in its regular position i.e. for driving in the straight direction and the switch operator takes it after placing the switch in the correct position and brings it to the block house and places it in the designated interface. The key marked with (+) remains locked in the "Robel" lock, which ensures that the switch is placed in the correct position for the train run. At 7:24 p.m., the switch operator of block 1 calls the train operator and the switch operator of block 2 and says them that he has secured the entrance route for the train 5202 and then the switch operator of block 2 confirms his presence with his control number and that the switches in the overlap are in their regular position. At 7:27 p.m., the station dispatcher gives an order to the switch operator of block 1 to give the movement authority for the train 5202 with the manual signal "Forward" at the entrance signal "A". At 7:32 p.m., train 5202 arrives at Sunja station, and at 7:33 p.m. the train dispatcher gives the signal to the train dispatcher of Hrvatska Dubica station, and the switch operator of block 1 returns switch number 5 to its regular position. At 7:20 p.m., a train dispatcher in the Majur railway station requests permission for train 5122 from the train dispatcher in Sunja which gives than his control number and permission to receive train 5122. Train 5122 leaves Majur station at 19:28. The running time of the passenger train between Majur and Sunja stations is 14 min [3]. After the arrival of train 5202 and the return of the block 1 switch operator to the block house, the train dispatcher calls switch operators of block 1 and 2 to set the switches for the passing route of train 5122 from the direction of the Majur station. At 7:35 p.m., the train operator gives the order to the switch operators of block 1 and 2, who take the necessary switch keys and go to place the switches in the correct position for the intended train run. Since train 5122 passes through the fourth station track, only switch number 27 have to be set from its regular position in the straight direction to the correct position for the intended train run on the diverging track. After receiving the order from the train dispatcher, the switch operator of block 2 on the exit side for train 5122 takes the (+) key of the switch number 27 and unlocks the switch from the (+) position for driving in the straight direction, places it in the (−) position and locks it with the (−) key. The (−) key of switch number 27 was in the "Robel" lock while the this switch was in the regular position for the straight run, The switch operator takes it after placing the switch in the correct position for the upcoming train run and brings it to the interface in the block house, a (+) key of this swich number 27 remains locked in the "Robel" lock that ensures the switch is placed in the correct position for the train route. At 7:37 p.m., the switch operator of block 2 on the exit side for train 5122 and the switch operator of block 1 on the entrance side of train 5122 inform station dispatcher that passage for train number 5122 through the track number four is safe. At 7:37 p.m., the station dispatcher gives an order to the switch operator of block 1 to give a movement authority for train 5122 with hand signal "Forward" at the entrance signal "B", and at 7:38 p.m. the station dispatcher gives an order to the switch operator of block 2 on the exit side of train 5122 that at the last exit switch, i.e. switch number 28, gives the manual signal "Forward". At 19:39, the station dispatcher of Sunja station requests permission for train 5122 from the station dispatcher of Sisak Caprag station, and the station dispatcher of Sisak Caprag station gives permission to receive this train. At 7:42 p.m., train 5122 arrives at Sunja station, and after passengers exchange, it leaves Sunja station at 7:43 p.m. After the train has passed, the switch operator of block 2 returns switch number 27 to its regular position.

4 An Analysis of the Train Control Method in the Hrvatska Dubica Railway Station

Hrvatska Dubica railway station is secured mechanically, but without a block device. Instead of mechanical or light signals, entry into the station area is ensured by red and yellow batons, whereby the red batons are placed at the beginning of the entrance route and about 200 m away from the blades of the first switches, while the yellow batons, whose purpose is to pre-signal the red batons, are placed at about 100 m in front of the red batons on both sides of the station area. In this station there are six switches and they are all operated manually and secured with mechanical "Robel" locks (Fig. 4).

The switches are installed on the spot because there is no station safety device, but the safety is carried out through a book of phonograms of train routes for station dispatcher and switch operators, which is located in the traffic office and in which the station dispatcher is giving orders to the switch operator with his signature, and the switch operator confirms with his signature that he understood or executed the received order. The crossing of passenger train 5200 and train 71,501 was analysed for the needs of railway infrastructure. At 1:30 p.m., the station dispatcher of Sunja station requests permission for train 71,501 from the station dispatcher of Hrvatska Dubica station, and the train dispatcher of Hrvatska Dubica station gives permission to receive train 71,501. Train 71,501 leaves Sunja station at 1:35 p.m. Train running time between stations Sunja and Hrvatska Dubica is 32 min [3]. As train 71,501

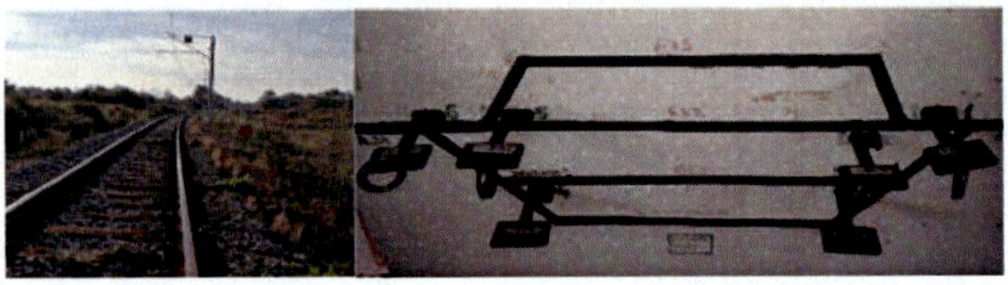

Fig. 4 Red baton at the home signal location "A" and station layout with switch keys in the Hrvatska Dubica railway station

arrives at the Hrvatska Dubica station to cross with passenger train 5200, the station dispatcher at 1:36 p.m. writes in a book of phonograms an order to the switch operator to secure the route for the train 71,501 on the fourth track and confirms it with his signature. The switch operator signs under the station dispatcher's signature and then dispatcher gives him the (+) key of switch number 5 so he can set it in the correct position for train 71,501 to enter the fourth track. The switchman goes to carry out the given order. At 1:50 p.m., he returns to the traffic office and shows the (−) switch number 5 key to the station dispatcher, which was in the switch while it was locked in the regular position, and puts it in the key slot and writes the time of arrival at the traffic station office in the a book of phonograms, i.e. that the entry route for the train 71,501 is secured, and confirms this with his signature. The station dispatcher than signs under the switch operator's signature and thereby confirms that the order to secure the track for the entry route of train 71,501 has been executed. At 1:51 p.m., the station dispatcher enters the time gave the order to switch operator to set the home signal for train 71,501 in the book and signs it. The switch operator signs then under the station dispatcher's signature and thereby confirms that he has received the order. After that he goes to the red signal pole located at the place of the entrance signal "B" and gives the movement authority to enter the station to the train 71,501 by manual signal. Train 71,501 arrives at Hrvatska Dubica station at 14.07 and at 14.08 the station dispatcher gives the information to the station dispatcher of the Sunja station that this train arrived in his station. At 2:12 p.m., the switch operator returns to the traffic office to get the (−) key of switch number 5 to return it to its regular position and at the same time to ensure the exit route for passenger train 5200 which has a passage through Hrvatska Dubica station. At 2:17 p.m., the switch operator returns the (+) switch number 5 key and thus proves that the exit route for passenger train 5200 is secured. At 2:00 p.m., the station dispatcher at Novska station requests permission for train 5200 from the train dispatcher at Hrvatska Dubica station. The station dispatcher at Hrvatska Dubica gives permission to receive train 5200 which departs from Novska station. The running time of this passenger train between stations Novska and Hrvatska Dubica is 24 min [4]. Passenger train 5200 needs to ensure passage through Hrvatska Dubica station, so the station dispatcher enters in the book the order time at 2:12 p.m. to ensure the passage of train 5200 through the third track and confirms this with a signature. The switch operator signs under the station dispatcher's signature and at 2:18 p.m. he enters the time that the track for the entry of passenger train 5200 is secured. At 2:19 p.m., the station dispatcher enters in the book the time of the order to give a movement authority signal for train 5200 to run into the station and signs it. The switch operator signs than under the station dispatcher's signature and thereby confirms that he has received the order and then he goes to the red signal pole located at the entrance signal "A" and gives the signal for the permitted run to train 5200. At 2:21 p.m., the station dispatcher in Hrvatska Dubica asks for permission for train 5200 from the station dispatcher in Sunja station which gives him this permission. Train 5200 than arrives at Hrvatska Dubica station at 2:26 p.m. and, after dwell stop for passenger exchange, leaves Hrvatska Dubica station at 2:27 p.m., and at 2:28 p.m. station dispatcher informs the station dispatcher in Novska station about this. At 2:34 p.m., the switch operator returns to the traffic office to get the (+) switch number 2 key to place it in the correct position and at the same time to ensure the exit for train 71,501 from the fourth track from Hrvatska Dubica station. At 2:35 p.m., the station dispatcher writes the order for train 71,501 to leave the fourth track. The switch operator signs under the signature of the station dispatcher and goes to carry out the given order. At 2:45 p.m., the switch operator returns (−) the switch number 2 key and thus proves that the exit for train 71,501 is secured. At 2:46 p.m., station dispatcher in Hrvatska Dubica requests and receives from the station dispatcher in Novska permission for train 71,501. Train 71,501 leaves Hrvatska Dubica at 2:47 p.m. After the departure of train 71,501 from Hrvatska Dubica station, the switch operator takes the (−) key of switch number 2 and goes to return it to its regular position. At 3:02 p.m., the switch operator returns to the traffic office with the (+) switch number 2 key and proves to the station dispatcher that switch number 2 is in its regular position.

5 Conclusion

The paper describes the train control systems of Popovača, Sunja and Hrvatska Dubica railway stations and analyses the methods of traffic management in these station areas. The analysis carried out showed that the trains crossing time is significantly shorter in station equipped with electro-relay interlocking system compared to stations without it and thus the capacity of the railway infrastructure is much higher. From the analysis, it can be seen that there is a big difference between the operation of a station with one switch operator and with two switch operators. On the example of a train crossing where the duration of crossing with one switch operator is about 40 min, while the duration of crossing with two switches is about 15 min, which is significantly shorter, but still significantly longer compared to the station with an electro-relay interlocking where the crossing time takes for about 3 min. Additionally, in comparison with train control methods in other two railway stations where these tasks are performed by station staff, train control method in the Popovača railway station due to application of electro-relay interlocking implies automated train route setting functions based on automatic supervision of track occupancy, which can reduce the probability of human errors related to traffic safety [2, 3]. The analysis showed that in the case of a lower degree of automation in a train control process, along with a decrease in traffic safety due to greater involvement, and thus human responsibility, at the same time the value of station intervals increases, which reduces the capacity of the network, i.e., the efficiency of using the railway infrastructure and vehicles.

References

1. Bădău, F.: Railway interlockings – a review of the current state of railway safety technology in Europe. Promet Traffic Transp. **34**(3), 443–445 (2022)
2. Janota, A., Pirník, R., Ždánsky, J., Nagy, P.: Human factor analysis of the railway traffic operators. Machines **10**(9), 820 (2022)
3. Seelmanna, H., Hromádkab, V., Korytárováb, J., Vítkovább, E., Funkb, T.: Influence of railway control and signaling equipment on reduction of occurrences in the state railway network of Czech Republic. Transp. Res. Procedia **44**, 47–52 (2020)
4. Hž Putnički prijevoz, Timetable 2021/2022. https://www.hzpp.hr. Accessed 21 May 2022

Regional Differences and Body Mass Index for Population of Croatian Boys

Renata Hrženjak and Ksenija Doležal

Abstract The World Health Organization has published results according to which 340 million children were obese in 2016, of which 19% were boys aged 5–19. Obesity is more prevalent among children than among adolescents. Research conducted in the Republic of Croatia indicates an increasing problem of obesity in the Croatian population. As childhood obesity can result in adult obesity and cause a variety of health problems such as cardiovascular disease, it is important to take care of diet and weight from an early age.

The paper presents the results of specific body differences of boys aged 6 to 12 years (BMI, body height, chest, waist and hip girth) in 5 Croatian regions (Goransko-primorska, Slavonia, Central Croatia, Dalmatia and City of Zagreb). In doing so, it was concluded that there are differences in the body structure of boys between regions. The biggest regional differences in girths were observed in the Goransko-primorska region, in 10 and 11-year-olds.

Keywords obesity · body mass index · regional differences · boys

1 Introduction

The majority of obesity in adulthood has its origins in childhood which makes obesity a pediatric concern and the period when interventions should be done. Both genetic and environmental factors play a role in the development of obesity [1]. According to data obtained by WHO, 340 million children were obese in 2016, of which 19% were boys aged 5–19 [2]. The results of new research into childhood obesity in Croatia indicate that 35% of the country's children aged 8–9 suffer from being overweight or obese. Nationally, results show the obesity problem is greater in boys, affecting 17.8% of boys compared to 11.9% of girls [3]. Research conducted on the obesity of children in 2011, conducted by the HZJZ on a sample of 11,000 children, showed

R. Hrženjak (✉) · K. Doležal
University of Zagreb Faculty of Textile Technology, Zagreb, Croatia
e-mail: renata.hrzenjak@ttf.unizg.hr

I. Salopek Čubrić et al. (eds.), *Proceedings of the 9th International Ergonomics Conference*, Lecture Notes in Networks and Systems 701,
https://doi.org/10.1007/978-3-031-33986-8_15

that obesity in children is increasing. Comparing these results, which found 24% of children overweight and 13% of children obese, with research from 2003, where only 5% of children had excessive weight and only 2–3% were considered obese, it can be concluded that if nothing is done, the obesity epidemic is yet to come [4].

Since 1980 obesity rates have risen three fold or more in some areas of North America, United Kingdom, Eastern Europe, Middle East, The Pacific Islands and Australia. In developing countries the prevalence of obesity also exhibited a dramatic increase, obesity now being present in 60% of women and 50% of men in many of these countries. The percentage of overweight children and adolescents also increased, by almost 50%, in the last two decades of the twentieth century. The prevalence and trends of obesity among US children and adolescents have continued to increase among all age groups; the increase has been more significant among African Americans and Hispanic individuals [2–4].

It has long been observed that about 40% of overweight children will continue to have increased weight during adolescence and 75–80% of obese adolescents will become obese adults. A child with a high BMI has a high risk of being overweight or obese at 35 years of life and this risk increases with age. The consequences of this disease starting in childhood may be more severe as the duration of obesity will be longer [4].

This paper presents the results of measuring the main body measurements by individual regions in the Republic of Croatia (Goransko-primorska, Slavonia, Central Croatia, Dalmatia and City of Zagreb). It also provides a representation of BMI values according to percentile values at the level of the entire country.

2 Subjects and Methods

A sample of 4037 boys aged 6 to 12 was included in this research. Given that the respondents were minors, the kindergartens and schools obtained the respondents' consent to participate in the body measurement research. The respondents were distributed in 5 Croatian regions: Goransko-primorska, Slavonia, Central Croatia, Dalmatia and City of Zagreb. The research was organized and conducted by University of Zagreb Faculty of Textile Technology in collaboration with several institutions (Institute of Anthropology of Zagreb, School of National Health "Andrija Štampar"—Faculty of Medicine of the University of Zagreb, Faculty of Medicine of the University of Split, Faculty of Philosophy of the University of Osijek and Faculty of Economics & Business of the University of Zagreb) in 2005 and 2006. Anthropometric measurements of boy's body (body height, chest girth, waist girth and hip girth) were measured with anthropometric instruments: measuring tape and one-arm anthropometer [5–12].

Body mass index (BMI) is a widely used method to define the relationship between weight and height. The BMI provides a practical clinical tool to classify individuals with normal and those with various degrees of obesity. The BMI system of classfication of obesity is important because it denotes the risk for medical complications of obese patients. Adult individuals with a BMI above 27 have a markedly increased risk for hypertension, hypercholesterolemia, and diabetes mellitus. In contrast, when the BMI index is less than 25, there are no apparent physical effects of obesity. However, the use of BMI has limited applications in the assessment of overweight children since its calculation is based primarily on a stable height, which is not applicable to growing children. Also, the BMI can underestimate the percentage of lean body mass since it does not account for variations in musculature [4, 13]. However plotting the growth and weight of children in age-specific growth charts allows a precise assessment of a child's status. It helps the clinician to evaluate a child's weight and its relationship to height and it provides a view of the growth patterns, which determine the gravity of the situation. The growth progression denotes if a child is gaining weight in excess of height over time, or if the body weight is increased, but it is maintained on the same percentile as the child grows older. These two different patterns of obesity in childhood need to be considered in the treatment to be provided [4, 14].

Body mass index (BMI) was calculated by Eq. (1) [15, 16]:

$$BMI\left(kg/m^2\right) = (Body\,mass\,in\,kg)/(Stature\,in\,meters)^2 \tag{1}$$

In children and adolescents aged 2 to 19 years, obesity was defined as a BMI at or above the 95th percentile of the CDC sex-specific BMI-for-age growth charts from

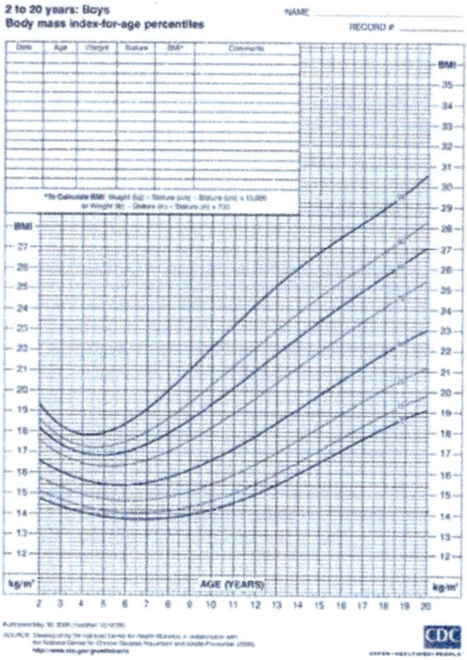

Fig. 1 2 to 20 years: boys body mass index-for-age percentiles [17]

2000, Fig. 1. This table includes the studied population of boys (aged 6 to 12 years) and is therefore suitable for comparison.

Overweight was defined as a BMI between the 85th and 95th percentiles. The estimates are presented as greater than or equal to both the 85th and 95th percentiles [18–20].

3 Results and Discussion

Changes of the main body measurements (body height, chest girth, waist girth and hip girth) by age and regions are presented in Figs. 2, 3, 4 and 5. BMI was calculated and presented in the Table 1. The statistical package SPSS was used for descriptive statistics and data analysis.

It can be seen from Fig. 2 that the growth of boys by age is linear, which is expected considering the investigated age groups. There are also visible regional differences, where growth in height is most pronounced in boys from the Dalmatia region (in all ages, except for seven-year-olds where Slavonians are only 0.2 cm taller, which is not significant). This is followed by boys from Slavonia (at 7, 8 and 9 years of age) and Central Croatia (at 6 years). Boys aged 12 years have the same average height in Slavonia as in Central Croatia (154.9 cm and 155 cm). Then there are the boys in the Goransko-primorska region, who grow the most at the age of 10 and 11 years compared to other regions. The shortest boys in height are in the City of Zagreb, of all ages, especially 9, 10 and 12 years old.

Chest girth size differences across regions are also noticeable (Fig. 3)

The boys from Central Croatia stand out the most here (ages 6 (60.6 cm), 7 (63.1 cm), 8 (65.2 cm) and 12 years (75.7 cm) together with Dalmatians). This is followed by boys from the Goransko-primorska region, ages 6, 8, 10 and 11 years.

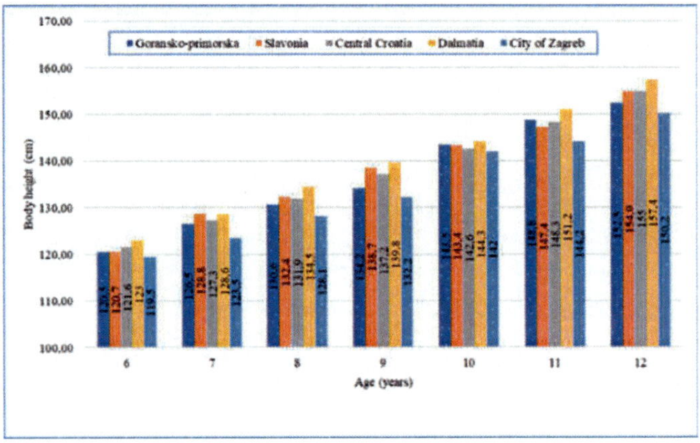

Fig. 2 Mean value of body height of boys distributed by age and regions

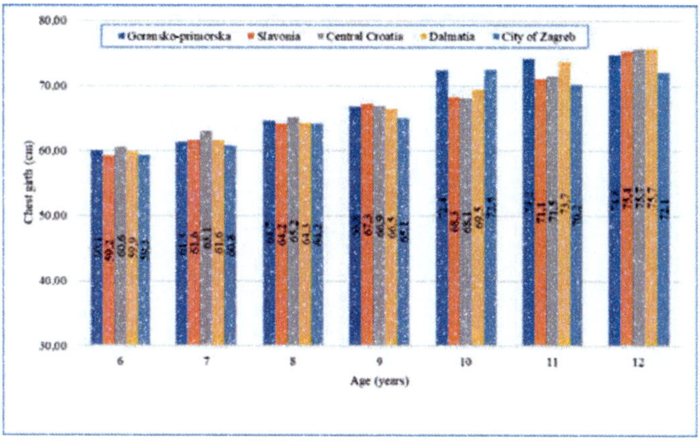

Fig. 3 Mean value of chest girth of boys distributed by age and regions

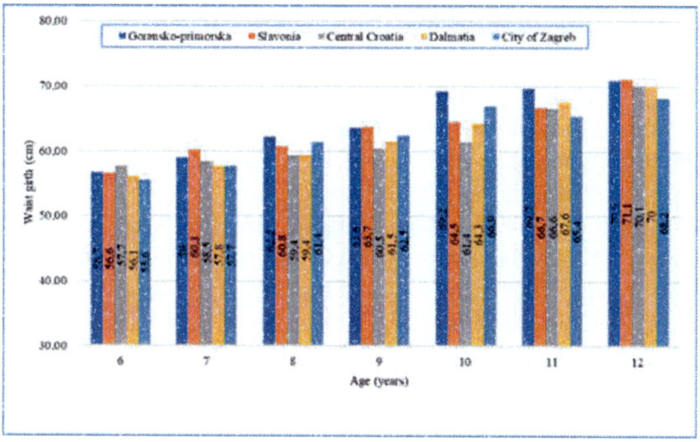

Fig. 4 Mean value of waist girth of boys distributed by age and regions

Boys in the City of Zagreb have the smallest chest girth in all age groups except ten-year-olds, who have the largest regional chest girth (72.5 cm).

For waist girth, regional changes are slightly different (Fig. 4). Boys from the Goransko-primorska region (ages 8, 10 and 11 years) have the largest girth, in the amounts of 62.2 cm, 69.2 cm and 69.7 cm. Then follow boys from Slavonia region, who have a larger girth at 7, 9 and 12 years. The boys from Zagreb do not have the lowest girth here, as with the chest, they have a larger girth than the boys from Slavonia region at the age of 8 and 10 years, but smaller than the boys from Goransko-primorska region.

In terms of hip girth, boys from Goransko-primorska region also stand out the most at the age of 8 (71.2 cm), 10 (80.8 cm) and 11 years (83.5 cm), Fig. 5. Dalmatians

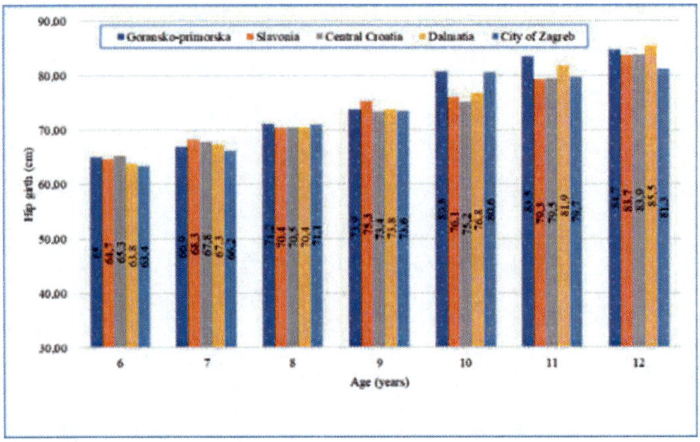

Fig. 5 Mean value of hip girth of boys distributed by age and regions

regionally have the largest hip girth at the age of 12 years. Boys from Zagreb aged 6, 7, and 12 years have the smallest hip girth. 8-year-olds from all regions have nearly the same hip girth (from 70.4 to 71.2 cm).

Table 1 shows physical parameters of boys aged 6 to 12 years. The investigated parameters are body height and body mass, which were used to calculate BMI values. Mean values and percentile values are shown.

By comparison of Table 1 with percentile curves (Fig. 1) it was concluded that average BMI for Croatian boys amounts between 50 and 85th percentile, representing a normal range. At the same time, 10% of the population of boys aged 6, 7, 9 and 11 years have a tendency to be overweight (between 85 and 95th percentile), and 5% of boys aged 8 and 10 years are obese ($\geq$95th percentile).

Table 1 Physical parameters of boys aged 6 to 12 years

Age (years)	N	Parameter	x	Percentiles						
				5	10	25	50	75	90	95
6	549	Body height (cm)	121.1	112.0	113.5	116.5	121.0	125.0	128.5	131.0
		Body mass (kg)	23.4	18.0	19.0	20.6	23.0	26.0	29.0	31.0
		BMI (kg/m^2)	**16.0**	**14.3**	**14.7**	**15.2**	**15.7**	**16.6**	**17.6**	**18.1**
7	608	Body height (cm)	126.9	117.0	119.0	123.0	126.6	130.5	135.1	138.0

(continued)

Table 1 (continued)

Age (years)	N	Parameter	x	Percentiles						
				5	10	25	50	75	90	95
		Body mass (kg) **BMI (kg/m²)**	26.2 **16.3**	20.0 **14.6**	21.0 **14.8**	23.0 **15.2**	25.0 **15.6**	29.0 **17.0**	32.6 **17.9**	36.0 **18.9**
8	629	Body height (cm)	131.8	121.0	123.5	127.4	132.0	136.0	140.0	142.8
		Body mass (kg) **BMI (kg/m²)**	29.7 **17.1**	21.0 **14.3**	22.5 **14.8**	25.0 **15.4**	29.0 **16.6**	33.0 **17.8**	39.0 **19.9**	42.0 **20.6**
9	576	Body height (cm)	136.8	125.3	128.0	132.5	136.5	141.0	145.5	148.1
		Body mass (kg) **BMI (kg/m²)**	32.9 **17.6**	24.0 **15.3**	25.0 **15.3**	28.0 **15.9**	32.0 **17.2**	36.0 **18.1**	41.0 **19.4**	46.1 **21.0**
10	652	Body height (cm)	143.2	131.5	134.0	138.5	143.0	147.9	152.0	155.7
		Body mass (kg) **BMI (kg/m²)**	37.6 **18.3**	26.0 **15.0**	28.0 **15.6**	31.0 **16.2**	35.2 **17.2**	42.0 **19.2**	50.0 **21.6**	56.0 **23.1**
11	507	Body height (cm)	148.1	136.0	138.4	143.0	148.0	153.0	158.0	161.0
		Body mass (kg) **BMI (kg/m²)**	41.0 **18.7**	29.5 **15.9**	31.0 **16.2**	34.0 **16.6**	39.0 **17.8**	47.0 **20.1**	52.5 **21.0**	60.0 **23.1**
12	516	Body height (cm)	154.3	140.0	143.0	147.0	153.6	160.0	167.7	172.1
		Body mass (kg)	45.9	31.0	34.0	38.0	44.0	52.0	60.3	67.0
		BMI (kg/m²)	**19.3**	**15.8**	**16.6**	**17.6**	**18.6**	**20.3**	**21.4**	**22.6**

4 Conclusion

It is necessary to continuously monitor the growth and development of children. Within this research it was confirmed that Croatian boys aged 6 to 12 years show positive secular changes, especially boys from the region of Dalmatia and then followed by Slavonia. In the City of Zagreb boys grew the least.

The biggest regional differences in girths: chest, waist and hip were observed in the Goransko-primorska region, in 10 and 11-year-olds. Boys aged 12 years regionally have similar girth values, only boys from City of Zagreb have lower values as for body height.

Further research and tracking the movement of BMI values of studied population is essential because there is a risk that obese children will become obese adults. The standards used to identify obesity and overweight in children and adolescents are in agreement with the standards for adults.

Acknowledgements The authors wish to thank all of the respondents and measurers who participated in this study.

Funding The publication of this paper was funded through the financial support for research at the University of Zagreb Development of clothing for specific purposes, TP2/2022.

References

1. https://www.ncbi.nlm.nih.gov/pmc/articles/PMC3005642/. Accessed 05 Sept 2022
2. https://www.who.int/news-room/fact-sheets/detail/obesity-and-overweight. Accessed 05 Sept 2022
3. https://www.who.int/europe/news/item/06-07-2021-overweight-or-obesity-affect-1-in-3-children-in-croatia-new-survey-shows. Accessed 05 Sept 2022
4. https://www.total-croatia-news.com/item/14572-croatia-ranked-eighth-in-obesity-of-the-28-eu-countries. Accessed 16 Sept 2022
5. Controlling the Global Obesity Epidemic. World Health Organization Obesity and Overweight Fact Sheet (2003)
6. Hedley. A.A., et al.: Prevalence of overweight and obesity among US children, adolescents and adults 1999–2002. JAMA **291**, 2847–2850 (2004). https://jamanetwork.com/journals/jama/fullarticle/198912. Accessed 07 Sept 2022
7. Ujević, D., et al.: Croatian antropometric system meeting the European union. Int. J. Clothing Sci. Technol. **18**(3), 200–218 (2006)
8. Hrženjak, R..: Influence of body proportions of children and youth as the basis for clothing construction. Ph.D. thesis, University of Zagreb, Zagreb (2012)
9. Hrženjak, R., Doležal, K., Ujević, D.: Development of sizing system for girls aged 6 to 12 years in Croatia. Coll. Antropol. **37**(4), 1095–1103 (2013)
10. Hrženjak, R., Ujević, D.: Hrvatski antropometrijski sustav - stanje. potrebe. perspektive. realizacija. Zbornik/Prvi kongres hrvatskih znanstvenika iz domovine i inozemstva. In: Barišić, P. (ed.) Zagreb-Vukovar. 15.-19. studenoga 2004. Ministarstvo znanosti obrazovanja i športa RH. Zagreb, pp. 218–221 (2004)
11. Buzina, R., et al: Morfološka i funkcionalna antropometrija. Praktikum biološke antropometrije. Zagreb (1984)

12. Hrženjak, R., Doležal, K., Brlobašić Šajatović, B.: Differences of body measurements for the youth population. In: Hodžić, A., Islamović, F., Hodžić, D. (eds.) Electronic Proceedings of 13th International Scientific Conference on Manufacturing Engineering - Development and Modernization of the Manufacturing. RIM 2021. Sarajevo. BIH. 29 September – 1 October 2021, pp. 352–357. Bihać. University oh Bihać. Faculty of Technical Engineering (2021)
13. Yajnik, C.S., Yudkin, J.S.: The Y–Y paradox. Lancet **363**, 163 (2004)
14. Alemzadeh, R., Rising, R., Lifshitz, F.: Obesity in children. In: Lifshitz, F. (ed.) Pediatric Endocrinology 5th edition, vol. 1, chap. 1. Informa NY (2007)
15. Meltzer, A., et al.: Weight history and hypertension. J. Clin. Epidemiol. **41**, 867–874 (1988)
16. Hrženjak, R., et al: Anthropometric measurements and analysis of body dimensions of Croatian female volleyball players. In: Rovero, G. (ed.) Proceedings of the 8th AUTEX Conference. Working Towards Change: Academy and Industry together. Città Studi. Biella. Italy. 24–26th June 2008, pp. 1–6. Torino. Politecnico di Torino. Poster session 1-Poster area 1 - (226) (2008)
17. https://www.cdc.gov/nchs/data/series/sr_11/sr11_246.pdf. Accessed 10 Sept 2022
18. Ogden, C.L., Flegal, K.M.: Changes in terminology for childhood overweight and obesity. National Health Statistics Reports (25), 1–5 (2010)
19. Kuczmarski, R.J., et al.: 2000 CDC growth charts for the United States: methods and development. Vital Health Stat. **11**(246), 1–190 (2002)
20. Ogden, L.C., et al.: Prevalence of childhood and adult obesity in the United States. 2011–2012. https://pubmed.ncbi.nlm.nih.gov/24570244/. Accessed 28 Aug 2022

The Formulation for Human Finger-Machine Contact Interaction Considering Nonlinear Flexibility and Non-smoothness Slipping by Multibody Dynamics Approach

Yuki Kitazawa, Yoshiki Sugawara, and Masakazu Takeda

Abstract In recent years, the opportunities for robot or machine to contact people directly are increasing in the field of nursing and rehabilitation. In such a field, it is very important to operate machines so that the person being assisted can feel at ease in the contact between the person and machine, such as gripping a person by a robot. In addition, recent years, technologies that reproduce sensations in metaverse space in the real world have been attracting attention. To reproduce the tactile sensation in the real space when touching an object in the Metaverse space, it is necessary to predict the deformation and stress of the skin around the contact area. The purpose of this study is to formulate interaction between the human finger and the machine. The human finger is a human-specific and human-like element that is dexterous and have very wide force range. Coefficient of friction between the skin and an object has been measured in past study. However, motion analysis considering the effects of contact phenomena has not been performed by considering the skin and the object as one system. The purpose of this study is to analyze the motion of human–machine system with phenomenon of contact motioned. This study proposes a modeling and numerical method suitable for the analysis of the motion of flexible bodies with contact. The numerical results are consistent with the empirical motion.

Keywords Multibody dynamics · Human–machine contact · Friction · Nonlinear flexibility · Non-smooth slipping

Y. Kitazawa (✉)
Graduate School of Aoyama Gakuin University, Sagamihara, Kanagawa, Japan
e-mail: d5622008@aoyama.jp

Y. Sugawara · M. Takeda
Aoyama Gakuin University, Sagamihara, Kanagawa, Japan

I. Salopek Čubrić et al. (eds.), *Proceedings of the 9th International Ergonomics Conference*, Lecture Notes in Networks and Systems 701,
https://doi.org/10.1007/978-3-031-33986-8_16

1 Introduction

In recent years, opportunities for direct contact between humans and machines have been increasing. In addition, the performance of these systems is sometimes evaluated based on human sensitivity. For example, human work support robot used in nursing care, medical care, rehabilitation, construction, etc. Mukai et al. developed ROBEAR that assists transfer assistance, which poses a large physical burden in nursing care [1]. In addition to covering the entire body with flexible materials, ROBEAR uses three types of haptic sensors to enable both contact that maintains softness and movements that require a large force.

In addition, Kobayashi et al. have developed and commercialized a wearable muscle strength assisting device called "Muscle Suit®" for the purpose of physically supporting all human movement [2]. This auxiliary device uses a high force McKibben type artificial muscle as an actuator and has a structure that can be easily attached and detached while reducing the physical burden on the operator. Machines that contact directly with people and assist their movements are required to have contact affinity to flexibly contact object (human). To increase the social acceptance of such a human–machine system, it is necessary to guarantee contact has high affinity and safety. By predicting the displacement and stresses that have occurred to humans as well as machines, affinity and safety can be guaranteed.

In addition, recent years, technological developments for reproducing real space tasks that involve contact in Metaverse space have carried out. If tactile sensation technology becomes more precise, the possibility of new communication using the Metaverse space can be expanded in a society where remote work is promoted. In order to reproduce the tactile sensation when touching an object in the Metaverse space, it is necessary to model the contact phenomenon between a person and an object and generate an appropriate task, for example output from the actuator. To predict the force and displacement generated by contact, it is necessary to predict the deformation and stress of the skin around the contact area and the accompanying change in stiffness and frictional force. But the technology is inadequate due to the non-linear stiffness of the skin and the difficulty of formulating contact with friction.

So far, several research have been conducted to measure the coefficient of friction between the skin and an object [3, 4]. However, motion analysis considering the effects of contact phenomena has not been performed by considering the skin and the object as one system. The purpose of this study is to analyze the motion of human–machine system with phenomenon of contact motioned (in the real space/metaverse space). This study deals with the contact between the machine and the finger at first. Fingers are the most used parts when humans intentionally touch objects, and they are capable of dexterous movements such as human-like grasping and manipulation. Also, fingers are highly capable of perceiving the texture of objects. In addition, since the subcutaneous tissue of the finger is thin, it is considered that behavior of contact part is susceptible to the effects of the human body structure (epidermis, subcutaneous tissue, and bone layer structure) compared to other parts. From these facts, it is considered that the derivation of the finger-machine contact model has high applicability in considering the contact problem of the human–machine system. By using a mathematical model of deformation and sliding of the human finger, it is possible to design machines and machine controls that can be operated intuitively and that can be used with confidence. It is expected that such a design will increase the receptivity to contact with machines and introduce machines that actively contact people in all aspects of daily life. In addition, it is possible to make the haptic feedback by contact in the Metaverse space more realistic, and it is expected that the Metaverse space will be used more effectively.

2 Analysis Target and Modeling

In this paper, we analyze fingertip sliding motion on a plane as a fundamental phenomenon of finger-machine contact. Figure 1 shows a schematic diagram of the operation. A pressing force F_N and a force F_T for sliding in the direction of movement are applied to the bones of the fingertips placed on the plane through muscles. The force applied to the bone is transmitted to the epidermis by deforming the subcutaneous tissue. On the other hand, friction occurs due to contact between the plane and the epidermal tissue. When the force applied to the epidermal tissue in the advancing direction exceeds the friction force, the epidermal tissue slides on the plane. When all epidermal tissue in contact with the plane slide, the whole finger slides on the plane. The Flores's method [5] is attracting attention as a method of introducing non-smooth contact forces such as Coulomb friction assuming rigid body contact into motion analysis. In this method, numerical analysis is performed by Moreau's time-stepping method [6] suitable for simulation of non-smooth systems. And non-smooth contact is formulated as an impulse in Linear Complementarity Problem (LCP). On the other hand, in this study, it is necessary to consider the non-smooth contact problem of flexible bodies. So, the spring mass model shown in Fig. 2 is introduced. For numerical analysis, the Moreau's time-stepping method is applied to the mass points with contact, while the Runge–Kutta method is applied to the mass points without contact. Combining the two numerical analysis methods is described in the next section. In this report, for simplicity, the model is modeled in a two-dimensional vertical plane (A-A cross section in Fig. 1(b)) to the longitudinal direction of the finger. The epidermal tissue is modeled with 5 mass points (m_{e1} ~ m_{e5}), and the subcutaneous tissue is also modeled with 5 mass point (m_{s1} ~ m_{s5}, where m_{s3} is the bone). Virtual springs that simulate the connection between tissues is introduced between each mass point (between epidermal tissue: k_{ee}, between subcutaneous tissue: k_{ss}, between epidermal tissue and subcutaneous tissue: k_{es}, between epidermal tissue and subcutaneous tissue: k_{diag}). Stiffness of skin tissue is nonlinear depending on internal stress. However, in this report, the virtual spring assumed to be linear spring for simplicity. For the same reason, the coefficient of friction μ between the plane and the epidermal tissue is also assumed to be constant.

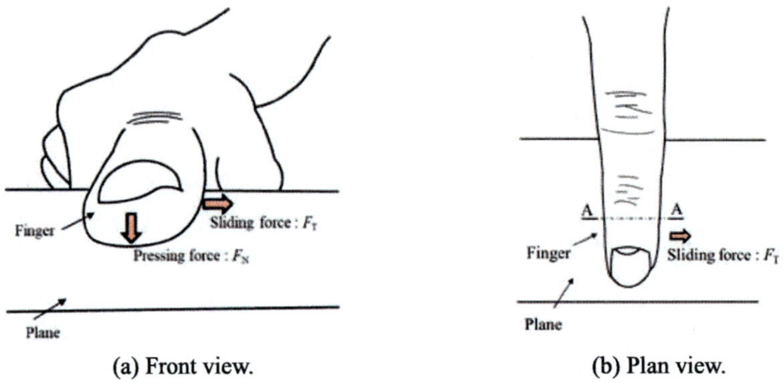

Fig. 1 Analysis target finger sliding motion on the plane

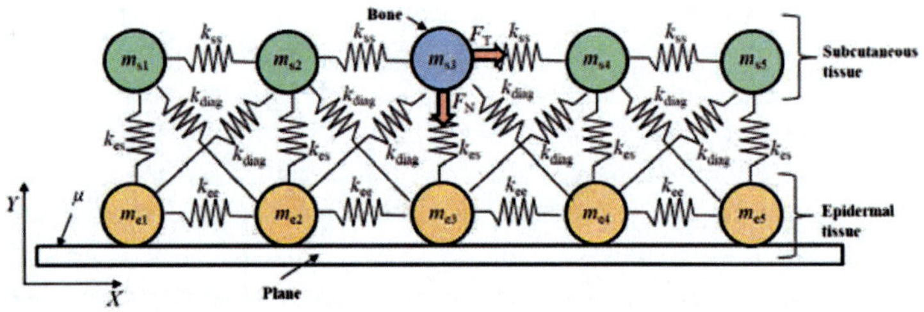

Fig. 2 Spring mass model expressing human finger-plane contact

3 Analysis Method

In this study, the Runge–Kutta method is used for motion analysis of subcutaneous tissue. On the other hand, Moreau's time-stepping method which is a discreate numerical analysis method suitable for non-smooth system simulation, is used for motion analysis of epidermal tissue that had contact phenomena with a plane. Figure 3 shows a flow chart of the analysis method that combines Runge–Kutta method and Moreau's time-stepping method. For explanation, Fig. 4 shows a schematic diagram corresponding to the analysis flow when two mass point for epidermal tissue and two mass point for subcutaneous tissue are used. The number I–XI in Fig. 3 and Fig. 4 correspond to the steps of the analysis procedure described later. For future explanation, the generalized coordinate vector $\boldsymbol{q}_{\mathrm{e}(i)}$ for the epidermal tissue, the generalized velocity vector $\dot{\boldsymbol{q}}_{\mathrm{e}(i)}$ for the subcutaneous tissue, and the generalized coordinate vector $\boldsymbol{q}_{\mathrm{s}(i)}$, and the generalized velocity vector $\dot{\boldsymbol{q}}_{\mathrm{s}(i)}$ are defined by Eqs. (1), (2), (3) and (4).

$$\boldsymbol{q}_{\mathrm{e}(i)} = [x_{\mathrm{e}1(i)} y_{\mathrm{e}1(i)} x_{\mathrm{e}2(i)} y_{\mathrm{e}2(i)} x_{\mathrm{e}3(i)} y_{\mathrm{e}3(i)} x_{\mathrm{e}4(i)} y_{\mathrm{e}4(i)} x_{\mathrm{e}5(i)} y_{\mathrm{e}5(i)}]^T \tag{1}$$

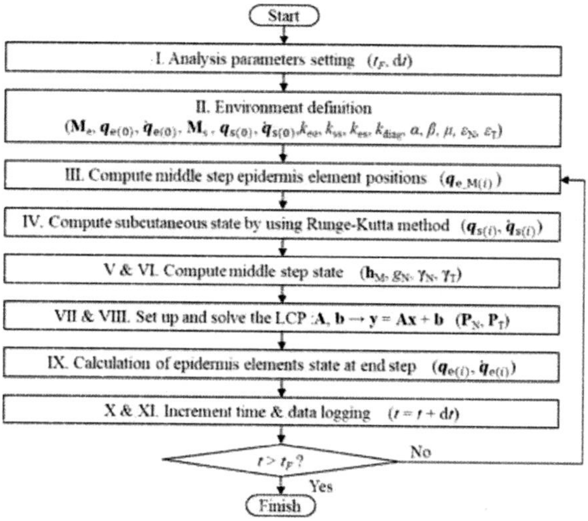

Fig. 3 Analysis flowchart

$$\dot{q}_{e(i)} = [\dot{x}_{e1(i)}\dot{y}_{e1(i)}\dot{x}_{e2(i)}\dot{y}_{e2(i)}\dot{x}_{e3(i)}\dot{y}_{e3(i)}\dot{x}_{e4(i)}\dot{y}_{e4(i)}\dot{x}_{e5(i)}\dot{y}_{e5(i)}]^T \tag{2}$$

$$q_{s(i)} = [x_{s1(i)}y_{s1(i)}x_{s2(i)}y_{s2(i)}x_{s3(i)}y_{s3(i)}x_{s4(i)}y_{s4(i)}x_{s5(i)}y_{s5(i)}]^T \tag{3}$$

$$\dot{q}_{s(i)} = [\dot{x}_{s1(i)}\dot{y}_{s1(i)}\dot{x}_{s2(i)}\dot{y}_{s2(i)}\dot{x}_{s3(i)}\dot{y}_{s3(i)}\dot{x}_{s4(i)}\dot{y}_{s4(i)}\dot{x}_{s5(i)}\dot{y}_{s5(i)}]^T \tag{4}$$

Numerical analysis is performed in the following 12 steps.

I. Set the conditions for numerical analysis. In this research, define the analysis end time t_F and the numerical integration step time dt.
II. Set the initial state and environmental conditions of the system to be analyzed. In this study, the mass matrix $\mathbf{M}_e$ of the epidermis element (5 × 5 diagonal matrix with all elements m_e), the initial position $q_{e(0)}$, the initial velocity $\dot{q}_{e(0)}$, the mass of the subcutaneous tissue element $\mathbf{M}_s$ (5 × 5 diagonal matrix with all elements in m_s), initial position $q_{s(0)}$, initial velocity $\dot{q}_{s(0)}$, , spring constant of the virtual spring connecting each element, dumping coefficient (Rayleigh damping, coefficient for mass α, the coefficient for stiffness β), the friction coefficient μ between the skin tissue and the object, the normal restitution coefficient ε_N, and the tangential restitution coefficient ε_T.
III. Following the rules of Moreau's time-stepping method, determine the position $q_{e\_M(i)}$ of the skin mass at the step time midpoint t_M.
IV. Numerical integration is performed using the Runge–Kutta method to obtain the position $q_{s(i)}$ and velocity $\dot{q}_{s(i)}$ of the subcutaneous tissue. In this step, the positions of the epidermal tissue are fixed on $q_{e\_M(i)}$ and the velocities are

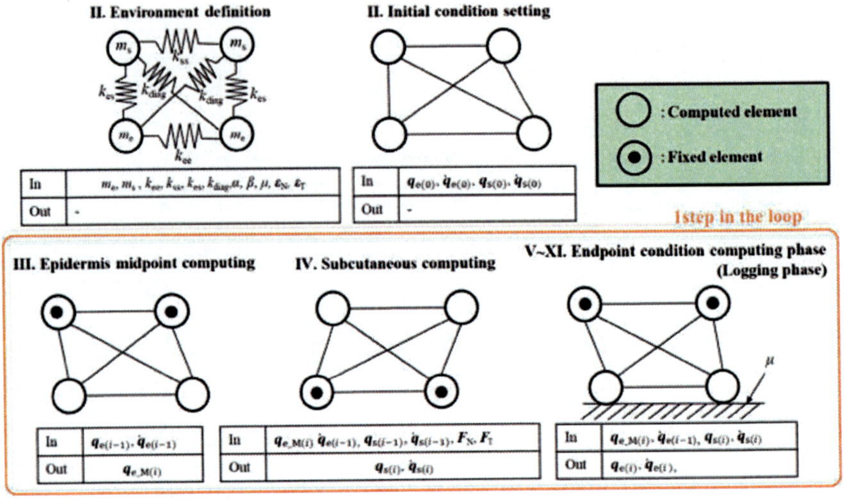

Fig. 4 Analysis overview diagram

fixed on $\dot{\boldsymbol{q}}_{\mathrm{e}(i-1)}$ because the epidermal tissue is analyzed by the Moreau's time-stepping method on other steps. In this study, the step size of the numerical integration using Runge–Kutta method is set to dt/10 compared to the step size dt of the Moreau's time-stepping method to obtain the analysis results with sufficient accuracy.

V. Position $\boldsymbol{q}_{\mathrm{e\_M}(i)}$ and velocity $\dot{\boldsymbol{q}}_{\mathrm{e}(i-1)}$ of the epidermal tissue obtained in III, and position $\boldsymbol{q}_{\mathrm{s}(i)}$ and velocity $\dot{\boldsymbol{q}}_{\mathrm{s}(i)}$, the external force $\mathbf{h}_{\mathrm{M}}$ (spring force) on the epidermal tissue when the position and velocity of the subcutaneous tissue is fixed is obtained.

VI. Calculate the vector $\boldsymbol{g}_{\mathrm{N}}$ $(= [g_{\mathrm{N1}}\ g_{\mathrm{N2}}\ g_{\mathrm{N3}}\ g_{\mathrm{N4}}\ g_{\mathrm{N5}}]^{\mathrm{T}})$ that aligns the gaps, which are relative coordinates between the epidermal tissue and the plane. Also, calculate the relative velocity vector $\boldsymbol{\gamma}_{\mathrm{N}}$, and the relative velocity vector $\boldsymbol{\gamma}_{\mathrm{T}}$ in the tangential direction. Since the position of the plane is fixed under the conditions of this paper, the relative velocity is equal to the skin mass velocity $\dot{\boldsymbol{q}}_{\mathrm{e}(i-1)}$.

VII. To account for non-smooth contact in the analysis, using the external force $\mathbf{h}_{\mathrm{M}}$ and the relative velocities $\boldsymbol{\gamma}_{\mathrm{N}}$ and $\boldsymbol{\gamma}_{\mathrm{T}}$ obtained in V and VI, derive the LCP of impulse in the i-th step [7].

VIII. Solve the LCP using Lemke's method [8] and obtain normal direction contact impulse $\mathbf{P}_{\mathrm{N}}$ and tangential direction contact impulse $\mathbf{P}_{\mathrm{T}}$.

IX. Find the velocity $\dot{\boldsymbol{q}}_{\mathrm{e}(i)}$ and the position $\boldsymbol{q}_{\mathrm{e}(i)}$ of the skin mass at the end of the i-th step.

X. Store the data and advance step.

XI. Repeat III to X until $t \geq t_F$.

By using this method, the effects of non-smooth contact such as friction and flexibility can be considered with sufficiently reduced time delay for motion analysis of systems including flexible parts such as fingers.

4 Analysis Conditions and Results

In this paper, the analysis conditions are set to the values shown in Table 1. The masses m_e and m_s of each element are 2.5 [g] by dividing the mass of a human index finger (approximately 25 [g]) into 10 equal parts. The coefficient of friction μ is set to 2 based on the results of actual measurements by Yamaba et al. [3]. The spring constants are given values that are considered appropriate from the analysis results, but in the future, it will be necessary to give appropriate values after careful examination by actual measurement experiments. The damping is given as Rayleigh damping, and the values proposed by Wu et al. [9] are used for the coefficient α for the inertia term and the coefficient β for the stiffness term. The step time was set to 0.0001 [s], which is enough to obtain analysis results with sufficient accuracy.

Figures 5 and 6 show the time series of the position of each mass point modeled on the X-axis and Y-axis, respectively. When $F_N = 1.5$ [N], the displacement to the epidermal tissue repeatedly starts and stops. This is because when a step-like

Table 1 Analysis parameters

Parameter		Symbol	Unit	Value
Mass of each element		m_e, m_s	kg	2.5*10^{-3}
Coefficient of friction		μ	-	2
Coefficient of restitution	Normal direction	ε_N	-	1
	Tangential direction	ε_T	-	1
Spring rate	Epidermis-Epidermis	k_{ee}	N/m	200
	Epidermis-Subcutaneous	k_{es}	N/m	200
	Subcutaneous-Subcutaneous	k_{ss}	N/m	200
	Epidermis-Subcutaneous(diagonal)	k_{diag}	N/m	100
Rayleigh damping $\mathbf{C} = \alpha\mathbf{M} + \beta\mathbf{K}$	Mass factor	α	N/m/s	1.0*10^{-7}
	Stiffness factor	β	N/m/s	0.4*10^{-3}
Input force on bone	X axis	F_T	N	2.5
	Y axis (Pressing force)	F_N	N	1.5
Step time		dt	s	0.0001
Analysis time		t_F	s	0.2

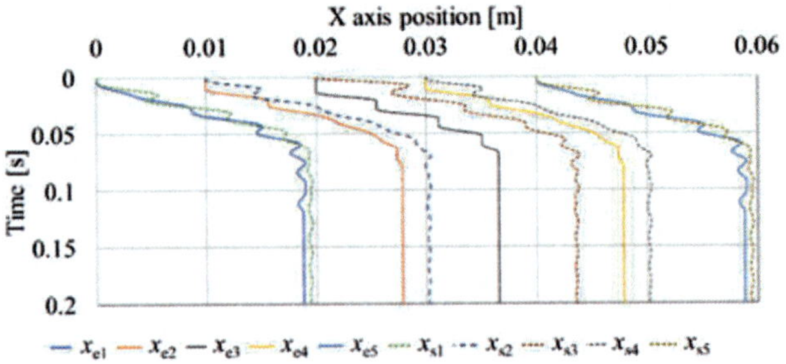

Fig. 5 Analysis result (X axis)

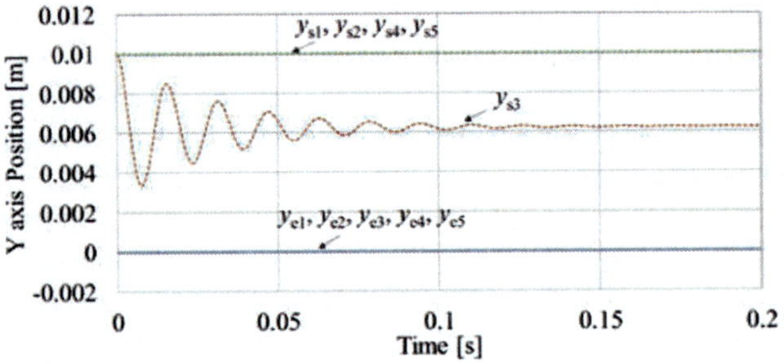

Fig. 6 Analysis result (Y axis)

force is transmitted from the bone to the epidermis, a vibration occurs in the normal contact force between the epidermis and the plane in a transient state. This vibration repeatedly exchanges the inequalities of the frictional force and the tangential contact force. Considering the movement of the whole finger, it represents the phenomenon of slipping accompanied by the stick–slip phenomenon. When the conditions were changed, some results showed deformation but no movement, while others showed iso-accelerative motion. These results are consistent with empirical results.

5 Conclusion

This study proposes modeling of human finger-plane contact at first. Contact model consists of mass points and virtual springs to introduce friction phenomenon. This study also proposes numerical analysis method suitable for prediction of motion that have contact phenomenon between soft material like a human finger and plane. This method consists with combination of Runge–Kutta method and Moreau's time-stepping method. When the sliding motion of the fingertip on the plane was analyzed by the proposed method, the sliding with stick–slip phenomenon, which occurs in actual motion, was observed. So, proposed method is suitable for motion analysis of human finger-plane contact. There are still parameters that have not been fully explored, such as the mass of the skin tissue and the spring constant between skin tissues, so it is needs to measure the motion to get a more appropriate value. In the future the correlation with the number of elements will be investigated in order to represent the motion of skin tissue with higher accuracy. In addition, it is needs to continue to examine the applicability of the proposed method to various parts of the body and movement.

Acknowledgements This work was supported by JST SPRING, Grant Number JPMJSP2103.

References

1. Szondy, D.: Robear robot care bear designed to serve Japan's aging population, from New Atlas. https://newatlas.com/robear-riken/36219/. Accessed 12 Sept 2022
2. Kobayashi, H., Uchimura, A., Ishida, Y., Shiiba, T., Hiramatsu, K., Konami, M., Matsushita, T., Sato, Y.: Development of muscle suit for upper body – realization of abduction motion. Adv. Robot. **18**(5), 497–513 (2004)
3. Yamaba, K., Takano, H., Ozaki, S.: Physical characteristics of the human finger. Biomechanisms **3**, 27–36 (1975). (in Japanese)
4. Shimada, A., Han, H., Kawamura, S.: Analysis of friction characteristics on human fingers. Trans. Soc. Instrum. Control Eng. **32**(12), 1581–1587 (1996). (in Japanese)
5. Flores, P., Leine, R., Glocker, C.: Modeling and analysis of planar rigid multibody systems with translational clearance joints based on the non-smooth dynamics approach. Multibody Syst. Dyn. **23**, 165–190 (2010)
6. Moreau, J.J.: Unilateral contact and dry friction in finite freedom dynamics. Nonsmooth Mech. Appl. **302**, 1–82 (1988)
7. Glocker, C., Studer, C.: Formulation and preparation for numerical evaluation of linear complementarity systems in dynamics. Multibody Syst. Dyn. **13**, 447–463 (2005)
8. Cottle, R.W., Dantzig, G.B.: Complementary pivot theory of mathematical programming. Linear Algebra Appl. **1**, 103–125 (1968)
9. Wu, J.Z., Krajnak, K., Welcome, D.E., Dong, R.G.: Analysis of the dynamic strains in a fingertip exposed to vibrations: correlation to the mechanical stimuli on mechanoreceptors. J. Biomech. **39**, 2445–2456 (2006)

Using Eye-Tracking Data to Investigate the Noticeability of Safety Pictogram on Transparent Packaging

Dorotea Kovačević, Maja Brozović, and Daria Mustić

Abstract The aim of the study was to investigate the influence of packaging transparency on the visual attention directed towards the information printed on the packaging surface. In order to get objective measures of participants' attention, their eye-movements were recorded by eye-tracking technology, and eye-fixation data for each visual element were analyzed. Given that visual signs play an important role in effective warning communication via product packaging, the study was mainly focused on the noticeability of the safety pictogram. The results showed that the packaging transparency did not influence the noticeability of the safety pictogram, but had a negative effect on the perception of less prominent information, such as the text displayed in small type. Additionally, the experimental manipulation of the pictogram's thickness demonstrated that the pictogram's visual appearance had a significant effect on attention capture. Taken together, the findings suggest that the noticeability of the small textual information is more sensitive to surface transparency than the pictorial information, and that thin safety pictograms capture more visual attention than the thick ones. Implications for an effective information presentation on transparent packaging were discussed.

Keywords eye-tracking · packaging · safety · pictogram · transparency

1 Introduction

Most products that people use every day include information on the packaging. If a product is potentially dangerous, it is necessary to provide the appropriate safety information. Displaying the information regarding product hazards can contribute to users' safety and, consequently, their satisfaction. The better the users perceive the risks of improper use of a product, the more likely they are to be protected from the unwanted consequences which could be dangerous for peoples' safety and health.

D. Kovačević (✉) · M. Brozović · D. Mustić
Faculty of Graphic Arts, University of Zagreb, Zagreb, Croatia
e-mail: dorotea.kovacevic@grf.unizg.hr

I. Salopek Čubrić et al. (eds.), *Proceedings of the 9th International Ergonomics Conference*, Lecture Notes in Networks and Systems 701,
https://doi.org/10.1007/978-3-031-33986-8_17

Previous research suggests that safety messages are more likely to be perceived if they are displayed in a form of a pictorial sign [1–3]. Still, using the pictograms is not an assurance that the user will definitely notice the safety message, especially if the pictograms are applied on packaging material which reduces the visibility of the printed information. Therefore, one of the aims of our study was to investigate the effect of packaging transparency on the noticeability of the safety pictogram.

Packaging transparency was selected for the investigation because of its successful use in product presentation. Many previous studies showed that material transparency is an advantageous attribute of packaging [4] and that consumer expectations of product quality can rely on transparency [5]. It was found that this can led to a higher willingness to purchase [6] and purchase decisions [7]. A recent study [8] showed that even a small part of packaging, such as a transparent window, can capture more attention compared to a completely opaque packaging which does not allow the consumer to see the product directly. However, despite the powerful role of packaging transparency in product presentation, the potentially negative effects on the information legibility should also be taken into consideration, especially when the packaging conveys safety messages.

In order to investigate the visual attention of users to safety messages on packaging, we used eye tracking technology in this study. This approach was selected because eye movements are considered as a key measure of attention to visual cues on product packaging [9]. Accordingly, eye tracking is a commonly used method for objective evaluation of the noticeability of visual information in a given surrounding [10], as well as for the measurement of consumer perception of warnings on product packaging [11, 12] and in other ergonomics studies [13, 14]. Our study aims to extend the existing research-based knowledge on users' visual perception of packaging information by employment of eye tracking equipment.

2 Methodology

In order to examine the influence of packaging and pictogram attributes on visual perception, two experiments were conducted. The first experiment investigated the effect of packaging transparency on the participants' fixations on different types of information displayed on the packaging. The second experiment examined the influence of the safety pictogram by manipulating its line weight. The participants' visual attention was measured directly using eye-tracking equipment in both experiments.

2.1 Experiment 1

Seventeen people (13 females and 4 males, mean age of 22.23 years, SD = 2.02) participated in the first experiment. The experiment was conducted in a research cabinet at the University of Zagreb Faculty of Graphic Arts. The eye-fixation data was recorded by Tobii Eye Tracker X60 (sampling rate of 60 Hz and an accuracy of 0.5 degree). The three-dimensional models of packaging were presented on a 17-inch computer display with a resolution of 1280 × 1024 pixels.

The packaging samples were pictures of three-dimensional product packaging in standard angular form and with typical graphic design. Both samples had same design elements and composition, but differed in the transparency of the packaging material. The material transparency was simulated by showing the product placed inside the packaging (an air-freshener container). The safety pictogram presented the warning message on prohibition of directing the air-freshener toward an open flame.

In order to avoid a goal directed visual attention process, the participants were instructed to view the packaging freely. For each participant the eye-tracking procedure started with a 5-point calibration. After the calibration, the participant's task was to fixate a white symbol "x" in the middle of the black computer screen for a duration of 1500 ms. Next, the picture of one packaging sample was displayed for 5 s. The same procedure was carried out for another packaging sample. The order of presenting the packaging samples was counterbalanced across the participants (Fig. 1).

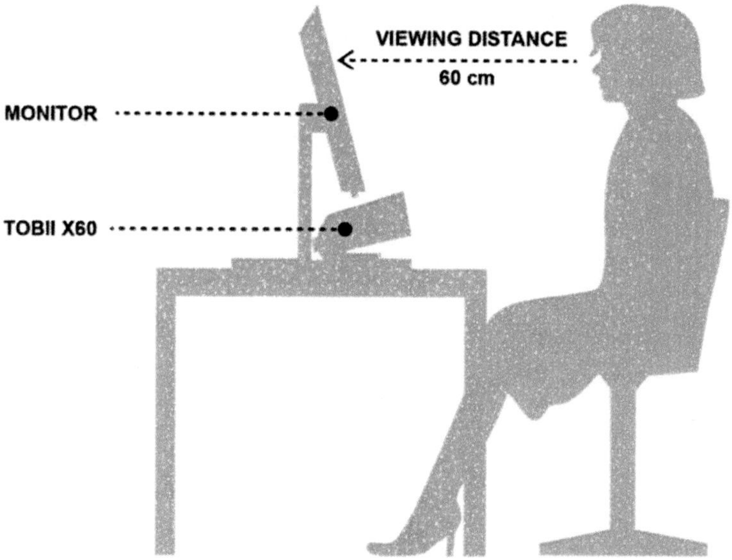

Fig. 1 Experiment setup

2.2 Experiment 2

A new group of seventeen persons participated in the second experiment. There were 10 females and 7 males (mean age of 21.88 years, SD = 2.96). In this experiment, the packaging samples differed in the level of the pictogram's thickness, so the participants were presented with one packaging containing a thin pictogram and with another one containing a thick pictogram. The experimental procedure was the same as in the first experiment.

2.3 Dependent Variables

During the experiments, the following eye-tracking data for each participant were measured: time to first fixation (TFF) on the area of interest and the first fixation duration (FFD). TFF was defined as the time measured from the packaging onset until the area of interest was fixated for the first time [15]. This measure was considered as an indicator of noticeability in previous studies [16–18]. FFD was defined as the time duration (in seconds) for the first fixation on the area of interest. The overall distribution of participants' visual attention was also recorded by the eye-tracker and presented by heat maps.

3 Results

3.1 Experiment 1

In the first experiment, the eye fixation data were calculated for four areas of interest on the packaging front panel. The areas of interest were: the safety pictogram, the imagery, the product name and the quantity information. The data were not normally distributed so nonparametric statistical tests were used. The median of the TFF and the FFD were taken as dependent variables. Comparisons between the groups which were performed using the Wilcoxon Signed Ranks Test (Tables 1 and 2) revealed the significant influence of packaging transparency on the visual perception of the quantity information. Specifically, the participants needed less time to detect the quantity information on the non-transparent packaging (Mdn = 0.01, range 0.00–3.57), than on the transparent packaging (Mdn = 2.09, range 0.00–4.25), $Z = -2.69$, $p < 0.01$. Furthermore, the quantity information held the participants' attention longer on the transparent packaging (Mdn = 0.20, range 0.00–0.97) than on the non-transparent packaging (Mdn = 0.00, range 0.00–0.45), $Z = -2.66$, $p < 0.01$. There was no statistical significant difference between the perception of the safety pictogram on the transparent and non-transparent packaging for both variables ($p = 0.33$ for TFF, $p = 0.23$ for FFD).

Table 1 Wilcoxon Signed Ranks Test for TFF (Transparent versus non-transparent packaging)

	Safety pictogram	Imagery	Product name	Quantity information
Z	−0.966	−0.970	−1.657	−2.691
Exact Sig. (2-tailed)	0. 334	0.332	0.097	0.007

Table 2 Wilcoxon Signed Ranks Test for FFD (Transparent versus non-transparent packaging)

	Safety pictogram	Imagery	Product name	Quantity information
Z	−1.194	−1.232	−0.517	−2.656
Exact Sig. (2-tailed)	0. 233	0.218	0.605	0.008

3.2 Experiment 2

In the second experiment, the noticeability of the safety pictogram was the object of the investigation and therefore the eye fixation data were calculated for one area of interest only (namely, the pictogram on the packaging front panel). The results showed no significant difference in TFF between the thick and the thin pictogram ($p = 0.25$). In contrast, there was a significant difference in FFD ($Z = -2.42$, $p < 0.05$), indicating that the thin pictogram held the participants' attention longer (Mdn = 0.29, range 0.00–0.88) than the thick pictogram (Mdn = 0.17, range 0.00–0.38). These results were in accordance with the heat maps generated by the eye-tracking tools used in the experiment (Fig. 2). Heat maps are commonly used in eye-tracking

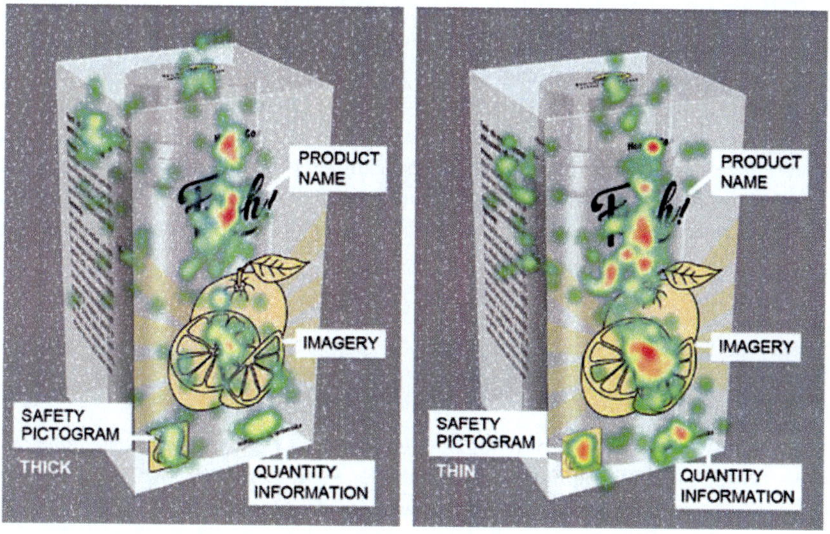

Fig. 2 Heat maps presenting the distribution of the participants' visual attention

research to present the parts of the stimuli that viewers pay attention to [19]. We used these eye-tracking visualizations to compare the fixations' distribution on the packaging with thick and thin pictogram.

4 Discussion

The results of our study showed that the packaging transparency did not influence the noticeability of the safety pictogram, but it had a negative effect on the perception of less prominent information, such as quantity information printed in small type size. Decreased noticeability of small textual information was reported in similar studies which investigated the size effects on eye-tracking data [12]. In our study, the visibility of the inner content (i.e. the product) was enabled by material transparency and this led to an even greater lack of visual prominence. This implies that the messages designed in textual form and printed in small size may result in decreased noticeability. This stresses the need for deliberate design of textual safety messages applied on the surface of transparent packaging.

In our study, the safety message was designed in the form of a pictorial sign which attracted and held the participants' visual attention easily. Similar results were found by Tórtora et al. who reported on the beneficial effects of the warning sign on people's attention [9]. By contributing to similar findings of previous studies on human factors and ergonomics [12, 20, 21], this result can be considered as another research-based support for the use of pictorial warnings in effective warning communication in the context of product packaging.

Another relevant finding of our study is that the weight of the pictogram's lines can influence the user's first fixation duration and prolong the attention maintenance. As stated by Wogalter, attention maintenance is an important part of effective warning communication [22]. On the other hand, we found no significant influence on the pictogram noticeability. This is in contrast with the results of the previous studies which demonstrated how the line thickness can be powerful in attracting people's visual attention within warning communication [12, 23]. It seems that more research is needed to reveal all the aspects of this particular design attribute in visual processing of the warning message.

5 Conclusions

The aim of the study was to investigate the influence of packaging transparency on users' visual attention to information on the packaging front panel. Therefore, the eye-fixation data for each visual element were analyzed.

It was found that the transparency of the packaging had a different impact depending on the format of visual information. Particularly, the transparent surface did not reduce the visibility of the safety pictogram, but it had an undesirable effect on the attention paid to the textual information in small size. In the light of these results, future research should focus on additional investigation of the visual perception of product-related information which could be relevant to the safety of the end user, but which cannot be displayed in optimal size due to the limited area on the packaging surface.

The main limitation of our study is the fact that the experiment was carried out in controlled viewing conditions. Due to the isolated experimental environment, it was not possible to get the same results as in realistic surroundings and everyday circumstances in which users observe the packaging before using the product. Regardless of the limitations, the findings of this study are based on objective and directly measured data which can contribute to a better understanding of packaging effects on the users' processing of safety messages.

References

1. Laughery, K.R., Young, S.L., Vaubel, K.P., Brelsford, J.W.: The noticeability of warnings on alcoholic beverage containers. J. Public Policy Mark. **12**(1), 38–56 (1993)
2. Young, S.L.: Increasing the noticeability of warnings: effects of pictorial, color, signal icon and border. Proc. Hum. Factors Soc. Ann. Meet. **35**(9), 580–584 (1991)
3. Lehto, M.R.: Designing warning signs and warning labels: Part I — guidelines for the practitioner. Int. J. Ind. Ergon. **10**(1–2), 105–113 (1992)
4. Eldesouky, A., Pulido, A.F., Mesias, F.J.: The role of packaging and presentation format in consumers' preferences for food: an application of projective techniques. J. Sens. Stud. **30**(5), 360–369 (2015)
5. Chandran, S., Batra, R., Lawrence, B.: Is seeing believing? Consumer responses to opacity of product packaging. In: McGill, A.L., Shavitt, S. (eds.) Advances in Consumer Research, pp. 970–971. Association for Consumer Research, Duluth (2009)
6. Simmonds, G., Woods, A.T., Spence, C.: 'Show me the goods': assessing the effectiveness of transparent packaging vs. product imagery on product evaluation. Food Qual. Prefer. **63**, 18–27 (2018)
7. Pal, S., Yevalkar, A., Bhattacharjee, A., Holkar, S.: Consumers' preference leading purchase intention toward manipulation of form and transparency for juice packaging design. J. Appl. Packag. Res. **10**(4), 79–88 (2018)
8. Ma, X., Zhuang, X., Ma, G.: Transparent windows on food packaging do not always capture attention and increase purchase intention. Front. Psychol. **11**(Nov) (2020)
9. Tórtora, G., Machín, L., Ares, G.: Influence of nutritional warnings and other label features on consumers' choice: results from an eye-tracking study. Food Res. Int. **119**(Oct), 605–611 (2019)
10. Kovačević, D., Brozović, M.: Pictograms on packaging: a comparison of objective and subjective measures of their noticeability, In: Pavlović, Ž. (eds.) Proceedings - The Eighth International Symposium GRID 2016, pp. 375–380. Faculty of Technical Sciences, Department of Graphic Engineering and Design, Novi Sad (2016)
11. Sillero-Rejon, C., et al.: Avoidance of tobacco health warnings? An eye-tracking approach. Addiction **116**(1), 126–138 (2021)
12. Kovačević, D., Brozović, M., Možina, K.: Do prominent warnings make packaging less attractive? Saf. Sci. **110**(Part A), 336–343 (2018)
13. Carr, D.B., Grover, P.: The role of eye tracking technology in assessing older driver safety. Geriatrics **5**(2), 1–14 (2020)

14. Han, Y., Yin, Z., Zhang, J., Jin, R., Yang, T.: Eye-tracking experimental study investigating the influence factors of construction safety hazard recognition. J. Constr. Eng. Manag. **146**(8), 1–13 (2020)
15. Lai, M.L., et al.: A review of using eye-tracking technology in exploring learning from 2000 to 2012. Educ. Res. Rev. **10**(88), 90–115 (2013)
16. Graham, D.J., Orquin, J.L., Visschers, V.H.M.: Eye tracking and nutrition label use: a review of the literature and recommendations for label enhancement. Food Policy **37**(4), 378–382 (2012)
17. Krugman, D.M., Fox, R.J., Fletcher, J.E., Rojas, T.H.: Do adolescents attend to warnings in cigarette advertising? An eye-tracking approach. J. Advert. Res. **34**(6), 39–52 (1994)
18. Roth, S.P., Tuch, A.N., Mekler, E.D., Bargas-Avila, J.A., Opwis, K.: Location matters, especially for non-salient features-an eye-tracking study on the effects of web object placement on different types of websites. Int. J. Hum. Comput. Stud. **71**(3), 228–235 (2013)
19. Zhou, C., Yuan, F., Huang, T., Zhang, Y., Kaner, J.: The impact of interface design element features on task performance in older adults: evidence from eye-tracking and EEG signals. Int. J. Environ. Res. Public Health **19**(15), 1–24 (2022)
20. Kovačević, D., Brozović, M., Možina, K.: Improving visual search in instruction manuals using pictograms. Ergonomics **59**(11), 1405–1419 (2016)
21. Laughery, K.R., Wogalter, M.S.: A three-stage model summarizes product warning and environmental sign research. Saf. Sci. **61**, 3–10 (2014)
22. Leonard, S.L., Wogalter, M.: Attention capture and maintenance. In: Wogalter, M.S., DeJoy, D.M., Laughery, K.R. (eds.) Warnings and Risk Communication, pp. 123–148. Taylor & Francis, London (1999)
23. Braun, C.C., Silver, N.C., Stock, B.R.: Likelihood of reading warnings: the effect of fonts and font sizes. Proc. Hum. Factors Ergon. Soc. Ann. Meet. **36**(13), 926–930 (1992)

Ergonomics in Ukraine

E. Lavrov

Abstract The article analyzes the development of ergonomics in Ukraine. It shows the history of the emergence of ergonomics, starting from the Soviet Union, and the succession of stages in the development of ergonomics, beginning with Professor Vladimir Bekhterev. The current state of the sciences that study the human factor is analyzed. The achievements of the scientific schools of Professors Anatoly Gubinsky, Vladimir Evgrafov, Akiva Asherov, Vladimir Voinenko, Alexander Rotstein, Pavel Chabanenko, Eduard Gavrilov, Viktor Dolia, Alexander Burov and other scientists are analyzed and described. The contribution of Ukrainian scientists to the theory of mathematical description of human–machine interaction, the methodology of evaluation and optimization of operator reliability, as well as to the theory of automation of ergonomic research is analyzed. Achievements in the field of ergonomics of information technologies, ergonomics of automated production control systems, ergonomics of transport systems, and ergonomics of e-learning are characterized. The state of the introduction of scientific achievements into university disciplines devoted to the study of the "human factor" is described.

Keywords Ergonomics · Man–Machine · Human Factor · Reliability · Ukraine

1 Introduction

The Fourth Industrial Revolution, comprehensive computerization, and the introduction of artificial intelligence not only do not exclude humans from modern automated systems, but also increase their role and aggravate the problem of the "human factor" [1–4]. We are seeing increased attention to [3–7]:

- ergonomic science,
- widespread use of the achievements of ergonomic science,

E. Lavrov (✉)
Sumy State University, Sumy, Ukraine
e-mail: prof_lavrov@hotmail.com

I. Salopek Čubrić et al. (eds.), *Proceedings of the 9th International Ergonomics Conference*, Lecture Notes in Networks and Systems 701,
https://doi.org/10.1007/978-3-031-33986-8_18

- ergonomic education,
- development and implementation of ergonomic standards, etc.

2 Problem Statement

The avalanche-like growth of publications and conferences devoted to ergonomics, for example [5–11], activation of national ergonomic associations, as well as the international ergonomic association, testify to the growing interest in the problems of "human factor". Due to the various problems in recent years the possibility of a wide presentation in the world scientific press of the state, achievements and problems of ergonomic research in such a large country as Ukraine, unfortunately, were limited.

Due to the crisis and war in Ukraine:

- new scientific results of Ukrainian scientists-ergonomists are poorly presented in English-language journals
- number of international projects in the field of ergonomics with the participation of scientists from Ukraine has decreased
- experience of teaching ergonomics in Ukraine is practically unknown in the world

Formulation of the Problem. Proceeding from this and in accordance with the existing interest of the world ergonomic scientific community in the present article we set the task to describe the history and state of ergonomic research in Ukraine.

3 Methods

To analyze the state of the issue we use the method of literary review, as well as personal experience of participation in various projects related to the "human factor" of the author of this article. The databases of the National Library of Ukraine named after Vernadsky, as well as Scopus and Web of Science were used to search for literature. To classify the results, we use a systematic approach to ergonomic studies [10–17]. This approach made it possible to systematize the presentation of ergonomics in Ukraine as follows (Fig. 1).

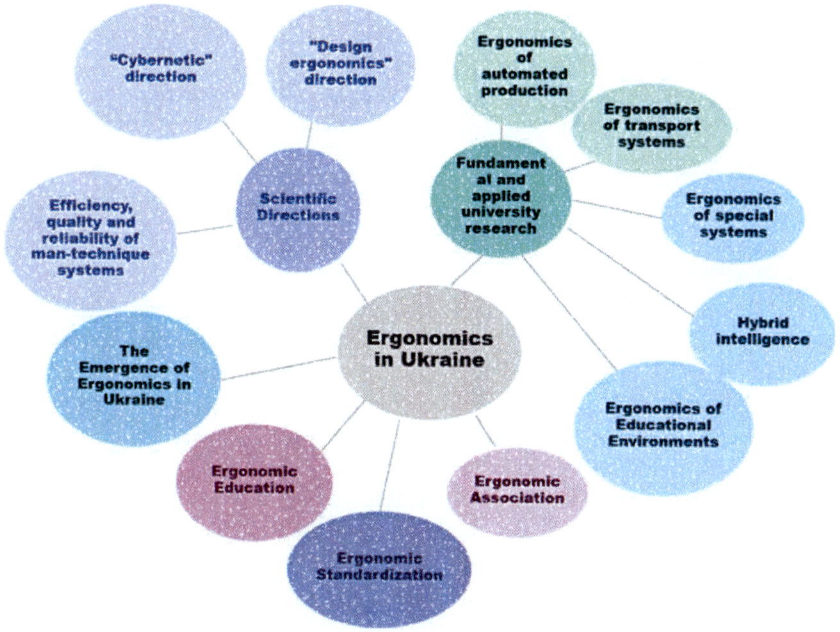

Fig. 1 The structure of the subject area "Ergonomics in Ukraine" (fragment)

4 Results

4.1 The Emergence of Ergonomics in Ukraine

Until 1991, Ukraine was a part of the Soviet Union, so the development of science in general and ergonomics in particular was determined by the development of Soviet science. As an important moment of the initial stage we should mention the ideas of academician V. M. Bekhterev, who in 1908 created the Psychoneurological Institute in St. Petersburg, and in 1918—The Institute of Brain and Mental Activity. In 1921, academician V. M. Bekhterev formulated a principally new approach to the problem of studying human labor: *"It is necessary that labor should yield maximum productivity at the optimum or maximum health in the absence not only of overwork, but also in the guarantee of full health and personal development of the worker"*. Although ergonomics had not yet formed into an independent scientific branch in the 20 s and 30 s, its goals and objectives were defined and organizational forms of research were outlined [12]. The period of the beginning of the intensive systematic development of ergonomics falls at the end of the 50 s. The first "unclassified" laboratory of industrial psychology was organized by B. F. Lomov at Leningrad State University in 1959. Boris F. Lomov played a key role in the development of engineering psychology in the USSR [12, 13] The main center of ergonomic research was located in Moscow at the All-Union Research Institute of Technical Aesthetics (VNIITE), and branches were located in several cities of the USSR, including Ukraine (Kiev, Kharkov, etc.). Thus, by the mid-60 s an organizational system for complex ergonomic research was created in the USSR, including Ukraine, and special "ergonomic" courses began to be introduced in universities.

4.2 *Scientific Directions*

Scientific Area "Efficiency, Quality and Reliability of Man-Technique Systems". One of the prominent organizers of ergonomic research was Professor Anatoly I. Gubinsky. In 1959, he took part in the first trip of a nuclear submarine through the Arctic Ocean to the North Pole. His researches in the field of ergonomic support were inspired by his experience in the Navy and practical acquaintance with real ergonomic problems. Together with Vladimir G. Evgrafov (who also served in the Navy), Anatoly I. Gubinsky developed the theory of description, evaluation and optimization of discrete algorithms for ship control systems, which he later extended to man–machine systems of various purposes [14]. In 1981, under the leadership of A. I. Gubinsky, the formation of the scientific direction "Efficiency, quality and reliability of "man-technique" systems" was completed [14]. From 1966 to 1990, A. I. Gubinsky organized regular symposia on "Efficiency, quality and reliability of "man-technique" systems", which allowed to create a scientific school involving scientists from almost all Union republics (Georgia, Latvia, Lithuania, Estonia, Armenia and others). The scientific school was represented by a large number of scientists from Ukraine (A. T. Asherov, P. P. Chabanenko, E. A. Pavlov, G. K. Kozhevnikov, A. P. Rotstein, etc.). As a result, a method for describing and evaluating the reliability of man–machine interaction using functional networks was developed as part of the scientific branch [14, 15]. During the same years a number of Ukrainian organizations, including sectoral research laboratory of industrial cybernetics (headed by Akiva T. Asherov) of the Ukrainian Correspondence Polytechnic Institute (Kharkov) faced the problem of low reliability of automated control systems of industrial processes (metallurgy, chemical production, machine building), which they were developing and implementing nationwide (Kharkov, Sumy, Magnitogorsk, Kalush, etc.). An urgent task arose—to find the "bottleneck" (man or machine?). The solution had to be based on the theory of Professor A. Gubinsky. The systemic application of functional networks made it possible to solve the problem on the national scale. Thus, the wide implementation and development of scientific direction "Efficiency, quality and reliability of man-technique systems" in Ukraine began.

The "Cybernetic" Direction. The Institute of Cybernetics of the Academy of Sciences of Ukraine becomes the center of research of human factor problems, in terms of cybernetics. It was here in 1950, under the leadership of Academician S. A. Lebedev, the first in the USSR and continental Europe Small Electronic Computing Machine "MESM" was created. In 1963, the director of the Institute Victor M.

Glushkov put forward the idea of uniting computer centers into a nation-wide network and on this basis building a nation-wide automated control system (it was historically the first approach to creating global networks; at that time no one even thought of the Internet anywhere in the world) [16]. The Institute carried out research on optimization theory, automata theory, artificial intelligence theory, etc. The methods and tools of cybernetics were extended to a number of sciences, including medicine, engineering psychology, and ergonomics [16].

"Design Ergonomics" Direction. The system of VNIITE branches carried out theoretical and applied research in the field of design. Ergonomics departments in these institutes developed normative documentation, methods of ergonomic expertise and organized the introduction of ergonomic methods into production and all spheres of human life. For example, the director of the Kiev institute, Vladlen M. Voinenko (who later immigrated to the United States), prepared a fundamental book on the principles of ergonomic design [17]. As an example of this institute's work, one can mention the work on the creation of an ergonomic data bank (international project).

Fundamental and Applied University Research. University science has become the most flexible form of ergonomic research (there are more than 300 universities in Ukraine). Here are just some examples of directions of ergonomic research:

Ergonomics of Automated Production and Information Technology. Traditionally it was and it is a subject of research of many scientists. The Ukrainian Correspondence Polytechnic Institute was one of the centers. The work was headed by A. Asherov, who became the organizer of the development of the scientific direction of A.Gubinsky "Efficiency, quality and reliability of man-technology systems" in Ukraine. He proved an increasing role of man in the development of information technology, he developed the mathematical foundations of the learning theory. His disciples, Evgeny A. Pavlov, Georgy K. Kozhevnikov, Sergey G. Isayenko, Tatiana V. Yashchun, and others, investigated the ergonomic support of control systems in aviation, flexible production systems, main gas pipelines, metallurgical enterprises, electron-beam welding, e-learning systems, etc. Today these studies of the human factor continue [18–24] and in a number of other universities—Zaporozhye Technical University (Sergiy N. Serdyuk and others), Sumy National Agrarian University (Nadia B. Pasko), etc.

Ergonomics of Transport Systems. In Ukraine, the research was initiated by Eduard V. Gavrilov at Kharkov Automobile and Road University and is being continued [25–28] by scientists at a number of other universities, such as the Kharkov National University of Municipal Economy (today it has the only Scientific Council in Ukraine for the defense of candidate and doctoral dissertations in ergonomics, created on the initiative of E. V. Gavrilov and A. T. Asherov). Among the scientists-ergonomists, who studied the questions of ergonomics on motor transport, there are Linnik Irina E. (daughter of E. V. Gavrilov), Victor K. Dolya, Yuri A. Davidich, Alexey O. Lobashev, Denis P. Ponkratov, etc. A number of scientists at the Ukrainian State University of Railway Transport (headed by Vitaly G. Brusentsov), the State

Scientific Research Center of Railway Transport (headed by Valery N. Samsonkin), the Lvov National Technical University (headed by Nikolay N. Zhuk) deal with ergonomics in the railway transport, and the National Aviation University in Kiev (headed by Alexey N. Reva) and National Aerospace University, Kharkiv (headed by Galina V. Migal) deal with ergonomics in the aviation.

Ergonomics of Special Systems. Pavel P. Chabanenko (Sevastopol, Kiev), Boris M. Gerasimov (Kiev), Maxim A. Pavlenko (Kharkov), Viktor M.Strelec (Kharkov). An example of a monograph devoted to the development of the ergo-network apparatus (modernized functional networks of Professor A. I. Gubinsky) is [29].

System Analysis and Reliability Design of Human–Machine Interaction. A. T. Asherov was the organizer of the research in this direction. Chabanenko Pavel P. [29] as well as scientists of Vinnitsa Technical University Alexander P. Rothstein and Sergey D. Shtovba (fuzzy reliability, fuzzy cognitive networks, modeling automation) made a great contribution to research of modeling issues [30].

Hybrid Intelligence, Adaptive Learning. Valery F. Venda was born in Ukraine, worked for a long time in a number of universities in the United States, and currently works at the Crimean University for the Humanities. He proposed a transformational theory of learning, studied the processes of mutual adaptation in the "human–machine-environment" systems [31].

Ergonomics of Educational Environments. The work of Ukrainian scientists in this direction is supervised by the Institute of Information Technologies and Learning Tools (Kiev). The head of the research is Alexandr Yu. Burov, who, in particular, proposed new methods for analyzing the work capacity of operators, mental activity, as well as issues of cybersecurity of educational environments [32, 33]. Developments in the field of educational ergonomics are underway in many other universities as well.

4.3 Ergonomic Education

In 1982 the Leningrad Electrotechnical University under the direction of A.I. Gubinsky began retraining specialists in "ergonomics" specialty. More than 20 candidates and 3 doctors of sciences graduated from the faculty, many of whom are now working in Ukraine. Preparation of bachelors and masters in "Ergonomics" specialty in Ukraine at the present time is not carried out. PhD and doctoral theses on specialty "ergonomics" are held in Kharkiv National University of Municipal Economy.Most universities have organized disciplines in ergonomics. The concept of ergonomic education in Ukraine was proposed by A.T. Asherov. The approach to the organization of ergonomic training of Masters of Computer Science in Sumy State University is described in [34].

4.4 *Ergonomic Standardization*

After the liquidation of VNIITE, the Ukrainian Scientific Research Institute of Design and Ergonomics at the National Aviation University in Kiev (Director—Vladimir A. Svirko) is developing standards and other normative documents in the field of design and ergonomics.

4.5 *Ergonomic Association*

Due to the known difficulties, the dynamic activity of the association has been temporarily suspended. Nevertheless, scientists follow the work of international ergonomic organizations and participate in a number of international projects. The work of Ukrainian ergonomists is coordinated by a group of scientists headed by A. Yu. Burov.

5 Conclusions

Ukraine has a large scientific potential in the field of research of reliability and efficiency of "human-technical environment" systems. There is a representative number of doctors and candidates of science in the specialty "Ergonomics". Ergonomic education is represented by the training of doctors and candidates of science, as well as special courses in the curricula of bachelor's and master's degree programs. The main areas of research are: ergonomics of automated manufacturing and information technology; ergonomics of transport systems; system analysis and reliable design of human–machine interaction; hybrid intelligence and adaptive learning; ergonomics of educational environments and several others. Research is concentrated in institutes of the Academy of Sciences and universities. Scientists consider as promising areas of research the directions related to cybersecurity, minimization of disaster risks and ergonomic norm violations, intelligent interfaces, and development of hybrid intelligence systems. It is planned to resume the work of the Ergonomic Association and expand international projects.

References

1. Zhong, R.Y., Xu, X., Klotz, E., et al.: Intelligent manufacturing in the context of industry 4.0: a review. Engineering **3**(5), 616–630 (2017)
2. Chicu, D.: Exploring the influence of the human factor on customer satisfaction in call centres. BRQ Bus. Res. Q. **22**(2), 83–95 (2019)

3. Meleschenko, E.E., et al.: Human factor management tool while performing high-risk work in thermal power engineering. AIP Conf. Proc. **2467**, 080009 (2022). https://doi.org/10.1063/5.0092470
4. Zhirabok, A.N., Kalinina, N.A., Shumskii, A.E.: Technique of monitoring a human operator's behavior in man-machine systems. J. Comput. Syst. Sci. Int. **57**(3), 443–452 (2018)
5. Cacciabue, P.C.: Human error risk management for engineering systems: a methodology for design, safety assessment, accident investigation and training. Reliab. Eng. Syst. Saf. **83**(2), 229–269 (2014)
6. Pisarev, A.S., et al.: Generalized model of cognitive activity taking into account uncertainty in an information-saturated environment. In: Proceedings of 2022 25th International Conference on SCM 2022, pp. 133–137 (2022). https://doi.org/10.1109/SCM55405.2022.9794831
7. Cao, Q., et al.: Extending axiomatic design theory to human-machine cooperative products. In: 2015 IEEE 10th Conference ICIEA, Auckland, pp. 329–334 (2015)
8. Bentley, T.A.: The role of organisational support in teleworker wellbeing: a socio-technical systems approach. Appl. Ergon. **52**, 207–215 (2016)
9. Zhang, Y., Sun, J., Jiang, T., Yang, Z.: Cognitive ergonomic evaluation metrics and methodology for interactive information system. In: AHFE 2019, AISC 965, pp. 559–570 (2019). https://doi.org/10.1007/978-3-030-20454-9_55
10. Bolčević, E., Tokić, S., Sumpor, D., Horina, J.L.: A preliminary study of the possible impact of non-related driving tasks on the car driver's performance using glance-based measures. In: Sumpor, D., Jambrošić, K., Jurčević Lulić, T., Milčić, D., Salopek Čubrić, I., Šabarić, I. (eds.) ERGONOMICS 2020. Advances in Intelligent Systems and Computing, vol. 1313, pp. 33–40. Springer, Cham (2021). https://doi.org/10.1007/978-3-030-66937-9_4
11. Havlikovaa, M., Jirglb, M., Bradac, Z.: Human reliability in man-machine systems. Procedia Eng. **100**, 1207–1214 (2015)
12. Sergeev, S.F.: Engineering psychology and ergonomics: history of development, concept and the conceptual basis. Educ. Technol. (1), 44–63 (2011). (in Russian)
13. Lomov, B.F.: Man and technology. Leningrad State University (1963). (in Russian)
14. Adamenko, A.N., et al.: Information controlling human-machine systems: research, design, testing. In: Gubinsky, A.I., Evgrafov, V.G. (eds.) Mashinostroenie, Moscow, Russia (1993). (in Russian)
15. Popovich, P.R., Gubinsky, A.I., Kolesnikov, G.M.: Ergonomic support of astronauts' activities. Mechanical Engineering, Moscow, Russia (1985). (in Russian)
16. Glushkov, V.M.: Fundamentals of paperless informatics. M. Nauka (1982). (in Russian)
17. Voinenko, V.M., Munipov, V.M.: Ergonomic design principles. Kyiv, Technika (1988). (in Russian)
18. Lavrov, E., Paderno, P., Siryk, O., Burkov, E., Kosianchuk, S., Bondarenko, N., Kyzenko, V.: Analysis of working conditions and modeling of activity algorithms for contact-center operators. In: Sumpor, D., Jambrošić, K., Jurčević Lulić, T., Milčić, D., Salopek Čubrić, I., Šabarić, I. (eds.) ERGONOMICS 2020. Advances in Intelligent Systems and Computing, vol. 1313, pp. 121–128. Springer, Cham (2021). https://doi.org/10.1007/978-3-030-66937-9_14
19. Lavrov, E., Pasko, N., Siryk, O.: Information technology for assessing the operators working environment as an element of the ensuring automated systems ergonomics and reliability. In: 2020 IEEE 15th International Conference Tcset, Lviv-Slavske, Ukraine, pp. 570–575 (2020). https://doi.org/10.1109/TCSET49122.2020.235497
20. Lavrov, E., Paderno, P., Siryk, O., Kyzenko, V., Kosianchuk, S., Bondarenko, N., Burkov, E.: Functional networks for modeling and optimization human-machine systems. In: Sumpor, D., Jambrošić, K., Jurčević Lulić, T., Milčić, D., Salopek Čubrić, I., Šabarić, I. (eds.) ERGONOMICS 2020. Advances in Intelligent Systems and Computing, vol. 1313, pp. 189–196. Springer, Cham (2021). https://doi.org/10.1007/978-3-030-66937-9_21
21. Lavrov, E.A., Siryk, O.E.: Functional networks for ergonomics and reliability tasks on the 90th anniversary of A. Gubinsky and V. Evgrafov. Sigurnost **64**(3), 245–257 (2022). https://doi.org/10.31306/s.64.3.3

22. Lavrov, E., Pasko, N.: Automation of assessing the reliability of operator's activities in contact centers that provide access to information resources. In: Proceedings of the 14th International Conference on ICT in Education, Research and Industrial Applications. Integration, Harmonization and Knowledge Transfer, Kyiv, Ukraine, vol. 1, pp. 445–448 (2018)
23. Lavrov, E.A., et al.: Information technology for modeling human-machine control systems and approach to integration of mathematical models for its improvement. In: 2020 XXIII International Conference SCM, St. Petersburg, Russia, pp. 117–120 (2020). https://doi.org/10.1109/SCM50615.2020.9198791
24. Lavrov, E., et al.: Ergonomics of IT outsourcing. Development of a mathematical model to distribute functions among operators. East.-Eur. J. Enterp. Technol. **2**(4), 32–42 (2016). https://doi.org/10.15587/1729-4061.2016.66021
25. Grigorova, T., Davidich, Y., Dolya, V.: Development of the model of the change in the passenger transport fatigue when approaching stopping points of suburban bus routes. East.-Eur. J. Enterp. Technol. **2**(3), 4–9 (2015). https://doi.org/10.15587/1729-4061.2015.38583
26. Davidich, N., et al.: Intelligent decision support system for modeling transport and passenger flows in human-centric urban transport systems. Energies **15**, 2495 (2022). https://doi.org/10.3390/en15072495
27. Gavrilov, E.V., Linnik, I.E., Sirota, V.M.: Ergonomic support of traffic management. Munic. Serv. Cities. K Tech. **2004**(58), 163–169 (2004)
28. Davidich, N., et al.: Monitoring of urban freight flows distribution considering the human factor. Sustain. Cities Soc. **75**, 103168 (2021)
29. Chabanenko, P.P.: Research of the safety and efficiency of the functioning of systems "human – technics" by ergonomic networks. Sevastopol, Academy of naval forces named after P. S. Nahimov (2012). (in Russian)
30. Rotshtein, A., Katielnikov, D., Kashkanov, A.: A fuzzy cognitive approach to ranking of factors affecting the reliability of man–machine systems. Cybern. Syst. Anal. **55**, 958–966 (2019). https://doi.org/10.1007/s10559-019-00206-8
31. Venda, V.F.: On the laws of mutual adaptation and transformations of the systems. Vopr. Filos. **2**, 94–105 (2017)
32. Burov, O., Lytvynova, S., Lavrov, E., Krylova-Grek, Y., Orliyk, O., Petrenko, S., Shevchenko, S., Tkachenko, O.M.: Cybersecurity in educational networks. In: Ahram, T., Karwowski, W., Vergnano, A., Leali, F., Taiar, R. (eds.) IHSI 2020. Advances in Intelligent Systems and Computing, vol. 1131, pp. 359–364. Springer, Cham (2020). https://doi.org/10.1007/978-3-030-39512-4_56
33. Burov, O., et al.: Cognitive performance degradation in high school students as the response to the psychophysiological changes. In: Ayaz, H., Asgher, U. (eds.) Advances in Neuroergonomics and Cognitive Engineering. AHFE 2020. Advances in Intelligent Systems and Computing, vol. 1201, pp. 83–88. Springer, Cham. https://doi.org/10.1007/978-3-030-51041-1_12
34. Lavrov, E., et al.: The method of teaching IT students computer analysis of ergonomic reserves of the effectiveness of automated control systems. In: E3S Web of Conferences, vol. 166 (2020). https://doi.org/10.1051/e3sconf/202016610017

Decision Support for Solving Problems of Ergonomic Provision of Contact Centers

E. Lavrov and O. Siryk

Abstract The article discusses the issues of decision support for engineers and managers to ensure the reliability, timeliness and convenience of the activities of information processing operators working in contact centers. A study of factors that reduce the effectiveness of call center operators was carried out. A method of formalized description of operators' activities based on functional networks of Professor A. Gubinsky is proposed. We propose a simulation model to analyze the processes of execution of all incoming requests. The use of this method makes it possible to assess reliability taking into account the structure of the operators' activities and working conditions. A model is proposed to justify the number of operators, design the structure of activities and working conditions. The study can be useful to justify the feasibility of improving ergonomics in contact centers of commercial and government organizations. The method ensures high reliability of application fulfillment processes and ensuring ergonomic standards and requirements.

Keywords Contact-center · reliability · simulation · ergonomics · man–machine · human factor

1 Introduction

Enterprises, the functioning of which is associated with the service of a large number of requests coming from a variety of sources (technical support and service departments, banks, process control systems; information services and communication systems, insurance companies, large trade organizations; trade networks; travel companies, cab companies, logistics process management systems, etc.) face the problem of efficient and timely processing of requests coming through the computer

E. Lavrov (✉)
Sumy State University, Sumy, Ukraine
e-mail: prof_lavrov@hotmail.com

O. Siryk
Taras Shevchenko National University, Kyiv, Ukraine

I. Salopek Čubrić et al. (eds.), *Proceedings of the 9th International Ergonomics Conference*, Lecture Notes in Networks and Systems 701,
https://doi.org/10.1007/978-3-031-33986-8_19

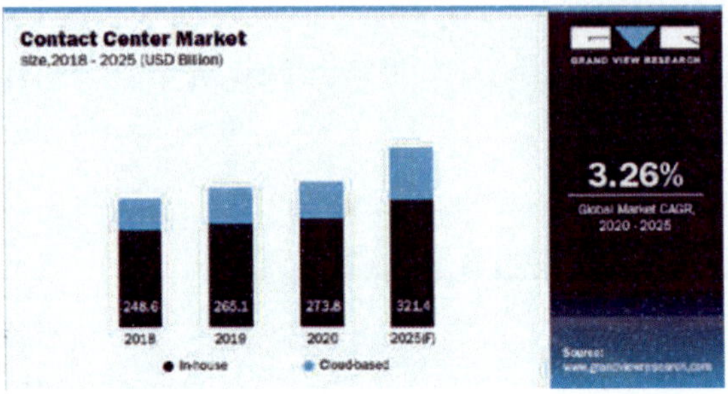

Fig. 1 Contact center market (status and forecast) [4]

or telephone [1, 2]. Special software systems are being developed to manage the work of contact center operators [3, 4].

The contact center market is quite voluminous [4] and is constantly growing (Fig. 1). At the same time the requirements to reliability and efficiency of contact centers are increasing, because every missed request or poorly executed request can lead to damages, accidents, fines, and risks to human health and the environment [1, 5].

2 Problem Statement

Based on the increasing contact center market (including pandemic and military-related problems), scientists around the world are working on the development of new software and methods for managing contact center processes, for example [2, 6], including the use of artificial intelligence [7]. Despite the widespread use of advanced technologies, the number of problematic situations is not only not decreasing, but also significantly increasing [1, 6, 8]. As a rule, this is due to ignoring the need to adapt management systems to the characteristics of people working in call centers. Very often operators work in conditions of information overload and even stress [8–10].

In this connection the number of researches, devoted to the consideration of the so-called "human factor" increases [9–15]. These studies are aimed at.

- Analysis of harmful environmental factors
- Operator training
- Improvement of psychological climate
- Operator incentives
- Etc.

For example, in [15] it is noted that many studies prove that one of the main ways in which operators can be more productive and, therefore, more satisfied, is to invest in training and development of employees (eventually, it will lead to an increase in customer satisfaction). Many techniques have been developed to manage "satisfaction" of contact center employees [15], which are quite successfully applied in various systems for managing requests for information processing. However, as a rule, these are heuristic approaches related to the specific features of particular systems.

Unfortunately, despite the large number of local developments, a unified comprehensive approach to the search for ergonomic reserves to improve the efficiency of contact centers has not been fully developed.

In this regard, the aim of this work is to develop a comprehensive approach to decision support for the ergonomic quality of contact centers based on the scientific achievements of ergonomics [16–23].

3 Methods

To make decisions on the design and operation of contact centers, we propose to move from the system of intuitive methods to the methodology of the functional-structural theory of ergatic systems by Professor A. Gubinsky [23, 24], which offers a system of numerical indicators to assess the environment, technology parameters, human characteristics, and predictable reliability characteristics of incoming requests and the states of operators working in the system. At the same time, the decision-making methodology should be based on the human-system approach, in which a human is "put in the forefront", and hardware and software are important, but auxiliary elements [24, 25]. Our methodology is based on the requirement to achieve specified characteristics of reliability and timeliness of fulfillment of requests while meeting ergonomic and health and minimizing stressful situations.

4 Results

4.1 Principles of the Decision Support System for the Ergonomic Quality of Contact Centers

Since the contact center usually has many operators, it should be classified as a "polyergatic system".

We single out 2 categories of operators:

- Executing operators (accept and process requests)

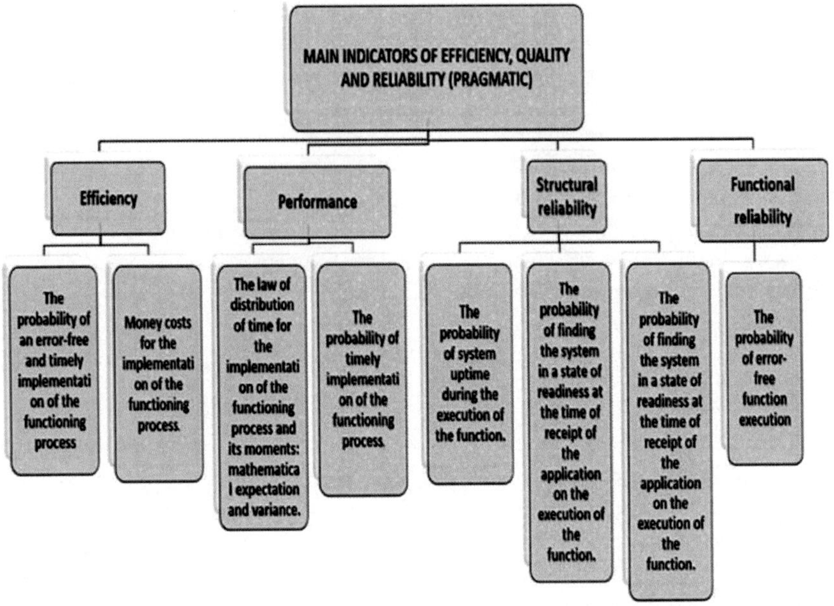

Fig. 2 Pragmatic indicators for assessing the functioning of the contact center

- Operators-managers or managers (manage the processes of execution of applications, control quality of processes, distribute functions between the operators-executors, and are responsible for ensuring comfort and ergonomics of workplaces).

It is proposed to make decisions based on a number of actual and predictable indicators of the characteristics of the system and the effectiveness of the execution of requests (Fig. 2), as well as the information load and performance of operators and environmental conditions (including the flow of requests and harmful factors).

The modeling is carried out according to:

- The activity approach to the analysis of contact center functioning, and we simulate the operators' activity with the help of functional networks (describe logical and functional connections between separate operations necessary for the execution of requests).
- Analysis of mass service systems we simulate, using a simulation model of the analysis of the processes of receipt and execution of requests, the processes of execution of the entire set of requests by all operators (which in this case are specific service "devices").

4.2 Basic Model for the Ergonomic Quality of the Contact Center

Based on the principle of maximizing the effectiveness of the contact center, subject to ergonomic norms and requirements, the task of operational management of the process of functioning can be defined in a simplified form as the task of selecting the necessary number of operators and technologies to fulfill requests

$$\begin{cases} \beta(X, N) \to \max & (1) \\ P\{T(X, N) < T_0\} > \alpha_0 & (2) \\ Z(X, N) \le z_0 & (3) \\ \lambda(X, N) \le \lambda_0 & (4) \\ K(X, N) \le k_0 K(X, N) \le k_0 & (5) \\ X \in X_0, \ \ N \in N_0, & (6) \end{cases}$$

where:

- $\beta(X, N)$ – the average probability of error-free implementation of the request;
- T – a random value of the time from receipt to the end of the request (the sum of the waiting time in the queue T_1 and the actual execution time of the request T_2);
- α_0– minimum acceptable probability of timely fulfillment of the request;
- T_0 – the directive time until the completion of the request;
- z_0 –operator workload factor;
- λ_0– maximum permissible value of the queue rate of requests waiting for operator's release;
- k_0 –maximum permissible value of the average queue of requests waiting for operator's release;
- X_0 –many alternative options for application technologies;
- N_0 – a set that defines the allowed number of operators.

In ergonomics, the following standards are accepted: $z_0 = 0.75$, $\lambda_0 = 0.4$, $k_0 = 3$ (contact centers can adjust these values based on the specific conditions of specific activities).

4.3 Modeling of Request Fulfillment Processes to Support the Decision-Making of Contact Center Managers

Our proposed approach to modeling the activity of contact center operators by functional networks is described in [9, 23], and the software for evaluating such activity is described in [26–28]. The availability of the results of the evaluation of alternative ways of execution of requests (taking into account the characteristics of operators and their current performance, the parameters of technical means and the external environment) allows us to simulate the processes of the whole complex of incoming requests at random times. Examples of modeling the processes of execution of applications, built in Matlab-Simulink (performed by Anna Krivodub) are shown in Figs. 3 and Fig. 4, and an example of the results of variant modeling is shown in Fig. 5.

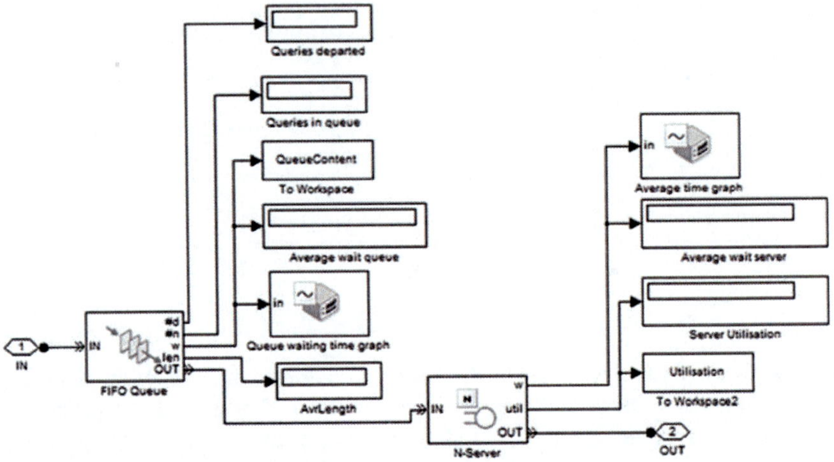

Fig. 3 A fragment of the simulation model of the process of execution of requests. Service channel subsystem

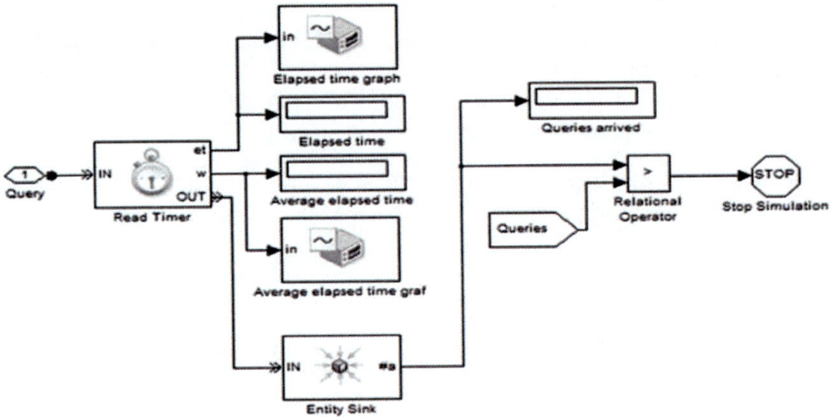

Fig. 4 A fragment of the simulation model of the process of execution of requests. Subsystem of the request receiver

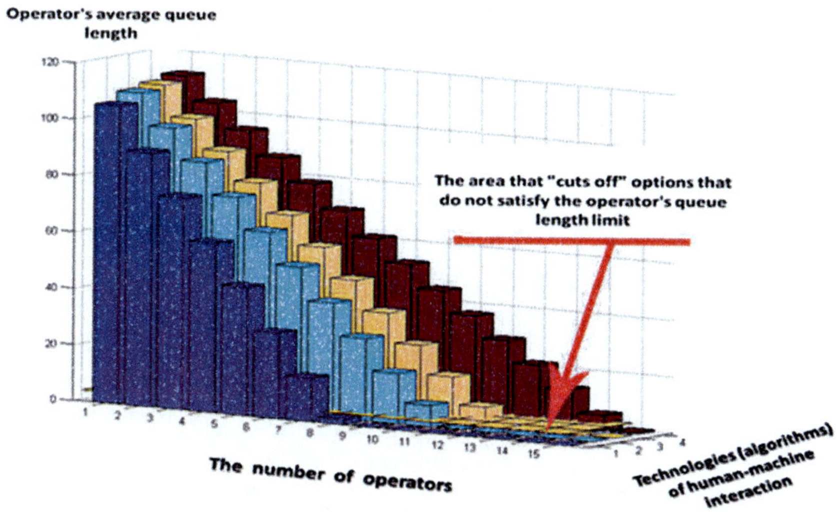

Fig. 5 A fragment of the results of variant modeling of the contact center functioning

5 Conclusions

The main problem of contact centers is the violation of ergonomic norms and requirements, which often leads to health problems of operators and stress, as well as to a decrease in the quality of service of incoming requests. Transition from intuitive methods of support of contact center employees to the method of ergonomic support based on qualimetric approach provides achievement of maximum levels of efficiency while maintaining the required values of operator workload. Combining the method of activity assessment with functional networks and method of analysis of mass service systems allows to solve the optimization task of contact center management subject to limitations on operators workload, queue coefficient, average queue value and probability of timely request fulfillment. Scientific novelty lies in the fact that, in contrast to the intuitive approaches to ergonomic quality management, we proposed a system of analytical-simulation models, based on the activity approach, which provide forecasting of pragmatic and ergonomic indicators of contact center functioning.

References

1. Sherstneva, A., Sherstneva, O.: The generalized model development for call center functioning. In: 2018 XIV International Scientific-Technical Conference APEIE, pp. 281–284 (2018). https://doi.org/10.1109/APEIE.2018.8545299

2. Zolkin, A.L., Burda, A.G., Avdeev, Y.M., Fakhertdinova, D.I.: The main areas of application of information and digital technologies in the agro-industrial complex. IOP Conf. Ser. Earth Environ. Sci. **677**(3), 032092 (2021). https://doi.org/10.1088/1755-1315/677/3/032092
3. Annisa, N.N., Sensuse, D.I., Wilarso, I.: Knowledge base model for call center department: a literature review. In: 2018 International Conference ICITSI, pp. 242–247 (2018). https://doi.org/10.1109/ICITSI.2018.8696042
4. https://www.grandviewresearch.com/pipeline/contact-center-market-procurement-intelligence-report. Accessed 25 Oct 2022
5. Bloomfield, R., Lala, J.: Safety-critical systems. The next generation. IEEE Secur. Priv. **11**(4), 11–13 (2013)
6. Aitchanov, B., et. al.: Contact centers management models: analysis and recommendations. In: 2021 IEEE International Conference SIST, 2021, pp. 1–5 (2021)
7. Al-Abbasi, L.M.S., et al.: An intelligent agent for E-government call center. In: 4th SMART CITIES Symposium (SCS 2021), pp. 558–565 (2021)
8. Mary, G.M.J., Munipriya, P.: Health monitoring of IT industry people. In: 2011 3RD International Conference on Electronics Computer Technology, Kanyakumari, pp. 144–148 (2011). https://doi.org/10.1109/ICECTECH.2011.5941875
9. Lavrov, E., et al.: Analysis of working conditions and modeling of activity algorithms for contact-center operators. In: Sumpor, D., Jambrosic, K., JurcevicLulic, D., Milcic, D., SalopekCubric, I., Sabaric, I. (eds) Proceedings of the 8th International Ergonomics Conference. AISC, vol. 1313, pp. 121–128. Springer, Cham (2021)https://doi.org/10.1007/978-3-030-66937-9_14
10. Lavrov, E., Pasko, N., Siryk, O.: Information technology for assessing the operators working environment as an element of the ensuring automated systems ergonomics and reliability. In: 2020 IEEE 15th International Conference TCSET, Lviv-Slavske, Ukraine, pp. 570–575 (2020). https://doi.org/10.1109/TCSET49122.2020.235497
11. Rashid, O., Qamar, A.M., Khan, S., Ambreen, S.: Intelligent decision making and planning for call center. In: 2019 International Conference on Computer and Information Sciences (ICCIS), Sakaka, Saudi Arabia, pp. 1–6 (2019)
12. Tao, D., Li, J.: Staff scheduling in call center considering global service quality. In: 2015 12th International Conference ICSSSM, Guangzhou, pp. 1–4 (2015). https://doi.org/10.1109/ICSSSM.2015.7170333
13. Cao, Q., et al.: Extending axiomatic design theory to human-machine cooperative products. In: 2015 IEEE 10th Conference on ICIEA, Auckland, pp. 329–334 (2015)
14. Bentley, T.A.: The role of organisational support in teleworker wellbeing: a socio-technical systems approach. Appl. Ergon. **52**, 207–215 (2016)
15. Chicu, D., del Mar, M., Pàmies, G.R., Cross, C.: Exploring the influence of the human factor on customer satisfaction in call centres. BRQ Bus. Res. Q. **22**(2), 83–95 (2019). https://doi.org/10.1016/j.brq.2018.08.004
16. Meleschenko, E.E., et al.: Human factor management tool while performing high-risk work in thermal power engineering. AIP Conf. Proc. **2467**, 080009 (2022). https://doi.org/10.1063/5.0092470
17. Zhirabok, A.N., Kalinina, N.A., Shumskii, A.E.: Technique of monitoring a human operator's behavior in man-machine systems. J. Comput. Syst. Sci. Int. **57**(3), 443–452 (2018)
18. Zhang, Y., Sun, J., Jiang, T., Yang, Z.: Cognitive ergonomic evaluation metrics and methodology for interactive information system. In: AHFE 2019, AISC 965, pp. 559–570 (2019).https://doi.org/10.1007/978-3-030-20454-9_55
19. Pisarev, A.S., et al.: Generalized model of cognitive activity taking into account uncertainty in an information-saturated environment. In: Proceedings of 2022 25th International Conference on SCM 2022, pp. 133–137 (2022). https://doi.org/10.1109/SCM55405.2022.9794831
20. Bolčević, E., Tokić, S., Sumpor, D., Horina, J. L.: A preliminary study of the possible impact of non-related driving tasks on the car driver's performance using glance-based measures. In: Proceedings of the 8th International Ergonomics Conference. Ergonomics 2020. Advances in Intelligent Systems and Computing, vol. 1313. Springer, Cham (2021).https://doi.org/10.1007/978-3-030-66937-9_4

21. Havlikovaa, M., Jirglb, M., Bradac, Z.: Human reliability in man-machine systems. Proc. Eng. **100**, 1207–1214 (2015)
22. Burov, O., et al.: Cybersecurity in educational networks. Adv. Intell. Syst. Comput. **1131**, 359–364 (2020). https://doi.org/10.1007/978-3-030-39512-4_56
23. Lavrov, E., et al.: Functional networks for modeling and optimization human-machine systems. In: Proceedings of the 8th International Ergonomics Conference. Ergonomics 2020. Advances in Intelligent Systems and Computing, vol. 1313. Springer, Cham (2021)https://doi.org/10.1007/978-3-030-66937-9_21
24. Lavrov, E.A., Siryk, O.E.: Functional networks for ergonomics and reliability tasks on the 90th anniversary of A. Gubinsky and V. Evgrafov. Sigurnost **64**(3), 245–257 (2022). https://doi.org/10.31306/s.64.3.3
25. Burov, O.Y.: ICT for performance assessment of emergent technologies operators. In: Proceedings of the 13th International Conference on ICT in Education, Research and Industrial Applications. Integration, Harmonization and Knowledge Transfer, Kyiv, vol. 1844, pp. 127–138. CEUR-WS (2017)
26. Lavrov, E., Pasko, N.: Automation of assessing the reliability of operator's activities in contact centers that provide access to information resources. In: Proceedings of the 14th International Conference on ICT in Education, Research and Industrial Applications. Integration, Harmonization and Knowledge Transfer, Kyiv, Ukraine, vol. 1, pp. 445–448 (2018)
27. Lavrov, E.A., et al.: Information technology for modeling human-machine control systems and approach to integration of mathematical models for its improvement. In: 2020 XXIII International Conference SCM, St. Petersburg, pp. 117–120 (2020). https://doi.org/10.1109/SCM50615.2020.9198791
28. Lavrov, E., Pasko, N., Krivodub, A., Barchenko, N., Kontsevich, V.: Ergonomics of IT outsourcing. Development of a mathematical model to distribute functions among operators. Eastern-Eur. J. Enterp. Technol. **2**(4(80)), 32 (2016). https://doi.org/10.15587/1729-4061.2016.66021

Recommendations for an Innovative Distance Continuing Education Device Intended for Orthodontic Practitioners

Mailloux Aurelie and Dinet Jérôme

Abstract The main objective of this multidisciplinary research (i.e., orthodontics and psycho-ergonomics) was to provide preliminary recommendations on designing an innovative distance continuing education device intended for French orthodontic practitioners. Before the COVID-19 crisis, the continuous training offer was reduced to face-to-face congresses. The COVID-19 crisis has changed the global organization of continuing education: all face-to-face congresses have been canceled and many videoconferences have emerged (i.e., without prior identification of orthodontists' expectations). These changes have modified the training needs and expectations of most orthodontic practitioners. To edit recommendations on creating an innovative useful, usable, and acceptable training device, a triangulation of qualitative methods was adopted. After an extensive literature review, in order to identify the existing remote devices, we conducted (i) four focus groups with novice orthodontists to generate innovative ideas, (ii) a survey by questionnaire to collect data about behaviors and expectations towards current and future education (from novice and expert practitioners), (iii) an ergonomic inspection of the eorthodontie.com website dedicated to training and information. Based on the combination of the three techniques (focus groups, a questionnaire, and an ergonomic inspection), several preliminary recommendations for designers of online distance environment in terms of interactions, contents and website architecture have been produced.

Keywords E-learning · Community of Practice · Psycho-Ergonomic · Innovative Device · Orthodontics · Continuing Education

---

M. Aurelie (✉)
Reims Hospital- Orthodontics, Reims, France
e-mail: aurelie.mailloux@univ-reims.fr

D. Jérôme
University of Lorraine, CNRS, INRIA, Nancy, Loria, France
e-mail: jerome.dinet@univ-lorraine.fr

I. Salopek Čubrić et al. (eds.), *Proceedings of the 9th International Ergonomics Conference*, Lecture Notes in Networks and Systems 701,
https://doi.org/10.1007/978-3-031-33986-8_20

1 Introduction

All medical practitioners must continue to train throughout their career to comply with legal, deontological, and ethical obligations. Before the COVID-19 crisis, the continuous training offer was mostly in the form of face-to-face congresses. Since the COVID-19 crisis, the training practices and needs of orthodontists have changed in connection with (i) the cancelation of all face-to-face continuing education congresses, and (ii) the emergence of numerous videoconferences. The replacement of congresses by videoconferences has also led to the loss of direct exchanges among practitioners (i.e., professional, and informal). These peer-to-peer interactions are, however, an important source of learning, cohesion and identification among them [1–6]. According to the state-of-the-art, only two websites were dedicated to the training and information of French orthodontists: the World Federation of Orthodontists (WFO), and the eorthodontie.com websites. However, the existing videoconferences and websites are not currently accredited by Continuing Professional Development organism and therefore cannot be recognized as official continuing education. In addition, none of the interviewed practitioners were registered with the WFO, probably because this is not adapted to the expectations of French orthodontic practitioners, that are depending, among others, on culture. A previous study has indeed already highlighted the importance of the cultural and/or social dimension for the acceptance of a new distance learning environment [7]. Concerning the eorthodontie.com website, created by a French orthodontist in 2007, the focus group participants reported its lack of updated content and scientific validity [8]. The purpose of this research was to edit recommendations on creating an innovative and complete distance continuing education environment dedicated to the French orthodontic practitioners. A complete, careful analysis of the needs, behaviors, and expectations of potential end-users was done in order to create an acceptable and usable distance learning device.

2 Methodology

The adoption of an innovative device by users is a complex process that involves technical factors related to the tool, organizational, and personal factors [8]. A user-centered design consists of identifying the characteristics of end users in terms of knowledge, skills, functions, habits. These data make it possible to represent users and to categorize them into homogenous subgroups in terms of needs and expectations [9] (Fig. 1).

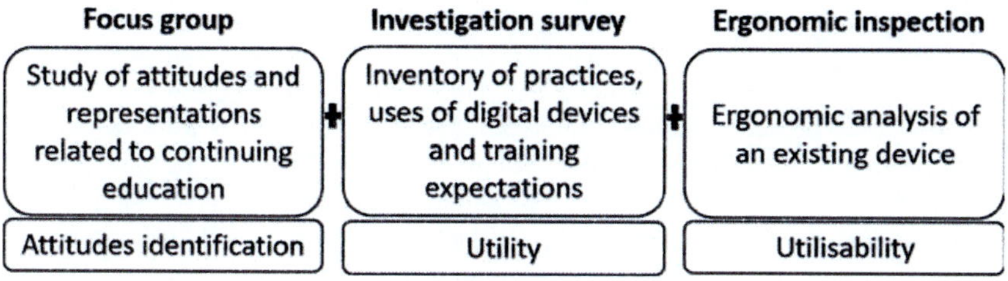

Fig. 1 Objectives of our triangulation of methods

2.1 Focus Group

Four focus groups of 4 to 6 novice practitioners were conducted. Three focus groups were handled before the COVID-19 and one after. The main objectives were to assess novices' representations and to generate innovative ideas related to current and future continuing training [10]. The focus group activities were carried out in three stages: (1) identification of the expectations, barriers, and perceptions of continuing education, (2) detailed presentation of the French training website: eorthodontie.com (sections of the home page, forum area, thematic classification), (3) and a collaborative construction of an optimal website architecture intended for orthodontists (i.e., based on card sorting methodology).

2.2 Questionnaire Survey

The questionnaire was administered during a professional conference (n = 42 interviewed practitioners, including 23 experts and 19 novices). The design of the questionnaire focused on the following items: current practices, digital uses, obstacles, the "successful" criteria of continuing education experience, opinions on distance continuing education? and the vision of continuing education in the future.

2.3 Ergonomic Inspection

The eorthodontie.com website (also presented during focus groups) was the subject of an ergonomic inspection, to determine if this interface was suitable for the practitioners [11].

2.4 Data Analysis

After a manual content analysis, some textual statistical analyzes were carried out using the free software IRAMUTEQ (http://www.IRAMUTEQ.org/) based on the R software and the Python language. Several automated analyses were applied such as, (i) the Reinert Descending Hierarchical Classification (DHC), (ii) the Factorial Analysis of Correspondences (FAC) to detect the proximity or the distance of the different classes, (iii) the similarity analysis when the number of segments was low. We compared the data collected before and after the COVID-19 crisis. In this research, novice orthodontists were defined as either orthodontic resident or practitioner with less than three years of liberal practice. Experts were characterized as orthodontists with more than three years of liberal practice. Data from novice orthodontists were compared with data from experts to identify their common or divergent expectations, needs and behaviors toward continuing education.

3 Main Results

For the interviewed practitioners and according to our extensive literature review, distance continuing training offered the following advantages: no trip, saving time and money, and flexibility (Fig. 2).

3.1 eorthodontie.com Website Evaluation and Ergonomic Inspection

The website low activity (i.e., very low discussion forum activity, and no recent publication of content) showed that this device did not match users' expectations.

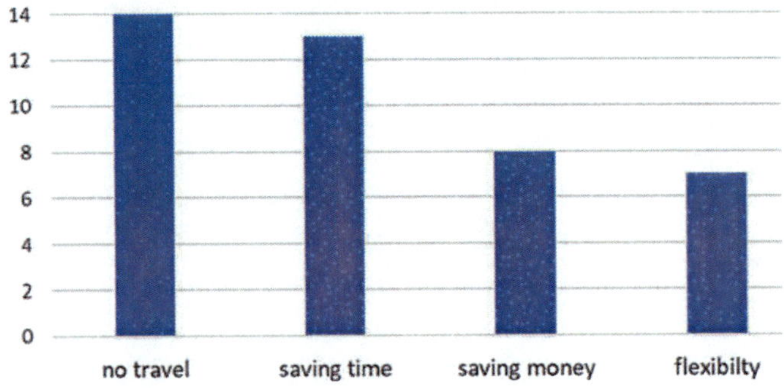

Fig. 2 Advantages of distance continuing education mentioned by practitioners and presented in descending order (number of occurrences after thematic grouping)

Thematic grouping	Focus group common remarks
content	lack of scientific validity
	lack of control
	lack of content (e.g., thesis, courses sections)
	participants, moderators and administrators unknow level of expertise
access	patients access should be limited
	no personal identification
general aspect	not pleasant
	videogame/advertising/streaming aspect
architecture	lack of organisation
	navigation "not instinctive"
	not keyword research
	no personal section

Fig. 3 Focus groups participants' common opinions and remarks about eorthodontie.com after manual thematic grouping

Figure 3 illustrates the practitioners' common opinions and remarks about several aspects of the eorthodontie.com website (i.e., content, access, general aspect, and architecture).

Concerning the content, the lack of scientific validity was a major drawback (e.g., content publication was unrestricted and unverified). The limitation of patient access was a recurring request, as well as the addition of a personal section, a personal identification, and a search bar. The ergonomic inspection revealed that the informational density and immediate feedback were well respected. But (i) the workload was increased (i.e., numerous unnecessary actions, lack of brevity and conciseness) (ii) some procedures were too heterogeneous, (iii) readability could be improved. In addition, the interface did not adapt to the user experience and did not handle his/her errors.

3.2 Interaction Needs and Expectations

According to the state-of-the-art, the loss of contact with peers, the lack of meetings and exchanges were the main distance learning disadvantage for interviewed novices and experts (based on the questionnaire survey collected data) [12] (Fig. 4). The different DHC and FAC produced from the focus group collected data, allowed us to identify their major expectations and needs in terms of interactions within an "ideal" device. Novices were very fearful of their peers' judgment and expressed their need (i) to interact with experts anonymously (to avoid their judgment), (ii) and to create restricted or extended online discussion groups (based on their professional experiences).

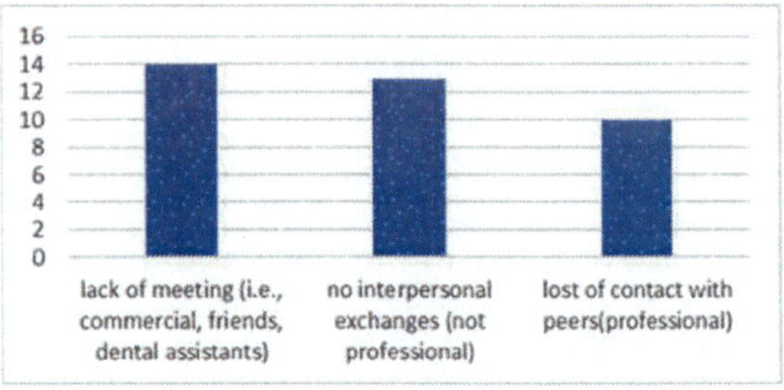

Fig. 4 Main disadvantages of distance learning mentioned by practitioners (in decreasing order of occurrence)

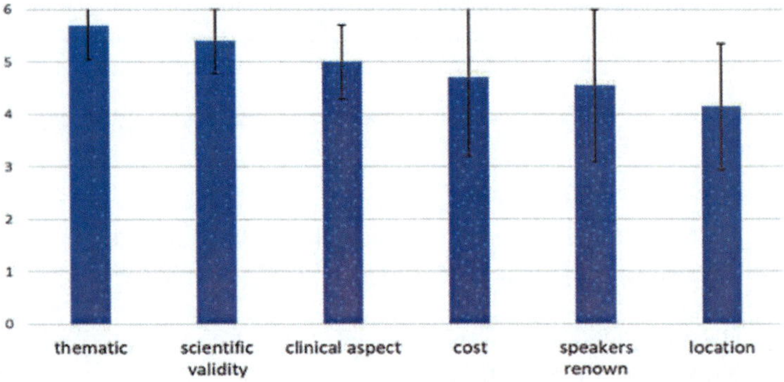

Fig. 5 Practitioners' expectations toward a continuing education experience after manual thematic grouping (average of scores from 0 to 6 and standard deviation, in descending order)

3.3 Content Expectations and Needs

Some needs were common before and after the COVID-19. The training should be delivered or translated into French. According to the questionnaire survey collected data, the main practitioners' expectations toward continuing education, shared by experts and novices, concerned: (1) content quality (i.e., thematic, and scientific validity), (2) clinical application, (3) cost (the lower the better), (4) the speaker renown, (4) the training location (close to the practice) (see Fig. 5).

3.4 Architecture of an "Ideal" Website Distance Continuing Education Device

This "ideal" website architecture (Fig. 6) is based on the last focus group collected data (conducted after the COVID-19). We considered that this architecture was closer to current practitioners' expectations (rather than data collected before the COVID-19 crisis). Nevertheless, we added several sections: "practice management" and "CPD account", absent from the post COVID-19 focus group, probably because none of

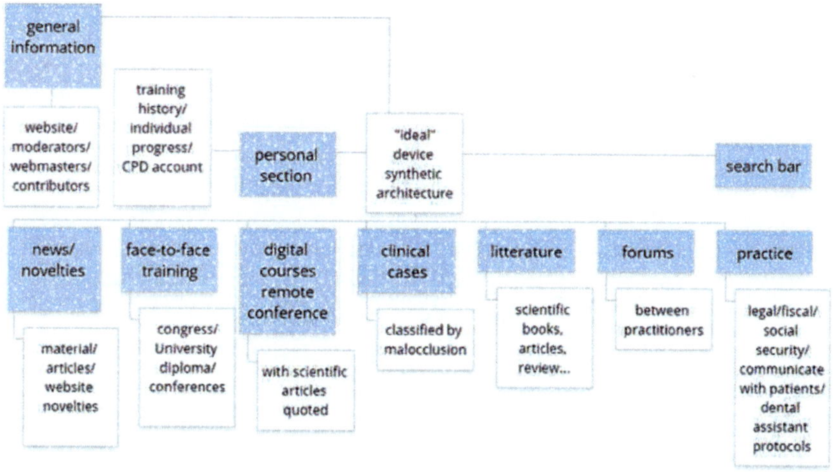

Fig. 6 "Ideal" device synthetic architecture constructed from the focus group collected data (blue square: home page section; white square: precisions on the section content)

the interviewed novices in the last focus group had a liberal experience. In addition, according to the eorthodontie.com evaluation, we also incorporated a search bar and a personal section and identification.

All architectures contained the following sections: clinical cases, literature, videoconferences, forums, and information about trainings (e.g., presentation, registration, cost, calendar). The clinical cases held a central place in all focus groups. This section should (i) contain treatment successes and failures, (ii) provide practitioners with the opportunity to discuss about clinical cases. The moderators and administrators' identities and level of expertise had to be specified (in the "general information" section). Moderators had to be either University Professor or "expert" practitioners to validate all the publications (e.g., articles, novelties). Replays, webinars, discussion forums and remote interactions were discussed at length in the post COVID-19 focus group, probably because participants had a huge experience of videoconferences during the confinement. The four focus groups agreed on the fact that only specialists (novices or experts) should be able to access the website (and therefore excluding patients and practitioners from other training). In addition, the possibility of an access on both computer and telephone was unanimous.

3.5 Recommendations

According to the state-of-the-art and our research, and in order to match with practitioners' needs and expectations, the future device should respect some criteria: (1) videoconferences direct translation into French, (2) content scientific validity should be ensured by international/European clinical experts, teachers from universities, or a mixed college of both, (3) the home page should contain the following sections: clinical cases, literature, digital learning and videoconferences, forums, information about trainings, practice management, a search bar (by keyword), and a personal section, (4) the device must be easy to access on both smartphone and computer, (5) and clinical cases should be accessible (success AND failure), and detailed step by step with the possibility of soliciting feedback from peers, (6) in the forum section, posting and participation could be anonymous (e.g., using a pseudonym), (7) however, the professional experience and diploma of each practitioner should be known, so that novice orthodontists could trust in the posted content.

4 Discussion

All the studies (focus group, questionnaire survey, and ergonomic inspection) were concordant and complementary [13]. However, professional barriers (e.g., solitary liberal practice, geographically dispersed liberal practice, time-consuming liberal practice) prevented us from conducting focus group of expert orthodontists.

This study revealed that French orthodontists needed: (i) scientifically validated content, (ii) discussions in small groups or extensive group (depending on the professional experience of the practitioners), (iii) anonymity to exchange on the discussion forum, (iv) and a section devoted to clinical cases. According to the state of the art and our user-centered research, an innovative distance continuing education environment could meet many needs of orthodontic practitioners [7].

Our collected data clarified why the WFO and the e-*orthodontie.com* websites were not used by the practitioners interviewed. Concerning the WFO website, cultural aspect apart [7], there was a strong language barrier (i.e., for 52% of experts and all the focus group novices).

Concerning the *e-orthodontie.com* website, apart from the opinions of the interviewed practitioners, the ergonomic inspection revealed some weak points (e.g., readability, workload, procedures heterogeneity, interface adaptation, users' error management), underlying the importance of a prior user-centered research to evaluate practitioners' needs and expectations [10, 14].

All the surveys revealed the significant tension among orthodontists related to their various types of academic achievements (part-time private education versus full time public education). The access of non-specialist orthodontists (i.e., from the part time training courses) to the future distance training environment should be carefully considered.

5 Conclusion

The collected data revealed French orthodontists' needs and expectations related to continuing education in terms of (i) scientific validity, (ii) sharing clinical experiences, (iii) content, (iv) and interactions between novice and expert orthodontists. On this basis, recommendations, and a website architecture for designing a future training device have been suggested. For the design process, the interactions between novice and expert orthodontists represented a major issue to be considered. The needs and attitudes of orthodontists in terms of interactions are the subject of another article [15]. Our user-centered approach must be extended during the design/redesign phases by empirical methodology, using personas or "real" end-users, to ensure compliance with the ergonomic criteria. To justify the usefulness of the proposed requirements, further experimentation should be carried out, especially to deepen the needs of practitioners after the-COVID-19 crisis. The future device should meet the previously identified needs of most practitioners, and comply with legal obligations, especially in terms of sharing medical data, which is not the case for existing websites.

References

1. Editorial board: Informal learning communities, communities of practice - Learning with, by and for others Interrogations ? review, vol. 32 (2019). Accessed May 2022
2. Mercieca, B.: What is a community of practice? In: McDonald, J., CaterSteel, A. (eds.) Communities of Practice, pp. 3–25. Springer, Singapore (2017). https://doi.org/10.1007/978-981-10-2879-3_1
3. Wenger, E.: Communities of practice and social learning systems: the career of a concept. In: Blackmore, C. (ed.) Social Learning Systems and Communities of Practice, pp. 179–198. Springer, London (2010). https://doi.org/10.1007/978-1-84996-133-2_11
4. Lave, J., Wenger, E.: Situated learning: legitimate peripheral participation, pp. 26–49. Cambridge University Press, New York (1991)
5. Wenger, E., McDermott, R.A., Snyder, W.: Cultivating Communities of Practice. Harvard Business Review Press, Boston (2002)
6. Lièvre, P., Bonnet, E., Laroche XXI, N.: Etienne Wenger. Community of practice and social theory of learning. EMS Editions, pp. 427–447 (2016)
7. Miller, K.T., Hannum, W.M., Morley, T., Proffit, W.R.: Use of recorded interactive seminars in orthodontic distance education. Am. J. Orthod. Dentofac. Orthop. **132**(3), 408–414 (2007)
8. Caro Dambreville, S.: Design of digital documents. Methodological path. Digital Documents. **12**(2), 7-22 (2009)
9. Brangier, É., Bastien, J.M.C.: The evolution of the ergonomics of computer products: accessibility, usability, emotionality and influenceability. Presses Universitaires de France, pp. 307–328 (2010). Accessed April 2022
10. Kitzinger, J., Markovà, I.: Centered groups (focus groups), p. 116 (2004)
11. Bastien, J.M.C., Scapin, D.L.: A validation of ergonomic criteria for the evaluation of human-computer interfaces. Int. J. Human-Comput. Interact. **4**(2), 183–196 (1992)
12. Spallek, H., Pilcher, E., Lee, J.-Y., Schleyer, T.: Evaluation of web-based dental CE courses. J. Dental Educ. **66**(3), 393–404 (2002). https://doi.org/10.1002/j.0022-0337.2002.66.3.tb03518.x
13. Apostolidis, T.: Social representations and triangulation: theoretical-methodological issues. In: Abric, J.C. (ed.) Methods of studying social representations, Ed Érès, pp. 13–35 (2003)

14. Gould, J.D., Lewis, C.: Designing for usability—key principles and what designers think. Communications of the ACM, pp. 50–53, December 1983. Accessed May 2022
15. Mailloux, A., Dinet, J.: Requirements analysis towards future design of an innovative distance learning device intended for French orthodontic practitioners contribution of a community of practice analysis. In: The Sixteenth International Conference on Digital Society, June 2022

Cognitive Ergonomics of the Integrated Collaborative, Interactive E-Learning and E-Training- upon Innovative Networks Empowered with AI and Big Data Analytics and Computing- of the Remote E-Multidisciplinary Team Board in Breast Cancer Decision Making and Treatment

Constantinos S. Mammas and Adamantia S. Mamma

Abstract Background and purpose: The project refers to the effect that interactive E-Learning among expert participants in our innovative Remote E-Multidisciplinary Board for Breast Cancer (PROMETHEUS II pn:1009078, 34931,34932,34933) is being integrated with the Hellenic Hyper-computer ARIS for Big Data analytics and computing. Design/methodology/Results: Integrated analysis of two research projects results showed that the E-Multidisciplinary Board in Breast Cancer may also be the substance of Innovative Networks for Interactive Collaborative E-Learning and Training among experts, residents and students for further and deep learning of the cognitive ergonomics of the E-Multidisciplinary Board incorporating AI and Big Data Analytics and Computing to optimize quality in Breast Cancer decision making, management and treatment in the peripheries. Conclusions.: Ergonomics of Innovating Training Networks for Interactive E-Learning in the context of the E-Multidisciplinary Board in cancer decision support and making, management and treatment in combination with the Mobile Clinical Unit ''Macedonia" for Breast Cancer prevention may be instructed remotely and interactively, while integration with AI and Big Data analytics interactive E-Learning may teach cognitive ergonomics of the E-Multidisciplinary Board and sustain an efficient surveillance system on a European Level and in the Balkans to optimize quality in Breast Cancer Screening, Prevention, Treatment and Survivorship especially for the peripheries.

C. S. Mammas (✉) · A. S. Mamma
Program of Excellence, Siemens Program for Greece, 2014-16 Athens, Hellas
e-mail: csmammas@med.uoa.gr

I. Salopek Čubrić et al. (eds.), *Proceedings of the 9th International Ergonomics Conference*, Lecture Notes in Networks and Systems 701,
https://doi.org/10.1007/978-3-031-33986-8_21

Keywords Cognitive Ergonomics · Remote e-multidisciplinary team board E-Learning and E-Training · AI and Big Data Analytics and Computing · Breast Cancer

1 Introduction

The optimal management of patients with breast cancer (BC) requires the expertise of specialists from different disciplines. This has led to the evolution of multidisciplinary team Board (MTB) in surgical oncology (SO), allowing all key professionals to jointly discuss individual patients and to contribute independently to clinical decisions. Also, it proved that MTBs for Multidisciplinary Oncology Conference in cooperation with Multidisciplinary Management of BC in high volume Breast Units (BU) optimize quality and have significantly better clinical outcomes for the benefit of the patients. However, MTBSO for BC decision making in different regions in the periphery of Greece and in other countries are scarce [1, 2]. Our project analyzes the E -learning and E -training technology, method, process, ergonomics and standards of the remote, specialized, multidisciplinary and personalized Remote E-Multidisciplinary Team Board in Surgical Oncology (E-MTBSO) for Breast Cancer (BC) integrating Prometheus II digital medical device with AI and Big Data analytics (PROMETHEUS II pn:1009078) technologies and methods to optimize decisions and outcomes for the benefit of the patients in the peripheries. In that context, it was given an emphasis on the remote digital macroscopic and microscopic diagnosis and BC TNM classification and accordingly to the optimum remote decision making and treatment integrated with AI empowered Big Data analytics [1–5]. Currently, within the experience of the international lockdowns in 2020 the whole project seems to be the right solution in health care and specifically in surgical oncology (SO) by providing not only: (a) experts' approach for continuous training for and from where ever, (b) acceleration of the diagnostic process and decision making, (c) evidenced based decisions, and (d) significantly lowering of the costs, but mainly by building up hands-on experience and skills in (e) the remote diagnosis and decision making and planning to optimize quality in health care services in the peripheries and improve BC patients outcomes then on a clinical level.

In the context of the collaborative e-training oriented session- using an experimental version of PROMETHEUS II pn:1009078 digital medical device the Exp-TS II (pn:100978, 34931, 34932, 34933) and its innovative training networks for collaborative E-learning and E-training- the invited specialists and the registered members used the official digital platform the program of Excellence 2014–16-Siemens program for Greece (priibite.wordpress.com), interactively, collaboratively and remotely, and trained by using the abovementioned digital medical device, technology and method aspects and functions (Telecommunications, Tele-radiology, Tele-pathology, Tele-cytology, Tele-microbiology, Tele-molecular biology). Also, they implemented them in the simulating clinical training practices and in the geographic and in the proper timing strategies for remote diagnosis and classification and decision support and making, in SO to: (1) narrow the quality differences in the health care services between the centers and the peripheries, (2) optimize decisions and treatments and improve outcomes in BC by remote accurate decisions and planning integrated with AI and Big Data analytics and computing [1, 3, 4, 6, 7].

On a clinical level, the PROMETHEUS II pn:1009078 digital medical device for BC, applied on a real clinical case from the GH of Kalymnos and the Department of Radiology of the Attikon University Hospital and an Oncologist for a patient who had suffered from a right breast cancer in 2011 that had been treated by right breast resection, chemotherapy and radiotherapy at that time and then suffered from an exacerbation with an advanced right breast cancer that needed further evaluation and treatment in 2021. The patient worked-out remotely by the Remote (E-MTBSO) for Breast Cancer (BC) integrating Prometheus II digital medical device with AI and Big Data analytics (PROMETHEUS II pn:1009078, priibite.wordpress.com of the program of Excellence 2014–16-Siemens program for Greece) from the GH of Kalymnos in June 2021 in which it was made the decision for a scheme of palliative chemotherapy and radiotherapy. The patient refused to undertake radiotherapy and that is why undertook only chemotherapy that monitored remotely in the periphery. Regarding the process of the Remote (E-MTBSO) for Breast Cancer (BC) participants applied tele-communications (TC) as a base for the Coordination Process (CP) and Tele-radiology (TRE) for the remote evaluation of the Breast and for BC TNM classification between the hospital in the periphery (PH) and the Breast Unit (BU) to make decisions and to optimize outcome [1–5].

On a E-training and on a E-learning Exp.TS II digital medical device based technology, method, process, service, clinical and training standards analysis and ergonomics integrating AI and Big Data analytics and computing assisted level for BC it has been applied a series of officially scheduled on-line training courses on 25.11.2021, on 29.01.2022 and on 19.02.2022 (priibite.wordpress.com of the program of Excellence 2014–16-Siemens program for Greece) already, and still goes on until the end of 2022 and on. On this level and regarding the process participants applied tele-communications (TC) for the Coordination Process (CP), Tele-radiology (TRE), Tele-pathology (TPE), Tele-cytology (TCE), Tele-microbiology (TME), Tele-genetic biology TGBE) for the remote evaluation of the breast and for BC TNM classification clinical process simulating interactively the hospital in the periphery (PH) and the Breast Unit (BU) to make also crucial decisions and optimize outcomes upon simulating BC cases [1–6].

According to the didactic Remote (E-MTBSO) for Breast Cancer (BC) on 25.11.2021 and on 29.01.2022 and on 19.02.2022 (priibite.wordpress.com, Breast Cancer section) the elaborated case referred to the previously mentioned clinical patient in the periphery suffering from an advanced breast cancer and cancer exacerbation and who needed further evaluation and palliative treatment. Regarding the E -training process specialists participants (nl = Surgeon, n2 = Radiologist, n3 = Pathologist, n4 = Cytologist, n5 = Microbiologist, n6 = Biologist) allocated in different points and connected remotely in the internet collaborating in the Exp.TS II digital medical device (pn:1009078, 34931,34932,34933) structure and AI and Big Data analytics by which interactively learn and train themselves by the above mentioned structure and process. In fact, PROMETHEUS II pn:1009078 digital medical device based technology, method, process, architecture and service integrated with AI and Big Data analytics, can be seen as a triangle (including the patient in the periphery, the coordinator in a central unit of the government and the Remote (E-MTBSO) for Breast Cancer (BC) from where it can be applied as the didactic base the remote evaluation of the oncologic patient and for instant remote macroscopic and microscopic evaluation and decision making. Also, for the preoperative planning, so as to empower trainees, students and other responsible professionals in SO, to make remotely valuable decisions by interactive E -learning and E -training of the up to date clinical guidelines and contemporary technology in SO. So, participants had the chance to learn and practice the right processes and develop skills for valuable clinical services in SO to confront clinical complexity at distance and in low cost and improve significantly therapeutic outcomes on a national and international level. Moreover, they developed the necessary clinical thoughts and hands on practice by using high technology to build-up the special skills in clinical and technological terms that needed to diagnose remotely the macroscopic and microscopic lesions of the Breast and make decisions upon TNM classification [1, 2, 4, 6–8]. The interactive collaborative E-training process started by a Remote (E-MTBSO) for Breast Cancer (BC) on 25.11.2021 and on 29.01.2022 and on 19.02.2022 had been archived in the innovative networks of PROMETHEUS II pn:1009078 digital medical device technology and method in the collaborative e-training for further Cloud computing AI and Big Data analytics integration (Fig. 1).

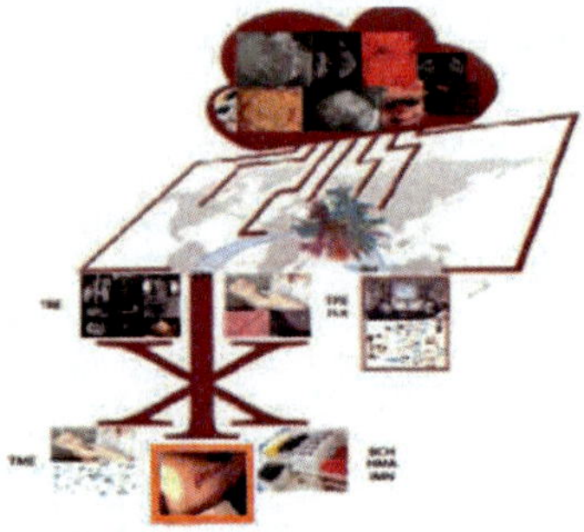

Fig. 1 The clinical Remote (E-MTBSO) for Breast Cancer (BC) on 25.11.2021 and on 29.01.2022 and on 19.02.2022 technology and process applied remotely by the specialists participants using the PROMETHEUS II pn:1009078 digital medical device technology and method directed the Exp.TS II assisted collaborative, interactive E-learning and E-training technology and method for Cloud computing of indicators, empowered by AI and Big Data analytics integration among remotely connected members of the Remote (E-MTBSO) for Breast Cancer (BC) on 25.11.2021 on 29.01.2022 and on 19.02.2022 and on, for instant decision making and treatment (n1 = Surgeon, n2 = Radiologist),(n1 = Surgeon, n2 = Radiologist, n3 = Pathologist, n4 = Cytologist, n5 = Microbiologist, n6 = Biologist)

2 Paper Content and Technical Requirements

Given that the ergonomic impact of the electronic space (ES) for the remote macro- and micro- examinations diagnosis of BC, had been already evaluated on the digital medical device Prometheus (pn: 2003016, 2013) in the context of previous experimental simulations, and in the collaborative e-training session the cognitive ergonomics of the PROMETHEUS II pn:1009078 digital medical device MTBSO for BC are based on the cognitive ergonomics of Digital Macroscopic and Microscopy (DM) in Tele-medicine, in TRE, in TPE, in TCE and in TME, for the remote macroscopic and microscopic diagnosis of injury/trauma, inflammatory, infectious and/ or malignant lesions of the solid abdominal and thoracic organs for remote decision making and treatment. The simulating experimentation for the collaborative e-training on 25.11.2021 and on 29.01.2022 and on 19.02.2022 among members of the E-MTBSO for BC included: (1) The development of an experimental digital medical device similar to Prometheus II pn:1009078 and to OTE-TS, the Exp.-TS II [5, 9, 10], (2) Integration of the Exp.-TS II with the Virtual Slide system (VS) for digitalization of images for TPE or TCE applying DM, AI analytical method in the process (Table 1), (3) Simulation of static or dynamic TPE and TCE for microscopic diagnosis of BC (Fig. 1) [11–13].

Table 1 Comparison of the modules between OTE-TS and Prometheus II Exp.-TS (pn:1009078, 34931,34932,34933) integrated with AI, Big Data analytics and Computing

Modules	OTE-TS	Prometheus II Exp.-TS
Medical record process	+	+
Examinations results	+	+
Capture/ imaging	+	+
DICOM and PACS	-	+
Real-time tele-conference	+	+
Chat and whiteboard	+	+
Tele-secretary facilities	+	+
Tele-Mentoring facilities	+	+
Telecommunication net	ISDN based	Internet based
Virtual Slide Integration	-	+
Cloud Computing AI	- -	+ +
Big Data Analytics	-	+

2.1 *Designing for Human Factors: The Technology for the Remote E-MTBSO for BC Collaborative and Interactive E-Learning and E-Training Among Specialists in SO*

According to the didactic protocol that members of the Remote (E-MTBSO) for Breast Cancer (BC) on 25.11.2021 and on 29.01.2022 and on 19.02.2022 by using the Prometheus II (pn:1009078) digital medical device integrated with AI and Big Data analytics and computing in the simulating clinical sessions, the data integrated with the Cloud archives of Prometheus II (pn: 1009078) in a concrete path for: (1) retrospective and/or real time remote diagnosis and classification of BC, applying telemedicine and remote visualization standards enhanced by AI to make instant decisions about the optimum management and treatment in each case, (2) computations of the requested clinical and financial indicators (number of BC cases, number of operative treatments, cost of BC management and treatment, total cost for each European country and in total in Europe applying AI supported Big Data analytics and computing and making decisions instantly. The results depicted, instantly, on tables and graphs for each (the same can be applied on a Balkan level under the condition that the Balkan Oncology is based on a common oncologic registry) (Fig. 2) [14].

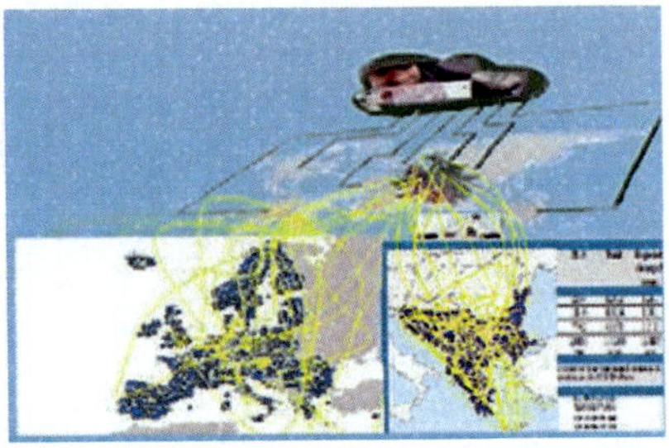

Fig. 2 Prometheus II (pn:1009078) digital medical device based AI and Big Data integrated collaborative E-training on line on the Remote (E-MTBSO) for Breast Cancer (BC) on 25.11.2021 and on 29.01.2022 and on 19.02.2022, connected with partners i.e. specialists for the E-MTBSO for BC -from which the digital images and the incidences of BC in each country in Europe computed by each member - after formation of structured Data in the European cancer registry- via Cloud computing of indicators (incidences, mortality, survival) empowered by AI retrospective and/or real time Big Data analytics and computing to make decisions for an optimum SO system on a national, Balkan, UK and European level

2.2 Designing for Human Factors: The Method of Prometheus II pn:1009078 Digital Medical Device Assisted Collaborative E-Learning and E-Training for the Remote E-MTBSO for BC, Integrating AI and Big Data Analytics for Computations and Mega Data Based Decisions in SO

The integrated TC, CP, TRE, TPE, TCE, TME and TGBE of BC in SO and the remote pre-operative planning and/or decision making and planning had shown high sensitivity, specificity and accuracy (ranging from 90% to 98%) in the remote diagnosis of traumatic, infectious, inflammatory, malignant lesions of our project that assessed specific thoracic and abdominal organs in organ transplantation integrating AI and Big Data analytics (Prometheus pn:2003016/2013 digital medical device of the program of Excellence 2014–16-Siemens program for Greece) [11, 14–18]. Taking that results into consideration and that a modern Hyper-computer is equipped with accelerating modules and with new technologies and that is the perfect choice for integrated AI and Big Data analytics and computing [17, 19], the corresponding SO data in Europe in 2020 had been archived in the working Cloud of the program of Excellence 2014–16-Siemens Program for Greece, as a structured material for real time Big Data analytics and computing, so that Prometheus II pn:1009078 digital medical device users could accumulate them for: (1) remote diagnosis collaborative interactive e-training, and for (2) instant AI and Big Data computations training on the Remote (E-MTBSO) for Breast Cancer (BC) on 25.11.2021 and on 29.01.2022 and on 19.02.2022 [14, 15, 19]. The participants had registered in the working site (priibite.wordpress.com) and actively implemented the best practices, high technology and geographic and timing strategies for Remote E-MTBSO for BC, integrating AI and Big Data analytics and decision support and making, to optimize decisions and improve outcomes in BC management and treatment by pre-operative planning according to the referred didactic and simulating interactively processed clinical cases on the Remote (E-MTBSO) for Breast Cancer (BC) on 25.11.2021 and on 29.01.2022 and on 19.02.2022 and on. As abovementioned, the results smoothed the path for initiating the clinical and training standards for the real time Big Data analytics and computing of clinical and financial indexes referring to the quality of SO in Europe and in the Balkans with a priority to the cost-effectiveness analysis for further decision making and on time interventions to improve the Health care systems [19]. The reliability and accuracy of a modern Hyper-computer considered by definition as the ideal [12, 14]. Prometheus II (pn 1009078) digital medical device assisted TRE, TPE, TCE, TME and TGBE Remote E-MTBSO for BC integrating AI and Big Data analytics and computing collaborative E -learning and E -training for pre- operative decision making and planning in BC management and treatment has already shown high sensitivity, specificity and accuracy upon remote diagnosis and classification of BC among specialists [3, 5, 19].

In the collaborative e-training process the material that used by the participants on the Remote (E-MTBSO) for Breast Cancer (BC) on 25.11.2021 and on 29.01.2022 and on 19.02.2022 and on based on the EU registry data from 2020 archived in the Cloud. The real time Big Data financial analysis with the technology and method of the digital medical device Prometheus II (pn: 1009078) completed with the real time computation of the indicators of interest (incidences, mortality, survival) of the BC treatment in Europe [1, 4]. These indicators and the cost-effectiveness analysis should have a priority in the computations that the members of the E-MTBSO for BC should apply and analyze and interpret their results instantly. In this context, the incidents of BC and the quality control of its management in Europe in terms of mortality and survival indicators computed by the participants of the for Remote E-MTBSO for BC on a real time Big Data analytics and computing framework simulating the proposed process as it has been suggested by the Program of Excellence 2014–16-Siemens program for Greece since 2016 and shaping our proposal for a realistic and effective leadership in SO nationally and on a European level to optimize quality in BC management and treatment. As far as the choice of the members of the proposed Remote E-MTBSO for BC integrating AI and Big Data analytics and computing is concerned, the qualitative E -training in new technologies and the quality control shapes a new fair system in SO that could be established in Europe for the benefit of the patients and for saving national and European and/or the Balkan economies. It is underlined that the accuracy of the computational analysis was really high and lasted for a half an hour maximum given that a software for mathematical analytics and computing had been incorporated in the system, forming the AI and Big Data analytics and computing part (Figs. 3, 4).

Fig. 3 Prometheus II (pn:1009078) digital medical device integrated with AI and Big Data for interactive collaborative e-learning and e-training on line on the Remote (E-MTBSO) for Breast Cancer (BC) decision making and treatment on 25.11.2021 and on 29.01.2022 and on 19.02.2022, connected with partners i.e. specialists for the Remote E-MTBSO for BC -from which the digital data as images and numbers of BC cases in each country in Europe sent to each member in the E-Learning process - after formation of structured Data- for AI supported Big Data analytics and computing of indicators via Cloud computing

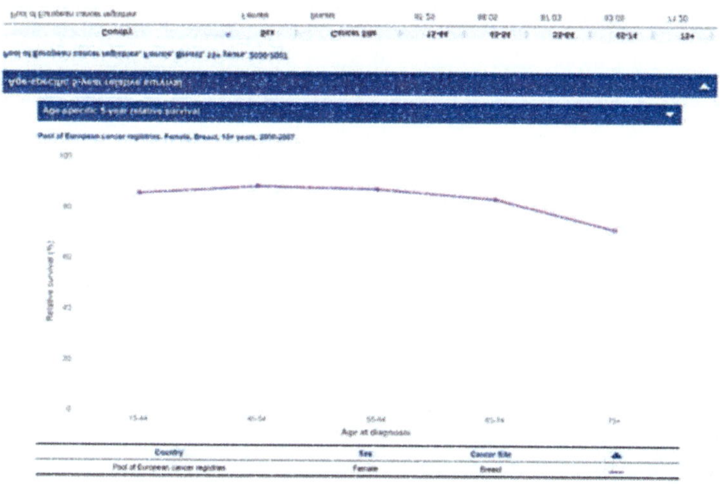

Fig. 4 Prometheus II (pn:1009078) digital medical device integrated with AI and Big Data for interactive collaborative E -learning and E -training on line on the Remote (E-MTBSO) for Breast Cancer (BC) decision making and treatment on 25.11.2021 and on 29.01.2022 and on 19.02.2022, connected with partners i.e. specialists for the Remote E-MTBSO for BC -from which the digital data as images and numbers of BC cases in each country in Europe had sent to each member in the E-Learning process- after formation of structured Data- for computation of indicators (number of cases, mortality, survival) by each participant in the E-Learning and E-Training process via Cloud computing, empowered by AI supported retrospective and/or real time Big Data analytics and computing to make decision for an optimum SO system on a multinational level Prometheus II (pn:1009078, 34931, 34932, 34933)

The abovementioned computational analysis results can be used by the Ministry of Health to intervene locally or nationally on time, so as to correct several problems referring to the cost-effectiveness balance and to the quality of BC management and treatment (in terms of the resources and morbidity-mortality rates) or to correct mistakes and/or destructive pricing policies of the biomedical products, on time [9–11, 17]. On the other hand they may be used as accurate and real time tools on a weekly base to make decisions about the biomedical technology products in SO expenses and pricing for the benefit [1, 3, 11]. The main drawback of this study is that integration of Big-Data computational analysis on 25.11.2021 and on 29.01.2022 and on 19.02.2022 based on a year framework of clinical SO on a European level simultaneously and not on a weekly level (2020). However, this drawback didn't have any negative impact on the participants collaborative E -learning and E -training given the aim of the training process was to be accustomed to the technology the method and the process to participate in the remote E-MTBSO for BC integrating AI and Big Data analytics and computing. Another drawback was that there was non-parallel comparative computational process using the Hyper-computer (HC) ARIS-IBM. However, the reliability and accuracy of a modern Hyper-computer considered by definition as the ideal one. In addition, we are developing new projects for the use of the ARIS-IBM HC as the source of mega data and the real computational center not only for training but also for clinical reasons in the direction of E-MTBSO for BC integrating AI and Big Data analytics and computing [12, 14].

3 Conclusion

Ergonomics of Innovating Training Networks for Interactive E-Learning and E-Training in the context of the Remote E-Multidisciplinary Board in Breast Cancer decision support and making, management and treatment in combination with the Mobile Clinical Unit ''Macedonia" for Breast Cancer prevention may be instructed remotely and interactively with a high clinical impact and optimum results in the outcomes of BC patients in the peripheries. The integration with AI and Big Data analytics interactive E-Learning and E-Training may teach cognitive ergonomics of the Remote E-Multidisciplinary Team Board in Surgical Oncology for Breast Cancer and sustain an efficient surveillance system on a European Level and in the Balkans to optimize quality in Breast Cancer Screening, Prevention, Treatment and Survivorship especially for the peripheries.

References

1. Saini, K.S., et al.: Role of the multidisciplinary team in breast cancer management: results from a large international survey involving 39 countries. Ann. Oncol. **23**(4), 853–859 (2012). https://doi.org/10.1093/annonc/mdr352
2. Shao, J., Rodrigues, M., Corter, A.L., Baxter, N.N.: Multidisciplinary care of breast cancer patients: a scoping review of multidisciplinary styles, processes, and outcomes. Current Oncol. **26**(3), 385–397 (2019)
3. Mammas, C.S., Mamma, A.S., Papaxoinis, G., Georgiou, I.: AI supported decision making and planning in 40 breast surgery operations. In:Aristoteleio University of Thessaloniki Editor. 32nd Panhellenic Congress of Surgery and International Forum, HSC Jun 9–12 Proceedings, p. 237, Hellenic Surgical Society, Thessaloniki (2021)
4. Mammas, C.S., Mamma, A.S.: E-multidisciplinary oncology conference for liver cancer as a technology and method to optimize personalization and outcomes in the peripheries. Stud Health Technol. Inform. **289**, 313–316 (2022)
5. Mammas, C.S., Mamma, A.S., Papaxoinis, G., Georgiou, I.: Remote AI supported e-multidisciplinary oncology conference in breast cancer as a technology and method to optimize outcomes in the peripheries. Stud Health Technol. Inform. **289**, 309–312 (2022)
6. Furness, P.N., Bamford, W.M.: Tele-pathology. Curr. Diag. Pathol. **7**, 281–291 (2001)
7. Coiera, E.: The internet and telemedicine. In: Coiera, E., (ed.) Guide to Medical Informatics. 2nd ed. CHM, New York (c2007)
8. Sandel, J.S.: What Money Can't Buy-The Moral Limits of Markets. Allen Lane, London (2012)
9. TO BHMA: 90 Years. In the Era of the Collapse of Illusions. Volume 9th 2002–2011,Labrakis PO, Athens (2012) (in Hellenics)
10. Kousoulis, L.: Day after day Hellas is sinking in a night without dawn. Pronews **2**, 62 (2016). (in Hellenics)

11. Friend, P.: Organization of Organ Retrieval-Bitter Lessons and Ambitious Plans Homepage. http://vincentbourquin.files.wordpress.com/2010/11/friend-Hammersmith-11-10-10.pdf. Accessed 17 July 2012
12. Raghupathi, W., Raghupathi, V.: Big data analytics in healthcare: promise and potential. Health Inf. Sci. Syst. **2**(3), 1 (2014)
13. Wilson, KM, Helton, WS ,Wiggins, MW.: Cognitive engineering. Wiley Interdisciplinary Reviews: Cognitive Science, 4 (1), 17(2013).
14. Aoun, M.A.: Advances in three hyper-computation models. EJTP **13**(36), 169 (2016)
15. Gualtieri, M.: True Streaming Analytics Platforms for Real-Time Everything Homepage. https://go.forrester.com/blogs/16-04-16. Accessed 03 Feb 2018
16. Panahiazar, M., Taslimitehrani, V., Jadhav, A., Pathak, J.: Empowering personalized medicine with big data and semantic web technology: promises, challenges, and use cases. In: Proceedings of the IEEE International Conference on Big Data, p. 790, IEEE (2014)
17. Karavatselou, E., Economou, G.P., Chassomeris, C., Daneli, V., Lymperopoulos, D.: OTE-TS-a new value-added telematics service for telemedicine applications. IEEE Trans. Inf. Technol. Biomed. **5**(3), 210–224 (2011)
18. Eng, J.: Computer network security for the radiology enterprise. Radiology **220**(2), 303–309 (2001). https://doi.org/10.1148/radiology.220.2.r01au32303
19. Mammas,C.S., Mamma, A.S.: Innovative training networks for interactive e- learning and application of ergonomics of the remote evaluation of the grafts in organ transplantation. In: Proceedings of the 49th Annual Conference of Nordic Ergonomics and Human Factors Society on Proceedings, p.114. Lund University, Lund (2017)

Interactive E-Learning and E-Training Enhanced by AI and Big Data Analytics of Cognitive Ergonomics of the Remote E- Multidisciplinary Board in Liver Cancer

Constantinos S. Mammas and Adamantia S. Mamma

Abstract **Background and purpose**. The project refers to the effect that interactive E-Learning among expert participants in our innovative Remote E-Multidisciplinary Board for Liver Cancer (PROMETHEUS II pn:1009078) is being integrated with the Hellenic Hyper-computer ARIS for Big Data analytics and computing. Design/ methodology/Results: Integrated analysis of three research projects results showed that the E-Multidisciplinary Board in Liver Cancer may also be the substance of Innovative Networks for Interactive Collaborative E-Learning and Training incorporating AI and Big Data Analytics and Computing among experts, residents and students for further and deep learning of the cognitive ergonomics of the E-Multidisciplinary Board to optimize quality in Liver Cancer decision making, management and treatment in the peripheries. Conclusions: Ergonomics of Innovating Training Networks for Interactive E-Learning for remote evaluation of the patients suffering from Liver Cancer in the context of the E-Multidisciplinary Board in cancer decision support and making, management and treatment in combination to the Mobile Clinical Unit ''Macedonia'' for Liver Cancer prevention may be instructed remotely and interactively, while integration with AI and Big Data analytics interactive E-Learning may teach cognitive ergonomics of the E-Multidisciplinary Board and sustain an efficient surveillance system on a European Level and in the Balkans to optimize quality in Liver Cancer Screening, Prevention, Treatment and Survivorship especially for the peripheries.

Keywords Cognitive Ergonomics · E-Learning and Training · AI and Big Data Analytics and Computing · Liver Cancer

C. S. Mammas (✉) · A. S. Mamma
Program of Excellence 2014-16, Siemens Program for Greece, Athens, Hellas
e-mail: csmammas@med.uoa.gr

I. Salopek Čubrić et al. (eds.), *Proceedings of the 9th International Ergonomics Conference*, Lecture Notes in Networks and Systems 701,
https://doi.org/10.1007/978-3-031-33986-8_22

1 Introduction

Our project analyzes the e-learning and e-training technology, method, process, ergonomics and standards of the remote, specialized, multidisciplinary and personalized remote e-Multidisciplinary Team Board in Surgical Oncology (E-MTBSO) for Liver Cancer (LC) integrating Prometheus II digital medical device with AI and Big Data analytics (PROMETHEUS II pn:1009078) technologies and methods to optimize decisions and outcomes for the benefit of the patients in the peripheries. **In that training context, was given an emphasis on the remote digital macroscopic and microscopic LC diagnosis and classification and accordingly to the optimum remote decision making and treatment integrated with AI empowered Big Data analytics** [1–5]. Currently, within the experience of the international lockdowns in 2020 and in 2021 as well as the continuous need to narrow the differences in the quality of health care services between the big civil centers and the peripheries and/or between the central and north Europe and the south Europe the whole project seems to be the right solution in health care services and specifically in surgical oncology (SO) by providing not only: (a) experts' approach with a continuous training from where ever, (b) acceleration of the diagnostic process and decision making, (c) evidenced based decisions, and (d) significantly lowering of the costs, but mainly with the capacity to build-up hands-on experience and skills in (e) the remote SO related diagnosis, classification, decision making and planning.

In the context of the collaborative e-training oriented session- using an experimental version of PROMETHEUS II pn:1009078 digital medical device technology and method the Exp-TS II and its innovative training networks for collaborative e-learning and e-training- the invited specialists and the registered members used the official digital platform of the program of Excellence 2014–16- Siemens program for Greece (www.priibite.wordpress.com) and trained remotely and interactively supported and using the abovementioned digital medical device, technology and method aspects and functions (Telecommunications, Tele-radiology, Tele-pathology, Tele-cytology, Tele-microbiology, Tele-molecular biology). Also, they implemented them in the simulating clinical training practices and in the geographic and in the proper timing strategies for remote LC diagnosis and classification and decision support and making so as to: (1) enhance LC related decision making accuracy and (2) optimize treatment efficacy and safety and (3) improve outcomes in LC by remote optimum decision and planning e-learning and e-training integrated with AI and Big Data analytics and computing [1, 3, 4, 6, 7].

On a clinical level, the PROMETHEUS II pn:1009078 digital medical device based technology, method, process, service, clinical and training standards analysis and ergonomics for LC, applied on a real clinical case from the GH of Kalymnos and a Department of Radiology in Athens and an Oncologist for a patient who had suffered from a renal cancer that had been treated by renal resection in the past but an exacerbation with a suspicious right lobe liver mass that needed further evaluation on June 2021. The patient worked-out remotely by the E-MTBSO for LC integrating Prometheus II digital medical device with AI and Big Data analytics

(PROMETHEUS II pn:1009078, priibite.wordpress.com of the program of Excellence 2014–16-Siemens program for Greece) in which made the decision for a surgical treatment in a Liver Unit of a tertiary hospital. Regarding the process participants applied tele-communications (TC) as a base for the Coordination Process (CP) and Tele-radiology (TRE) upon upper abdominal US for the remote diagnosis and classification of the liver mass and then for LC TNM classification upon MRI between the hospital in the periphery (PH) and the Liver Unit (LU) to make decisions and optimize outcome [5, 8–10].

On a e-training and on a e-learning the Exp.TS II digital medical device based technology, method, process, service, clinical and training standards analysis and ergonomics integrating AI and Big Data analytics and computing assisted level for LC it applied on a series of officially scheduled on-line training courses on 25.11.2021 and on 18.03.2022 (priibite.wordpress.com of the program of Excellence 2014–16-Siemens program for Greece) already, that still goes on until the end of 2022 and on for LC. On this level and regarding the process participants applied tele-communications (TC) for the Coordination Process (CP), Tele-radiology (TRE), Tele-pathology (TPE), Tele-cytology (TCE), Tele-microbiology (TME), Tele-genetic biology TGBE) for the remote evaluation of the Liver and for LC TNM classification clinical process simulating collaboratively and interactively the hospital in the periphery (PH) and the Liver Unit (LU) to make remotely accurate decisions and optimize outcomes upon simulating LC cases [11–14]. According to the didactic LC case on the 25.11.2021 and on 18.03.2022 (priibite.wordpress.com, Hepatobiliary and Pancreatic Cancer section) the elaborated case referred to the previously mentioned clinical patient who was a young woman of 58 y.o. in the periphery suffering from a suspicious mass and who needed further evaluation and treatment. Regarding the e-training process specialists participants (n1 = Surgeon, n2 = Radiologist, n3 = Pathologist, n4 = Cytologist, n5 = Microbiologist, n6 = Biologist) allocated in different points and connected remotely in the internet collaborating in the Exp.TS II digital medical device structure and AI and Big Data analytics by which interactively learnt and trained themselves by the above mentioned structure and process [15–18].

In fact, PROMETHEUS II pn:1009078 digital medical device based technology, method, process, architecture and service integrated with AI and Big Data analytics, can be seen as a triangle (including the patient in the periphery, the coordinator in a central unit of the government and the E-MTBSO for LC from where it can be applied as the didactic base the remote evaluation of the oncologic patient and for instant remote macroscopic and microscopic evaluation and decision making. Also, for preoperative planning, so as to empower trainees, students and other responsible professionals in SO, to make remotely valuable decisions by interactive e-learning and e-training of the up to date clinical guidelines and contemporary technology in SO and especially in LC. So, participants had the chance to learn and practice the right processes and develop skills for valuable clinical services in SO to confront clinical complexity at distance and in low cost and improve significantly therapeutic outcomes on a national and international level. Moreover, they developed the necessary clinical thoughts and hands on practice by using high technology to

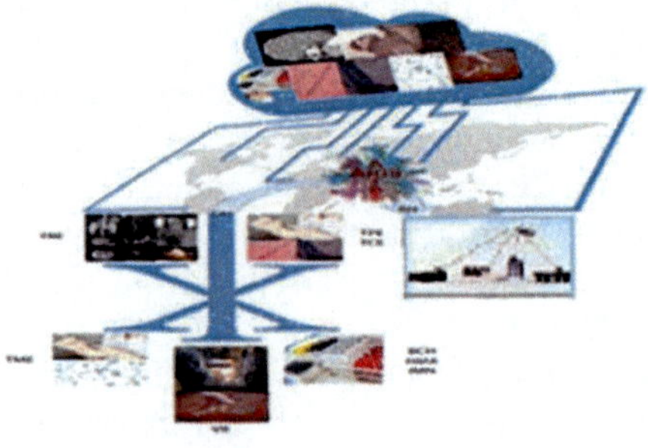

Fig. 1 The clinical Remote E-MTBSO for LC technology and process applied remotely by the specialists participants using the PROMETHEUS II pn:1009078 digital medical device technology and method on June 2021 directed the Exp.TS II assisted collaborative, interactive E-learning and E-training technology and method for Cloud computing of indicators, empowered by AI and Big Data analytics integration among remotely connected participants of the Remote (E-MTBSO) for LC on 25.11.2021 and on 18.03.2022, for instant decision making and treatment (nl = Surgeon, n2 = Radiologist),(nl = Surgeon, n2 = Radiologist, n3 = Pathologist, n4 = Cytologist, n5 = Microbiologist, n6 = Biologist)

build-up the special skills in clinical and technological terms that needed to diagnose remotely the macroscopic and microscopic lesions of the Liver and make decisions for further treatment upon TNM classification [1, 2, 4, 19, 20]. The interactive collaborative e-training process on started by a E-MTBSO for LC on 25.11.2021 and on 18.03.2022 upon data that had been archived in the innovative networks of PROMETHEUS II pn:1009078 digital medical device technology and method in the collaborative e-training for further Cloud computing AI and Big Data analytics integration (priibite.wordpress.com) (Fig. 1).

2 Paper Content and Technical Requirements

Given that the ergonomic impact of the electronic space (ES) for the remote macro- and micro- examinations diagnosis of LC, had been already evaluated on the digital medical device Prometheus (pn: 2003016, 2013) in the context of previous experimental simulations, and in the collaborative e-training session on 25.11.2018, the cognitive ergonomics of the PROMETHEUS II pn:1009078 digital medical device in the context of MTBSO for LC are based on the cognitive ergonomics of Digital Macroscopic and Microscopy (DM) in Tele-medicine, in TRE, in TPE, in TCE and in TME, for the remote macroscopic and microscopic diagnosis of injury/trauma, inflammatory, infectious and/or malignant lesions of the solid abdominal and thoracic organs for remote decision making and treatment. The simulating experimentation for the collaborative e-training on 25.11.2021 and on 18.03.2022 among members of the E-MTBSO for LC also included: (1) The development of an experimental digital medical device similar to Prometheus II pn:1009078 and to OTE-TS, the

Table 1 Comparison of the modules between OTE-TS and Prometheus II Exp.-TS (pn:1009078) integrated with AI, Big Data analytics and Computing

Modules	OTE-TS	Exp.-TS II
Medical record process	+	+
Examinations results	+	+
Capture/ imaging	+	+
DICOM and PACS	-	+
Real-time tele-conference	+	+
Chat and whiteboard	+	+
Tele-secretary facilities	+	+
Tele-Mentoring facilities	+	+
Telecommunication net	ISDN based	Internet based
Virtual Slide Integration	-	+
Cloud Computing AI	- -	+ +
Big Data Analytics	-	+

Prometheus II Exp.-TS II [5, 9, 10], (2) Integration of the Exp.-TS II with the Virtual Slide system (VS) for digitalization of images for TPE or TCE applying DM, AI analytical method in the process (Table 1), (3) Simulation of static or dynamic TPE and TCE for microscopic diagnosis of LC (Fig. 1) [11–13].

2.1 *Designing for Human Factors: The Technology for the Remote E-MTBSO for LC Collaborative and Interactive E-Learning and E-Training Among Specialists in SO*

According to the didactic protocol that members of the E-MTBSO for LC should follow on 25.11.2021 and on 18.03.2022 by using the Prometheus II (pn 1009078) digital medical device integrated with AI and Big Data analytics and computing in the simulating clinical sessions, the data integrated with the Cloud archives in a concrete path for: (1) retrospective and/or real time remote diagnosis and TNM classification of LC, applying telemedicine and remote visualization standards enhanced by AI to make instant computations and decisions about the optimum management and treatment in each case, (2) computations of the requested clinical and financial indicators (LC cases in each country and in Europe, mortality caused by LC and survival in each country and in Europe) applying AI supported Big Data analytics and computing and making decisions instantly on a national and/or international level. The results depicted, instantly, on tables and graphs for each European country and EU in total (the same can be applied on a Balkan level under the condition that the Balkan Oncology is based on a common oncologic registry) (Fig. 2).

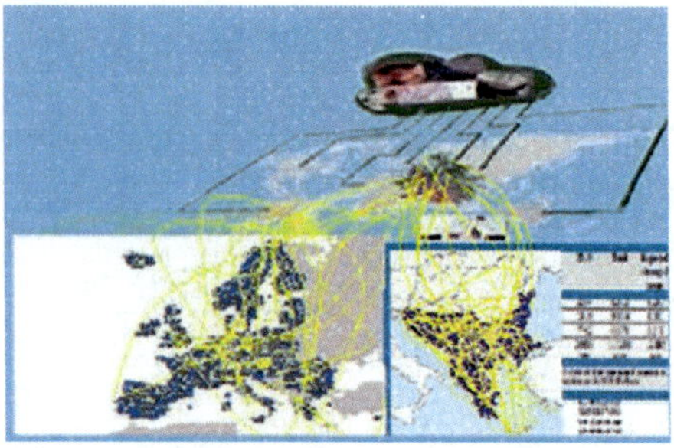

Fig. 2 Prometheus II (pn:1009078) digital medical device based AI and Big Data integrated collaborative E-training on line on 25.11.2021 and on 18.03.2022, connected with partners i.e. specialists for the E-MTBSO for LC -from which the digital images and the incidences of LC in each country in Europe computed by each member - after formation of structured Data in the European cancer registry- via Cloud computing of indicators (incidences, mortality, survival) empowered by AI retrospective and/or real time Big Data analytics and computing to make decisions for an optimum SO system on a national, Balkan, UK and European level

2.2 Designing for Human Factors: The Method of Prometheus II pn:1009078 Digital Medical Device Assisted Collaborative E-Learning and E-Training for the Remote E-MTBSO for LC, Integrating AI and Big Data Analytics for Computations and Mega Data Based Decisions in SO

The integrated TC, CP, TRE, TPE, TCE, TME and TGBE of LC in SO and the remote pre-operative planning and/or decision making and planning for further adjuvant and/or neoadjuvant therapy had shown high sensitivity, specificity and accuracy (ranging from 90% to 98%) in the remote diagnosis of traumatic, infectious, inflammatory, malignant lesions of our project that had assessed specific thoracic and abdominal organs in organ transplantation integrating AI and Big Data analytics (Prometheus 2003016/2013 digital medical device of the program of Excellence 2014–16-Siemens program for Greece) as abovementioned. Taking that results into consideration and that a modern Hyper-computer is equipped with accelerating modules and with new technologies and that is the perfect choice for integrated AI and Big Data analytics and computing [17, 19, 20], the corresponding SO data in Europe in 2020 had been archived in the working Cloud of the program of Excellence 2014–16-Siemens Program for Greece, as a structured material for real time Big Data analytics and computing, so that Prometheus II pn:1009078 digital medical device users could accumulate them for: (1) remote diagnosis collaborative interactive e-training, and for (2) instant AI and Big Data computations training on 25.11.2021 and on 18.03.2022 [20].

The participants had registered in the working site (priibite.wordpress.com) and actively implemented the best practices, high technology and geographic and timing strategies for Remote E-MTBSO for LC, integrating AI and Big Data analytics and decision support and making, to optimize decisions and improve outcomes in LC management and treatment by pre-operative planning according to the referred didactic and simulating interactively processed clinical cases in LC on 25.11.2021, on 18.03.2022 and on. As abovementioned, the results smoothed the path for initiating the clinical and training standards for the real time Big Data analytics and computing of clinical and financial indexes referring to the quality of SO in Europe and in the Balkans with a priority to the cost-effectiveness analysis for further decision making and on time interventions to improve the Health care systems [20].

Prometheus II (pn:1009078*) based Collaborative E-Training for Real Time AI and Big-Data analytics and computing, of LC management and treatment and its quality control.*

The reliability and accuracy of a modern Hyper-computer considered by definition as the ideal [12, 14]. Prometheus II (pn 1009078) digital medical device assisted TRE, TPE, TCE, TME and TGBE remote E-MTBSO for LC integrating AI and Big Data analytics and computing collaborative e-learning and e-training for pre-operative decision making and planning in LC SO management and treatment has already shown high sensitivity, specificity and accuracy (ranging from 90% to 98%) upon remote diagnosis and classification of LC among specialists [1, 4, 20]. In the collaborative e-training process the material that used by the participants on 25.11.2021 and on 18.03.2022 and on based on the EU registry data from 2020 archived in the Cloud. The real time Big Data financial analysis with the technology and method of the digital medical device Prometheus II (pn: 1009078) completed with the real time computation of the indicators of interest (incidences, mortality, survival) of the LC treatment in Europe [1, 4]. These indicators and the cost-effectiveness analysis should have a priority in the computations that the members of the E-MTBSO for LC should apply and analyze and interpret their results instantly.

In this context, the incidents of LC and the quality control of its management in Europe in terms of mortality and survival indicators computed by the participants of the for Remote E-MTBSO for LC on a real time Big Data analytics and computing framework simulating the proposed process as it has been suggested by the Program of Excellence 2014–16-Siemens program for Greece since 2016 and shaping our proposal for a realistic and effective leadership in SO nationally and on a European level to optimize quality in LC management and treatment. As far as the choice of the members of the proposed Remote E-MTBSO for LC integrating AI and Big Data analytics and computing is concerned, the qualitative training in new technologies and the quality control shapes a new fair system in SO that could be established in Europe for the benefit of the patients and for saving national and European and/or the Balkan economies. It is underlined that the accuracy of the computational analysis was really high and lasted for a half an hour maximum given that a software for mathematical analytics and computing had been incorporated in the system structuring the AI and Big Data analytics (Figs. 3, 4).

Fig. 3 Prometheus II (pn:1009078) digital medical device based AI and Big Data integrated collaborative e-training on line on 25.11.2021 and on 18.03.2022, connected with partners i.e. specialists for the Remote E-MTBSO for LC -from which the digital data as images and numbers of LC cases in each country in Europe had sent to each member in the E-Learning process- after formation of structured Data- for computation of indicators (number of cases, mortality, survival) by each participant in the E-Learning and E-Training process via Cloud computing, empowered by AI supported retrospective and/or real time Big Data analytics and computing to make decision for an optimum SO system on a multinational level (34931)

The abovementioned computational analysis results can be used by the Ministry of Health to intervene locally or nationally on time, so as to correct several problems referring to the cost-effectiveness balance and to the quality of LC management and treatment (in terms of the resources and morbidity-mortality rates) or to correct mistakes and/or destructive pricing policies of the biomedical products, on time [9–11, 17, 20]. On the other hand they may be used as accurate and real time tools on a weekly base to make decisions about the biomedical technology products in SO expenses and pricing for the benefit [1, 3, 11].

The main drawback of this experimental study is that integration of Big-Data computational analysis on 25.11.2021 and on 18.03.2022 based on a year framework of clinical SO on a European level simultaneously and not on a weekly level (2020). However, this drawback didn't have any negative effect to the participants collaborative e-learning and e-training given the aim of the training process was to be accustomed to the technology the method and the process to participate in the remote

Age-specific 5-year relative survival

Pool of European cancer registries, Both sexes, Liver, 15+ years, 2000-2007

Country	Sex	Cancer Site	15-44	45-54	55-64	65-74	75+
Pool of European cancer registries	Female	Liver	25.60	17.74	13.68	9.99	5.13
Pool of European cancer registries	Male	Liver	20.90	16.73	13.54	10.24	5.74

Fig. 4 Prometheus II (pn:1009078) digital medical device based AI and Big Data integrated collaborative e-training on line on 25.11.2021 and on 18.03.2022, connected with partners i.e. specialists for the Remote E-MTBSO for LC -from which the digital images and the numbers of LC cases in each country in Europe sent to each member - after formation of structured Data- via Cloud computing applying AI empowered retrospective and/or real time Big Data analytics and computing to make decision for an optimum SO system on a multinational level

E-MTBSO for LC integrating AI and Big Data analytics and computing. Another drawback was that there was non-parallel comparative computational process using the Hyper-computer (HC) ARIS-IBM. However, the reliability and accuracy of a modern Hyper-computer considered by definition as the ideal one. In addition, we are developing new projects for the use of the ARIS-IBM HC as the source of mega data and the real computational center not only for training but also for clinical reasons in the direction of E-MTBSO for LC integrating AI and Big Data analytics and computing [12, 14].

3 Conclusion

Ergonomics of Innovating Training Networks for Interactive E-Learning for remote evaluation of the patients suffering from Liver Cancer in the context of the Remote E-Multidisciplinary Board in Liver Cancer decision support and making, management and treatment may be instructed remotely, collaboratively and interactively. The integration with AI and Big Data analytics interactive E-Learning and E-Training among specialsts in SO may teach cognitive ergonomics of the Remote E-MTBSO for LC and sustain an efficient surveillance system on a European Level and in the Balkans to optimize quality in Liver Cancer Decsions Making and Treatment as well as for the Screening, Prevention and Survivorship especially for the peripheries.

References

1. Siddique, O., Yoo, E.R., Perumpail, R.B., et al.: The importance of a multidisciplinary approach to hepatocellular carcinoma. J. Multidiscip. Healthc. **10**, 95–100 (2017)
2. Couri, T., Pillai, A.: Goals and targets for personalized therapy for HCC. Hepatol Int. **13**(2), 125–137 (2019)
3. Mammas, C.S., Mamma, A.S., Papaxoinis, G., Georgiou, I.: AI supported decision making and planning in 40 breast surgery operations. In: Aristoteleio University of Thessaloniki Editor. 32nd Panhellenic Congress of Surgery and International Forum, HSC Jun 9–12 Proceedings, p. 237, Hellenic Surgical Society, Thessaloniki (2021)
4. Mammas, C.S., Mamma, A.S.: E-multidisciplinary oncology conference for liver cancer as a technology and method to optimize personalization and outcomes in the peripheries. Stud. Health Technol. Inform. **289**, 313–316 (2022)
5. Mammas, C.S., Mamma, A.S., Papaxoinis, G., Georgiou, I.: Remote AI supported e-multidisciplinary oncology conference in breast cancer as a technology and method to optimize outcomes in the peripheries. Stud. Health Technol. Inform. **289**, 309–312 (2022)
6. Furness, P.N., Bamford, W.M.: Tele-pathology. Curr. Diag. Pathol. **7**, 281–291 (2001)
7. Coiera, E.: The Internet and Telemedicine. In: Coiera, E., editor. Guide to Medical Informatics. 2nd ed. CHM, New York, (c2007)
8. Priester, R.: Distributing Limited Health Care Resources University of Minnesota, Minessota-Center for Biomedical Ethics (1991)
9. Sandel, J.S.: What Money Can't Buy-The Moral Limits of Markets. Allen Lane, London (2012)
10. TO BHMA: 90 Years. In the Era of the Collapse of Illusions. Volume 9th 2002–2011,Labrakis PO, Athens (2012) (in Hellenics)
11. Friend, P.: Organization of organ retrieval-bitter lessons and ambitious plans. http://vincentbourquin.files.wordpress.com/2010/11/friend-Hammersmith-11-10-10.pdf. Accessed 17 July 2012
12. Raghupathi, W., Raghupathi, V.: Big data analytics in healthcare: promise and potential. Health Inf. Sci. Syst. **2**(3), 1 (2014)
13. Wilson, K.M., Helton, W.S., Wiggins, M.W.: Cognitive engineering. Wiley Interdiscip. Rev. Cogn. Sci. **4**(1), 17–31 (2013). https://doi.org/10.1002/wcs.1204
14. Aoun, M.A.: Advances in three hyper-computation models. EJTP **13**(36), 169 (2016)
15. Gualtieri, M.: True streaming analytics platforms for real-time everything. https://go.forrester.com/blogs/16-04-16. Accessed 03 Feb 2018
16. Panahiazar, M., Taslimitehrani, V., Jadhav, A., Pathak, J.: Empowering personalized medicine with big data and semantic web technology: promises, challenges, and use cases. In: Proceedings of the IEEE International Conference Big Data, p.790, IEEE (2014)

17. Karavatselou, E., Economou, G.P., Chassomeris, C., Daneli, V., Lymperopoulos, D.: OTE-TS-A new value-added telematics service for telemedicine applications. IEEE Trans. Inf. Technol. Biomed. **5**(3), 210–224 (2011)
18. Morgan, RH.: Computer network security for the radiology enterprise computer network. Radiology. **220**(2), 303 (2001)
19. Mammas, C.S.: The Remote evaluation of pap-test applying tele-cytology in the context of uterus graft remote evaluation and in the remote prevention of cervical cancer. http://www.lib.uoa.gr/fileadmin/user_upload/I.K.Y_Papanikolaou.pdf. Accessed 11 Apr 2015
20. Mammas, C.S., Mamma, A.S.: Innovative training networks for interactive e- learning and application of ergonomics of the remote evaluation of the grafts in organ transplantation. In: Proceedings of the 49th Annual Conference of Nordic Ergonomics and Human Factors Society on Proceedings, p. 114. Lund University, Lund (2017)

Effect of Gaming Controllers on Wrist Angles

Lizbeth A. Mariano, Ilham Priadythama, Ping Yeap Loh, and Satoshi Muraki

Abstract The study aimed to distinguish wrist movement characteristics when playing with different controllers. Five healthy university students played an esports game with three types of controllers/devices: a keyboard and mouse, a smartphone (iPhone 12), and a PS5 DualSense controller. Participants used the three controllers to play a first-person shooter (FPS) game for five minutes each. A 3D motion capture system was used to record and analyze variations in the wrist angle. Our results suggest that participants presented the highest wrist extension angles when using a PS5 DualSense controller for the left wrist and using a mouse for the right wrist. In contrast, for the ulnar deviation angle, both the keyboard and PS5 controllers yielded the highest angles for the left wrist. The PS5 controller yielded the highest ulnar deviation angles for the right wrist among the three devices. For all controller types, a combination of wrist extension and ulnar deviation angles was predominant.

Keywords esports · esports ergonomics · gaming controllers · wrist angle

1 Introduction

In recent years, competitive gaming or esports has seen enormous growth in terms of audience and revenue [1]. Esports place the same physical and mental demands on athletes as traditional sports [2]. As a result, esports athletes' health concerns must be addressed appropriately. To remain competitive, they must practice for extended periods of time. A study in 2019 found that collegiate esports players usually practice

---

L. A. Mariano (✉) · I. Priadythama
Graduate School of Design, Kyushu University, Fukuoka 815-8540, Japan
e-mail: lamariano@up.edu.ph

L. A. Mariano
Department of Industrial Engineering, University of the Philippines, 4031 Los Baños, Laguna, Philippines

P. Y. Loh · S. Muraki
Faculty of Design, Kyushu University, Fukuoka 815-8540, Japan

I. Salopek Čubrić et al. (eds.), *Proceedings of the 9th International Ergonomics Conference*, Lecture Notes in Networks and Systems 701,
https://doi.org/10.1007/978-3-031-33986-8_23

three to ten hours per day [3]. Competitive gamers practice for more than 12 h per day [1]. Long practice hours lead to several health hazards for these players, especially overuse injuries, that require attention [2, 3].

Esports are mostly played using computers and consoles [1], but in recent years, mobile/handheld devices have been favored [4]. According to Migliore [1], an understanding of input devices is necessary for healthcare providers. The mobile/handheld devices can be used in conjunction with a wide range of commercially available controllers. A variety of games are also available to play, such as first-person shooters (FPS), multiplayer online battle arenas (MOBA), and role-playing games (RPG). Gaming movements entail approximately 400–600 actions per minute [1]. A few studies have been conducted on the mechanisms of overuse injuries in esports.

There are many available tools and equipment that can be used in motion analysis. One such tool is motion capture, a range of technologies that record the movement of people and objects and sends the results to another application for postprocessing of data [5]. Motion capture systems have been used by several studies for the postural analysis of office-based and computer tasks [6]. It is a good tool for analyzing kinematic parameters in 3D, time-based parameters, and capturing small movements because of its high resolution. These capabilities of the motion capture system were needed in our study since we must observe small movements of the wrists in 3D space and the trend of these movements over time. This study investigated the wrist kinematic characteristics such as average wrist angles and their fluctuations resulting from using different gaming controllers/devices. Musculoskeletal disorders have been linked to awkward joint posture (exceeding normal range) and repetition [7] so we chose to measure the wrist angles and their fluctuations over time. The results of this study will be helpful for succeeding research on the prevention of wrist overuse injuries.

2 Methods

2.1 Participants

Five right-handed participants with no upper-extremity musculoskeletal disorders were recruited (two males and three females, 26.0 $\pm$ 4.1 years). Phalen's test and the Boston Carpal Tunnel Syndrome Questionnaire were administered to exclude participants with symptoms of carpal tunnel syndrome (CTS). All the participants provided written informed consent. The experiment was approved by the Ethics Committee of the Faculty of Design, Kyushu University, Japan.

(a) Mouse and keyboard (b) PS 5 Controller (c) iPhone 12

Fig. 1 Controllers/devices used

2.2 Games and Gaming Controllers/Devices

The experiment used FPS games and three controllers: a keyboard and mouse, a gamepad (PS5 DualSense Controller), and a smartphone (iPhone 12). The game used for the keyboard and mouse, and the PS5 DualSense was the Counter-Strike: Global Offensive (CS: GO) - Deathmatch mode. For the smartphone (iPhone 12), the call of duty mobile (CODM) multiplayer mode was used. Additionally, the participants practiced playing the FPS games with the devices until they were confident enough to use them. Because the participants were all right-handed, the left hand was used to operate the keyboard (movement and action keys) and the right hand for the mouse (aiming and shooting). For the PS5 Controller and the iPhone, the left hand was used to control the movement direction with an optional trigger for shooting, while the right hand was used to control aiming, shooting (main), and other action buttons. Figure 1 shows the controllers used in the experiment.

2.3 Experiment Setup, Protocol, and Measurements

The gaming station setup includes an office chair with adjustable height and a table where the computer monitor was placed 60 cm from the participant, and the controller (Fig. 1) to be used was placed on the table. The monitor height was adjusted to participant eye level and participants adjusted their preferred seat height prior to the experiment tasks. In addition, the participants wore black upper clothing throughout the experiment.

The motion capture system, Cortex v7.0 (Motion Analysis, Rohnert Park, CA), with seven near-infrared (NIR) cameras (Motion Analysis, Rohnert Park, CA) was used to collect wrist kinematics data. Seven 9.5-mm retroreflective markers (14 markers in total) were attached to the forearms and hands of each side of the participant, these were: (1) at the third metacarpal head, (2) at the second metacarpal head, (3) a reference marker at the middle of 1 and 2 horizontally, and 1 and the wrist vertically, (4) at the ulnar styloid, (5) at the radial styloid, (6) at 1/3 of forearm length, distal from the medial epicondyle, and (7) at 1/3 of forearm length, distal from the lateral epicondyle.

The experiment task was to play CS:GO Deathmatch for 5 min each for the mouse and keyboard and PS5 controller (Fig. 1a and 1b) while for the iPhone 12 (Fig. 1c) was to play CODM multiplayer game also for 5 min. The sequence of using the controllers differs among the participants. The participant was signaled at the start and the end of each recording. After each trial, the controller is changed and there is a 5-min break before the start of the next trial. Motion data recording was started after 10-s when the participants initiated the game.

2.4 Data Extraction and Processing

The wrist kinematics data recorded during the experimental tasks were extracted from Cortex v7.0 and analyzed with KineAnalyzer (Kissei Comtec Co., Ltd, Japan). The wrist kinematic measurements were performed continuously for each trial. The five-minute general average and the moving average of 255 s of both wrist flexion–extension and wrist radial-ulnar deviation data were analyzed. The wrist flexion and radial deviation angles are presented as negative angles. Subsequently, the percentage occurrence of each wrist angle was calculated.

2.5 Statistical Analysis

The Kruskal–Wallis test was conducted to examine the wrist angle differences of the three controllers among the participants. All statistical analyses were performed using SPSS version 28.0 (IBM Corp., Armonk, NY). All data are reported as means ± standard deviation (SD).

3 Results

3.1 Average of Wrist Kinematics

The average wrist angles with standard deviations were computed and summarized for each participant (ID1–ID5) in Figs. 2 and 3 for the left and right wrists, respectively. The distributions of all wrist angles were found to be significantly different between the controllers ($p < 0.05$). Based on Fig. 2, the wrist extension angle was the highest on average for the left wrist when using a PS5 controller. It also exhibited the greatest variation in the angles. On the right side, the highest wrist extension angles on average were reported when using a mouse, but it had the least variation in angles.

In contrast, Fig. 3 shows that for the ulnar deviation angle of the left wrist, keyboard use had the highest average angle, surpassing that for the PS5 controller. However,

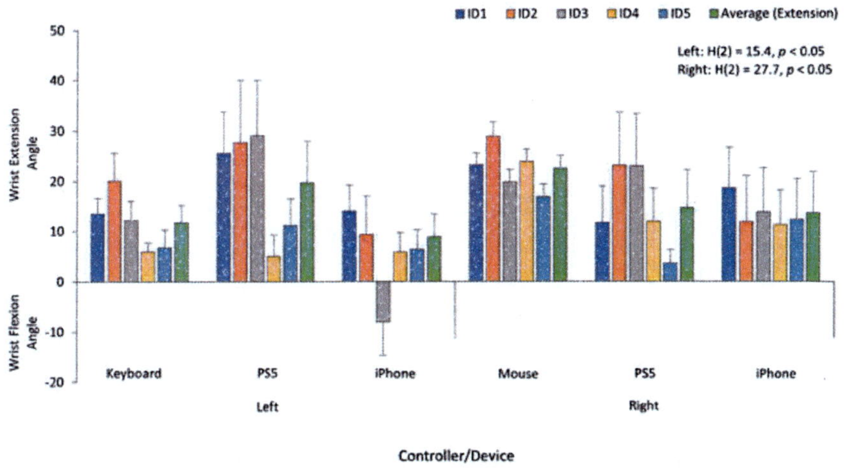

Fig. 2 Average wrist extension/flexion angles of each participant with the overall average

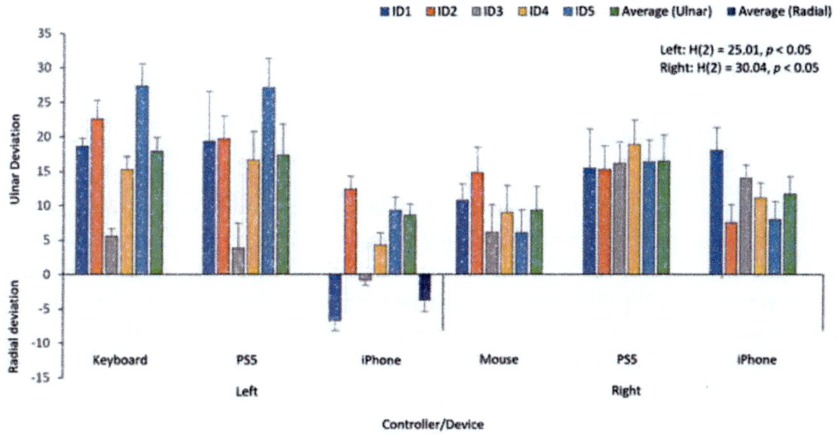

Fig. 3 Average wrist ulnar/radial deviation angles of each participant with the overall average

the PS5 controller had a higher variation in angles. On the right side, the wrist ulnar deviation angles were highest on average for the PS5 controller use. It also exhibited the highest variation in the angles. Not much wrist flexion and radial deviation angles were encountered during the experiment, but the largest average angle was observed when using iPhone 12 for both wrists. Relative to the other devices, the wrist flexion variation is medium and smallest at the radial deviation angle.

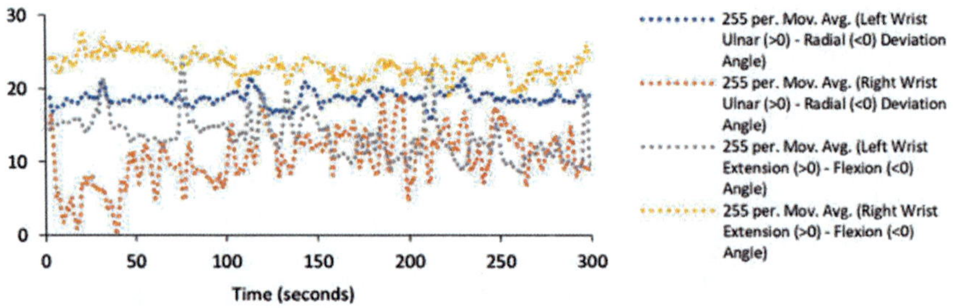

Fig. 4 Trendlines for wrist angle change when using a mouse and keyboard (example of ID1)

Table 1 The occurrence of wrist angles (percentage of total)

	Left				Right			
Device	Ulnar (%)	Radial (%)	Ext (%)	Flex (%)	Ulnar (%)	Radial (%)	Ext (%)	Flex (%)
Mouse & keyboard	100	0	99	1	97	3	100	0
PS5 Controller	98	2	88	12	100	0	94	6
Smartphone	89	11	67	33	100	0	79	21

3.2 Moving Average of Wrist Kinematics

A 255-s moving average was applied to smoothen the data. Figure 4 shows a sample graph for participant 1 using a mouse and keyboard during the game. All angles for the right wrist fluctuated more than those for the left wrist, according to the trend lines. When using the PS5 controller, the observed fluctuations in the wrist angles were similar for both the left and right wrists. Finally, when using a smartphone, the ulnar-radial deviation angles exhibited similar fluctuations for both sides. However, in terms of extension-flexion angles, the right wrist was observed to have more fluctuations than did the left wrist. The frequency of occurrence for each angle was also computed using the percentage of wrist time in flexion/extension and ulnar/ radial deviation (Table 1). For all controller types, a combination of wrist extension and ulnar deviation angles was predominant.

4 Discussion

When using the keyboard, most controls used are located on the left side; therefore, the ulnar deviation angle increased for the left wrist (Fig. 3). However, the movements were more consistent as the keys were fixed. Using the mouse with the right hand resulted in the highest extension angles (Fig. 2), which was expected because of its shape. Some participants also lifted their hands slightly when repositioning the mouse. In most cases, when using the PS5 Controller, a high variation in angles was encountered. This could be due to the motion sensor features on the PS5 controller. The participants could use it to direct their aim or to change the view angle. They also had more freedom in movement, as the controller was not fixed on the table. Lastly, when using the smartphone, a high occurrence of wrist flexion could be owing to the variations in holding the device among the participants. Because there is no definitive way to hold it, each user chose whichever was most comfortable for them. Overall, the wrist extension angles were greater than the wrist flexion angles on both the left and right sides. This could be because devices are easier to use when the wrists are extended.

Wrist deviation from the neutral position is considered a risk factor for CTS, especially when combined with repetition and forceful exertion [8]. Based on previous research [9], the normal values of wrist angles are as follows: 71° extension, 73° flexion, 33° ulnar deviation, and 19° radial deviation. Most of the measured angles in our study were within normal limits, except in a few instances, such as for the left wrist ulnar deviation when using a keyboard at 42° and the right wrist ulnar deviation at 36° when using a smartphone. We also observed considerable repetition in the movements of the participants. Further studies are needed to measure the impact of esports on wrists, such as acute muscle fatigue and soft tissue changes.

5 Conclusion

To better understand player movements during esports, it is imperative to determine the wrist angles when controllers are used. The study identified that the PS5 DualSense controller yielded the largest extension angles among the three controllers for the left wrist. On the right wrist, mouse use yielded the highest extension angles, while the other two controllers yielded almost similar angles. Although the study did not point directly to a factor causing injury, it is an important step to further our knowledge of musculoskeletal injury prevention in esports. Gaming devices and controller designers can draw insights from this study for designing products that are more ergonomic for players. Esports teams' medical supports can also use the results of this study in designing their injury prevention programs.

Acknowledgements The authors would like to express their sincere gratitude to JICA Innovative Asia for funding this research and to Mr. Keneth Sedilla for helping during the experimental trials.

References

1. Migliore, L., McGee, C., Moore, M. N. (eds.): Handbook of esports medicine. 1st edn. Springer Nature, Switzerland (2021)

2. Zwibel, H., DiFranciscoDonoghue, J., DeFeo, A., Yao, S.: An osteopathic physician's approach to the esports athlete. J. Am. Osteopat. Assoc. **119**(11), 756–762 (2019)
3. DiFranciscoDonoghue, J., Balentine, J., Schmidt, G., Zwibel, H.: Managing the health of the eSport athlete: an integrated health management model. BMJ Open Sport Exerc. Med. **5**, e000467 (2019)
4. Esports Insider. https://esportsinsider.com/2021/12/esports-on-smartphones-the-future-of-the-industry/. Accessed 10 June 2022
5. Mo-sys. https://www.mo-sys.com/what-is-motion-capture-and-how-does-it-work/. Accessed 17 Aug 2022
6. Maheronaghsh, S., Vaz, M., Santos, J.: Methods of posture analysis for computer workers. Int. J. Occup. Environ. Saf. **2**(2), 41–49 (2018)
7. CDC. https://www.cdc.gov/niosh/topics/ergonomics/ergoprimer/step1.html. Accessed 17 Aug 2022
8. Anderson, D.A., Agur, A.M., Oliver, M.L., Gordon, K.D.: Effects of slight flexion-extension and radial-ulnar deviation postures on carpal tunnel volume. Clin. Biomech. **92**, 105575 (2022)
9. Kim, T.S., et al.: A study on the measurement of wrist motion range using the iPhone 4 gyroscope application. Ann. Plast. Surg. **73**(2), 215–218 (2014)

Assessment of Acoustical Ergonomics Within Cessna Citation CJ2 Aircraft

P. K. Mikačić and T. Bucak

Abstract This paper provides an interior noise analysis of the light business jet Cessna Citation CJ2, with experimentally measured noise levels during normal flight operations. The study is made to collect acoustical information which provide in-depth view of human acoustical comfort and further increase the faculty's information database with aircraft acoustical information which can be later used by other authors for further study or be used as a reference value for individual use. Measured noise data is presented in a tabular and graphical way, giving precise and simple insight during all segments of the flight. The main results consist of Equivalent noise levels, Minimum and Maximum noise levels, and Octave band noise levels, showing that noise levels are fluctuating around high level margins hence appropriate noise protection options are recommended for long-exposed personnel, primarily pilots. Lastly, the results are processed through objective metrics, yielding the actual influence of Cessna Citation CJ2 aircraft noise levels on the crew and passengers. The results show that passengers are provided with a sufficient level of speech intelligibility while pilots are expected to wear the headset during the critical flight phases to maintain the highest possible flight safety level.

Keywords Aircraft interior noise · Cessna CJ2 · Human comfort · Psychoacoustic ergonomics

1 Introduction

With everyday exposure to noise, one would make sure to familiarize itself with noise related information. But not every data is easily accessible or available. The biggest concern of most aircraft manufacturers is external noise and how to comply

P. K. Mikačić (✉) · T. Bucak
Department of Aeronautics, Faculty of Transport and Traffic Sciences, University of Zagreb, Zagreb, Croatia
e-mail: mikacicpk@gmail.com

T. Bucak
e-mail: tino.bucak@fpz.unizg.hr

I. Salopek Čubrić et al. (eds.), *Proceedings of the 9th International Ergonomics Conference*, Lecture Notes in Networks and Systems 701,
https://doi.org/10.1007/978-3-031-33986-8_24

with noise abatement regulations, but interior noise is not regulated and is purely a matter of ergonomics and human comfort. Therefore, to find an answer to how much noise crew and passengers are exposed to in Cessna Citation CJ2, it was deemed necessary to conduct own experimental study.

The research was conducted in situ during multiple flights using a Class 1 Sound Analyzer according to ISO 5129:2001. Measurements were taken over the course of 2 days. The flight deck noise measurements were measured in position between pilots at the head level with the correctly placed seat, while the cabin measurements were measured on seat 6 at the head level of the upright passenger seat. Multiple takes were made for each flight phase to ensure constant result levels and brought into a graphical and tabular format for easy insight. Results are divided into the flight deck and cabin segment for a more precise picture of overall cabin ergonomics, therefore, focusing directly on crew noise exposure as well as passengers' acoustical comfort.

Aircraft crew spends significant amounts of time inside the flight deck carrying out everyday flight operations, so acquiring information on the interior noise gives a clearer picture of the noise exposure magnitude on human beings and provides valuable information regarding normal communications and overall ergonomics. Passengers are mostly not as exposed to long term interior noise effects as aircraft crew, but passengers' comfort regarding the acoustical impact plays a major role in ergonomics and as such, aircraft choice for future flights. Recent studies show that modern, large, long range business jets have interior noise levels values of 50–60 dBA while smaller business jets have higher noise levels of up to 80 dBA which is seen as ergonomically satisfactory as smaller jets are used for shorter flights hence noise exposure is shorter.

2 Interior Noise Sources

Sound is a part of mechanical wave motion, whose frequency lies in the audible range of the human ear, i.e., in the range of 20 Hz – 20,000 kHz. The formation of waves is caused by the movement of particles in the environment. Depending on the movement of the particles regarding the direction of propagation, there is a difference between transverse and longitudinal wave motion. Sound propagates through the air as longitudinal wave motion and can also spread through solids, liquids and other gases [1].

Having in mind the basic definition of sound, it is defined that noise is any sound that is subjectively perceived as uncomfortable or distracting. Aircraft as a source produces exterior and interior noise. Exterior noise is primarily due to high pressure differences at the engine exhaust and is regulated by limiting produced noise levels and forming special procedures that have to be followed to reduce the impact on the environment and local population around the airport, i.e., noise footprint. Interior noise is the combined product of engine operations and the aircraft's movement through the air that penetrate into the aircraft cabin.

High pressure changes during engine combustion and exhaust process are transmitted via the engine mounts to the aircraft fuselage and further on into the interior

Fig. 1 Cessna Citation CJ2 interior noise reduction methods

of the aircraft. To reduce the effect of all processes happening around the passengers and crew during the normal flight of the aircraft, different measures are introduced to reduce interior noise to a comfortable level.

Figure 1. represents technical methods used to reduce interior noise. Soundproofing methods that involve window geometry, pane thickness, seals and furniture placement all make a barrier to engine noise while using soundproofing materials such as fiberglass insulation or closed-cell vinyl, act as a great insulator of noise and outside conditions.

3 Interior Noise Measurements

Experimental measurements of interior noise are divided into the flight deck section and cabin section. As there are different levels of noise produced during different phases of flight, measurements are made in seven flight phases, as shown in Table 1.

Results ae graphically and tabularly displayed, as shown in Tables, 2, 3, 4 and 5 both flight deck and cabin are displayed in the same graph, set apart by different

Table 1 Phases of flight with respective power settings and Indicated airspeed

	Flight phase	Power setting [%]	Airspeed
1	(Engine) Start	Idle	0 kn
2	Taxi	30.0	20 kn
3	Take off	101.0	110 kn
4	Climb (5000 ft)	97.3	230 kn
5	Cruise (34,000 ft)	100.0	Ma 0.70
6	Descent (29,000 ft)	98.0	260 kn
7	Approach (3000ft)	65.0	140 kn

Table 2 Interior Noise Results: Equivalent Noise Levels [dBA].

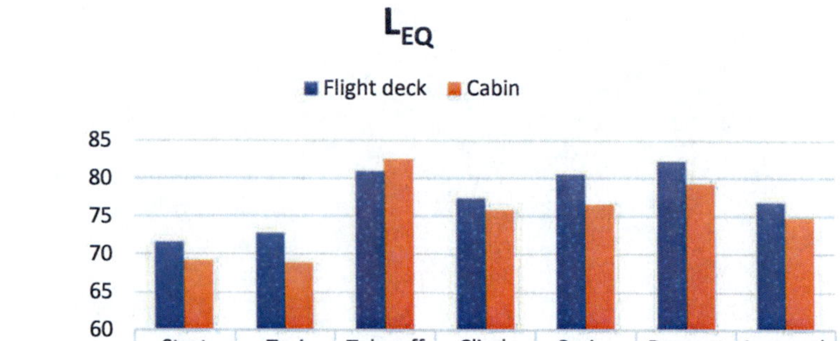

	Start	Taxi	Take off	Climb	Cruise	Descent	Approach
Flight deck	71.6	72.8	81	77.4	80.6	82.3	76.9
Cabin	69.2	69	82.6	75.8	76.6	79.3	74.8

colors. The Y axis represents noise levels in dBA and the X axis represents the flight phase. Table 2 provides Equivalent noise levels of A-pondered 10-s segment noise measurements. It is seen that take off phase causes the highest noise levels in the cabin because of the rear tail-mounted engines, while the descent phase causes the highest levels of noise in the flight deck due to the high level of aerodynamic noise caused by frontal placement into the high-speed flowing airstream. Table 3 provides the lowest and highest recorded noise levels in the 10-s segment hence providing insight into noise level fluctuation amplitudes.

Tables 4 and 5 represent octave band noise levels in the flight deck and cabin thus providing detailed acoustical pictures for each flight phase. Overall, it is seen that low and middle range octave frequencies dominate the acoustical picture so it can be concluded that noise does not cause direct discomfort during flight but significantly reduces level of speech intelligibility and increases speech interference level.

4 Objective Metrics Analysis

Results have been processed through objective metrics algorithms to determine the actual impact of noise on Speech Interference Level and Speech Intelligibility Index.

4.1 SIL Analyses

SIL (Speech Interference Level) is defined as an ISO standard that computes the arithmetic mean of sound pressure levels in 0.5, 1, 2 and 4 kHz octave bands. Results

Table 3 Interior Noise Results: Maximum and Minimum Noise Levels [dBA].

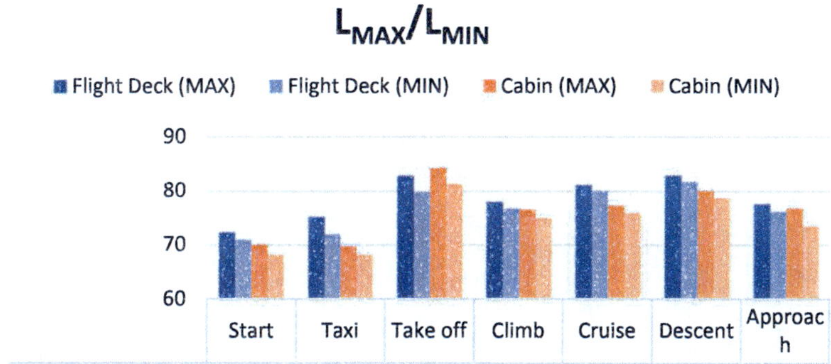

	Start	Taxi	Take off	Climb	Cruise	Descent	Approach
Flight Deck (MAX)	72.3	75.1	82.8	78	81.1	82.9	77.6
Flight Deck (MIN)	71	71.9	79.8	76.8	80	81.7	76.2
Cabin (MAX)	70	69.7	84.2	76.6	77.3	80	76.8
Cabin (MIN)	68.2	68.2	81.3	74.9	75.9	78.7	73.5

Table 4 Flight deck Noise Results: Noise Levels in Octave Band [dB].

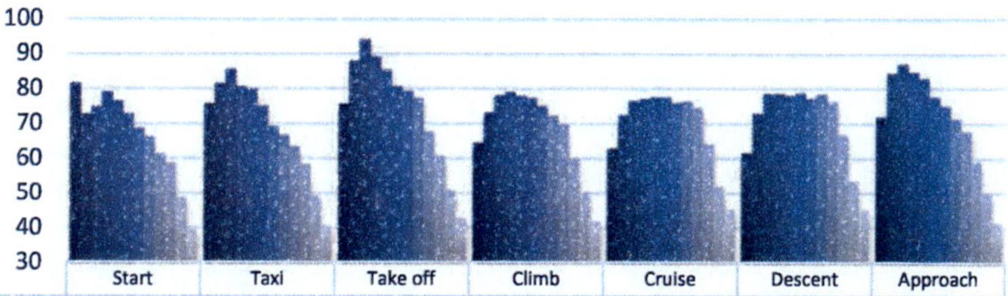

	Start	Taxi	Take off	Climb	Cruise	Descent	Approach
8.0 Hz	81.6	75.7	75.6	64.5	62.8	61.6	71.9
16 Hz	72.7	81.5	87.9	73.1	72.4	72.9	84.4
31.5 Hz	74.8	85.5	94	78	76.5	78.5	87.2
63 Hz	79	80.5	89.2	79	77.1	78.6	84.8
125 Hz	76.5	79.8	85.3	77.9	77.6	78	82.9
250 Hz	72.8	75	80.6	77.3	77.5	78.7	77.6
500 Hz	68.6	69.3	79.3	75	75.9	77.2	75
1.0 kHz	66.3	66.7	77.3	72.2	76.4	78.4	71.3
2.0 kHz	61.4	63.3	67.6	69.7	74.5	76.3	67.7
4.0 kHz	58.7	58.2	60.6	60.1	64.1	66.8	59.1
8.0 kHz	48.9	49.3	50.9	50.4	51.9	53.5	50.2
16.0 kHz	39.5	40.5	42.8	41.9	45.4	45.3	41.4

Table 5 Cabin Noise Results: Noise Levels in Octave Band [dB].

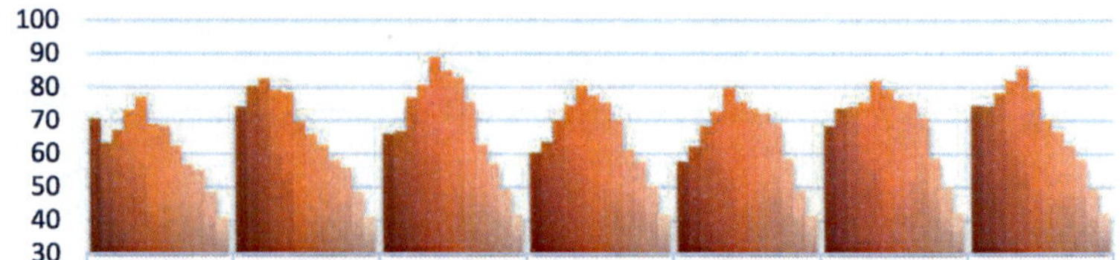

	Start	Taxi	Take off	Climb	Cruise	Descent	Approach
8.0 Hz	70.7	74.3	66	60.5	57.7	68.4	74.6
16 Hz	63.4	80.5	66.9	63.6	62.3	73.7	74.4
31.5 Hz	67.2	82.6	76.8	70.1	68.3	74.2	78.1
63 Hz	73.3	79.3	80.7	74.6	72.8	75.4	82
125 Hz	77.1	78.7	89	80.4	79.7	81.8	85.6
250 Hz	68.9	69.7	85	77.4	75.6	79.3	79
500 Hz	68.4	65.9	83	75.3	73.5	76.1	70.3
1.0 kHz	62.4	62.7	75.5	70.2	72.2	75.5	66.7
2.0 kHz	56.8	58.1	62.8	61.3	69.2	70.8	62.5
4.0 kHz	55.2	56	57.1	57.7	58.5	58.6	58.3
8.0 kHz	48.7	48.7	49.5	50.5	49.2	50	49.7
16.0 kHz	41	41.3	41.9	42	41.7	42.3	42.2

Table 6 SIL results

Flight phase	1	2	3	4	5	6	7
Flight deck SIL [dB]	63.8	64.4	71.2	69.3	72.7	74.7	68.3
Cabin SIL [dB]	60.7	60.7	69.6	66.1	68.8	70.3	64.5

are to be compared with SIL and distance relation table (Table 7), to provide the maximum distance between two subjects on which the listening is at a satisfactory level. Measurements from Tables 4 and 5 are used in Eq. 1. to calculate SIL [2].

$$SIL = \frac{L_{P500}+L_{P1000}+L_{P2000}+L_{P4000}}{4} [dB] \quad (1)$$

SIL results (shown in Table 6) are referenced to Table 7 to provide information regarding the maximum distance between subjects, while 'A' represents the normal voice and 'B' represents the louder voice. For example, during descent (flight phase 6), in the cabin, the distance between two subjects may be a maximum of 0.13 m if normal voice communication is desired.

4.2 *SII Analyses*

SII (Speech Intelligibility Index) is an objective metric used to describe voice intelligibility in unfavorable conditions. For SII calculations, the computer algorithm '*SII*

Table 7 SIL and Distance Relation Table

SIL [dB]	A [m]	B [m]
35	7.5	15
40	4.2	8.4
45	2.3	4.6
50	1.3	2.6
55	0.75	1.5
60	0.42	0.85
65	0.25	0.5
70	0.13	0.26

Table 8 SII results

Flight Phase	SII – Flight deck I Cabin							
	Normal		Raised		Loud		Shout	
Start	0.13	0.24	0.37	0.48	0.60	0.71	0.81	0.84
Taxi	0.11	0.23	0.35	0.47	0.58	0.70	0.79	0.85
Take off	0.01	0.06	0.15	0.21	0.38	0.44	0.60	0.65
Climb	0.02	0.07	0.20	0.30	0.43	0.53	0.65	0.74
Cruise	0.01	0.04	0.10	0.23	0.32	0.46	0.55	0.68
Descent	0.00	0.02	0.05	0.16	0.26	0.40	0.49	0.62
Approach	0.03	0.11	0.23	0.34	0.46	0.57	0.68	0.78

Table 9 SII description

Incomprehensible	Understandable	Highly understandable
0–0.45	0.45–0.75	0.75–1

Calculation 1.0 for ANSI S3-5 – 1997.' made by J. Hou is used. Octave band noise results, in the range of 0.25 kHz to 8 kHz, are inserted and desired voice loudness is selected [3].

Table 8 shows objective metrics results divided in columns for each voice loudness and further divided into the flight deck and cabin sections. Results are to be referenced in Table 9. For example, during engine start (flight phase 1), cabin SII for loud voice loudness is 0.71 which referenced to Table 8 shows that intelligibility is Understandable.

5 Conclusion

Noise can have a huge impact on passengers' acoustical comfort as well as disturb the normal flow of communications. Results prove that Cessna Citation CJ2 has quite high levels of noise as it closely follows the upper limit of comfort with L_{EQ} values passing 80 dBA.

The lowest degree of human comfort is during the phases of take off and descent. During the take off phase, aircraft engines produce the highest levels of noise, especially in the cabin which is situated close to the engines. During the descent phase, the highest noise levels are recorded in the flight deck, which are caused by the high airspeed of the aircraft through the air, causing high amounts of aerodynamical noise. High levels of noise, followed by vibrations, cause a decrease in the level of acoustical ergonomics, so therefore, choosing new engine options and improving the technical noise reduction methods may highly contribute to the overall comfort during the flight.

By observing octave frequency band noise levels, it is seen that low and middle-end frequencies make a resounding part of an acoustical picture. Octave frequency band values are further used to provide objective metrics in determining the impact on everyday communication and overall acoustical comfort. It is determined that SIL results are quite lower in the cabin compared to the flight deck, thus enabling normal communications with less strain, and deeming necessary use of a headset in the flight deck, as some vital information may be influenced by high levels of noise. SII results are trending towards the lower side of the scale hence presenting that additional vocal effort must be used during most of the flight to provide an acceptable level of communication.

The collection of interior noise results provides a great amount of data regarding the everyday acoustical comfort and ergonomics of Cessna CJ2 aircraft. The gathered results may be used as a future reference to the updated noise reduction methods.

References

1. Malcolm, C.: Encyclopedia of Acoustics. John Wiley & Sons, New York (1997)
2. Acoustical Society of America, ASA Standards. https://asastandards.org/terms/speech-interference-level/. Accessed 09 Oct 2022
3. Acoustical Society of America, Methods For Calculation of the Speech Intelligibility Index, New York (1997). (Revised 2017)

Ergonomic Design of the Hand Saw Handle

Diana Milcic, Adisa Vucina, and Marija Bosnjak

Abstract In order to design the handle of a hand tool as ergonomically acceptable, it is necessary to know all aspects of the human hand and define ergonomic design guidelines. Defining requirements and technical specifications ensures the quality of ergonomic design. Following the results of current research in tool handle design, the possibility of better interaction between the tool and the user is created. The handle of the hand tool was created based on defined ergonomic guidelines in order to reduce disorders of the musculoskeletal system and increase functionality, comfort and safety. The research was conducted on the example of a hand saw handle.

Keywords ergonomic design · hand saw handle

1 Introduction

In a large number of occupations many employees have to use hand tools to accomplish their duties. The interaction between the hand tool and the user should be such that it does not cause discomfort and acute trauma caused by a single use or cumulative trauma as a result of repeated long-term overexertion. Improper use or poor design of the handle of hand tools contributes to various injuries or musculoskeletal disorders affecting upper limb and neck pain [1]. The prevention and reduction of worked injuries depends of the design of hand tool, but designer also should consider the working environment and the way of the task performing. The most critical part of the hand tool is its handle which design depends on the purpose of the hand tool. The anthropometric data of the human hand is very important in the process of designing the handle of a hand tool [2, 3].

D. Milcic (✉)
Faculty of Graphic Art, University of Zagreb, Getaldiceva 2, Zagreb, Croatia
e-mail: diana.milcic@grf.hr

A. Vucina · M. Bosnjak
Faculty of Mechanical Engineering, Computing and Electrical Engineering, University of Mostar, Matice Hrvatske bb, Mostar, Bosnia and Herzegovina

I. Salopek Čubrić et al. (eds.), *Proceedings of the 9th International Ergonomics Conference*, Lecture Notes in Networks and Systems 701,
https://doi.org/10.1007/978-3-031-33986-8_25

In this study, authors are dealing with the design of the hand saw handle in order to reduce discomfort and injuries that may occur during use. The hand saw is one of the frequently used hand tools and vary in their function and design. It is a basic tool that is used for different purposes and different materials using your own strength. Because of that there are plenty of different types of manual saws that can be used, some of them are: basic hand saw, wallboard saw, pull saw, coping saw, hacksaw, crosscut saw, rip saw etc. [4].

2 Ergonomic Design Guidelines

By collecting and analyzing data from the available literature, the guidelines that should be followed in order to create an ergonomically acceptable hand saw handle will be listed below. By carefully collecting data and defining design guidelines (GD), the potential for error in designing an ergonomically correct tool is reduced.

GD1: Neutral Hand Position. One of the main rules that should be followed when creating any ergonomically acceptable hand tool is to ensure a neutral position of the operator's hand. Tools and handles should be designed so that, when performing the necessary operation, the hands and joints can be in a comfortable position. The arms have more strength and are less susceptible to injury when the joints are straightened. The strength of the human hand decreases by up to 55% in bent hand positions, depending on the position [5]. In order to ensure a neutral position of the forearm when handling a hand saw, it is necessary to bend the handle so that the joint suffers minimal stress. For a hand tool with a normal linear handle, the center line of the cross-section of the handle and the forearm should form an angle of 110° so that the joint is in a neutral position during the operation (Fig. 1) [6].

GD2: Choice of Grip Type. Ergonomic analyzes divide grips into three basic types, defined by the strength and precision of the grip, which are distinguished by those parts of the hand that are in direct contact with the object: strong grip - includes the entire palm with all five fingers, uses relatively strong muscles in the forearm; middle grip - performed with all fingers, and the palm is mainly used for support; tweezers

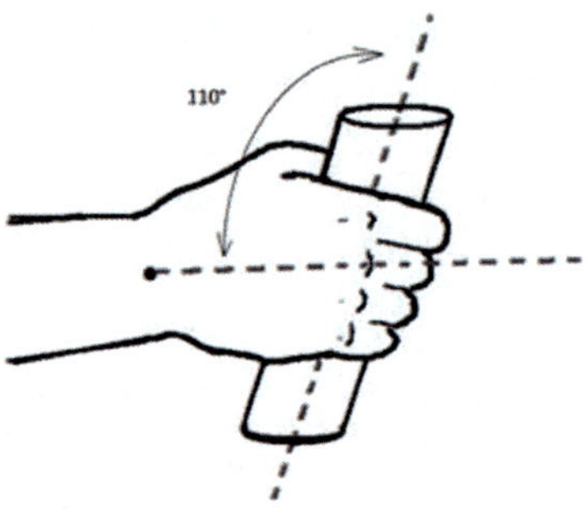

Fig. 1 The center line of the cross section of the handle and the forearm should form an angle of 110° Source: Taken from Justin G. Y., Jia-Hua L. et al. (2013) [6]

grip - the object is held only with the tips of the thumb and forefinger, it uses smaller and weaker finger muscles and ensures that only small objects are held. Operating a handsaw requires a strong grip involving the entire hand with all fingers. The thumb, index finger and middle finger are the strongest fingers and use them to produce the greatest grip strength. The ring finger and little finger have less power and serve to stabilize the handle rather than being the primary force contributors.

GD3: Cross Section of the Handle. The shape of the handle should be an elliptical shape with a width to length ratio of 1:1.25 which can be used to exert a push force and a pull force [7]. The cross-section of the handle in relation to the length of the palm: 17.3% and 21.5% of the length of the palm (ratio of the width and length of the elliptical cross-section) [7]. It is necessary to ensure an optimal cross-section of the handle, taking into account the diameter along the entire length of the handle. Inadequate handle diameter can limit the operator's ability to produce the required force. Figure 2 shows the definitions of the measurement of the human hand, and the basic measurements of the handle cross-section.

GD4: Handle Length. Research has shown that the palm width of 95% of workers is less than 100 mm, so the length of the handle should not be less than 100 mm [8]. The length of the handle should be increased if it is planned to wear gloves when working with the tool. By allowing more contact between the handle and the hand, bending of the joint, stretching of the tendons, stretching of the elbow and limitation of shoulder movement are avoided. If the handles are too short, forces can be concentrated at the base of the palm.

GD5: Handle Without Sharp Edges and Finger Grooves. The gripping surfaces must not have finger recesses or sharp edges. Finger recesses should be avoided because their distance may not correspond to different hand sizes and gripping a

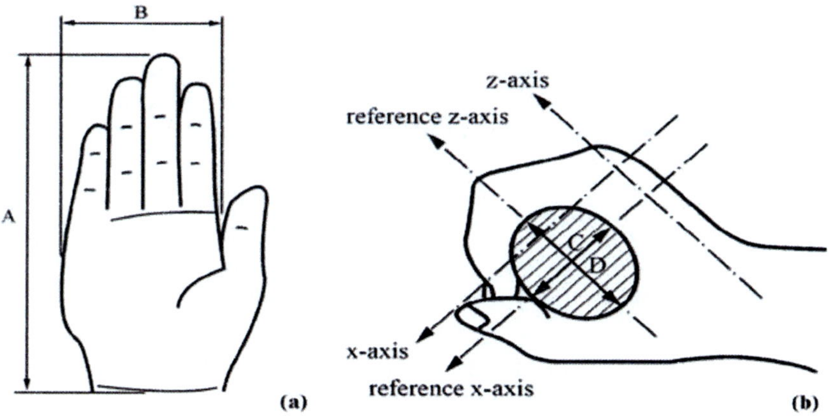

Fig. 2 **a** A – hand length, B – hand breadth, **b** C – grip width diameter, D – grip length diameter
Source: Taken from Wang, C., Cai D. (2017) [7]

handle that does not fit the hand may create compressive forces on the soft tissues under the fingers. The force should be spread over a larger area or directed to less sensitive areas of the hand. Sharp edges on tool handles can also injure the underlying tendons on the volar sides of the fingers, contributing to a condition known as trigger finger - stenosing tenosynovitis of the underlying tendons [9].

GD6: Increasing the Contact Area Between the Hand and the Handle of the Tool. In order to produce as much force as possible when handling a hand tool with as little expenditure of operator power as possible, it is necessary to design the handle in such a way that the contact surface between the hand and the handle of the tool is as large as possible. This is achieved by a handle design that follows the contours of the palm along the entire length of the handle.

GD7: The Material of the Tool Handle. There must be sufficient friction between the handle and the hand to provide a secure grip and prevent the tool from slipping. The friction characteristics of the handle affect the grip force required to maintain control over the tool and the ability to apply force or torque. The amount of friction depends on the coefficient of friction between the hand and the handle material. Some materials have higher coefficients of friction and, consequently, better friction characteristics than others. On handles where large forces must be applied, the surface texture of the material is useful because it increases friction. This makes the handle slip-resistant, maintains a better grip and contributes to better performance of tasks. The material of the handle should be a poor conductor of heat and electricity and should not be porous so as not to absorb and retain oil or other liquids. The materials should be strong enough not to chip or crack and injure the hand.

GD8: Handling Safety. To provide safety for the hand from the saw blades and to prevent the hand from sliding forward, it is useful to place a guard on the front of the handle. Adequate clearance should be made between the handle of the hand saw and the guard so that the handle can be easily used by providing enough room for the fingers, even if the operator is wearing gloves.

GD9: Handle Suitable for Both Right-Handed and Left-Handed Users. The handle should be equally suitable for both left-handed and right-handed users. This can be achieved with a symmetrical handle design that will fit both right and left hands equally.

3 Design of the Ergonomic Hand Saw Handle

Collecting the data that operators require from a hand saw handle, the following requirements are listed that a hand saw handle should fulfill: neutral joint position, comfortable grip, suitable handle length, non-slip surface of the handle, handle suitable for both right-handed and left-handed users, fits the hand well, safe use, ease of use, can offer high task performance.

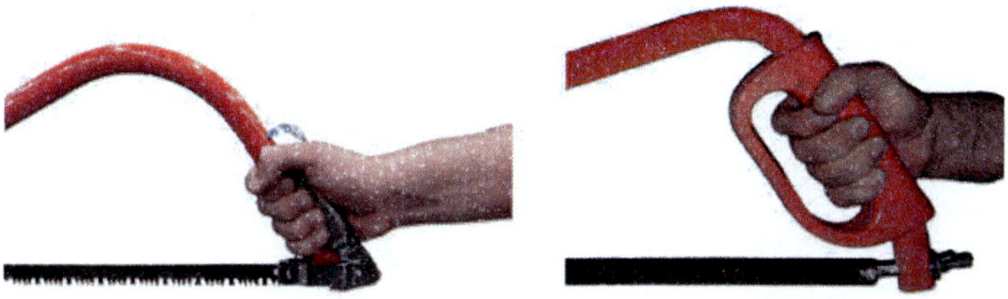

Fig. 3 Existing hand saws left: specimen 1, right: specimen 2

3.1 Evaluation of the Design of Existing Hand Saw Handles

In order to see to what extent the existing handles meet the previously mentioned requirements, two typical examples of hand saw handles will be analyzed from the aspect of ergonomics (Fig. 3). A survey was conducted on a sample of 10 subjects. These are people who use a saw in household chores and workers who use a hand saw for a certain period of time.

Both handles are angled, have a suitable length and suitable design for both right- and left-handed users. Handle 1 does not provide a comfortable grip due to the inadequate cross-section and the presence of sharp edges. Due to the metal material from which the handle is made, it does not have a non-slip surface. The handle poorly follows the contours of the human hand, so it fits the hand poorly. It does not contain a finger guard, so the hand is not protected from saw blades.

Handle 2 provides a somewhat comfortable grip due to its elliptical cross-section. It does not contain sharp edges, and it is made of plastic material with a texture, which makes it a less slippery surface. It contains slots for the fingers, which are not recommended from an ergonomic point of view. The guard on the front provides some protection to the hand from saw blades.

3.2 Shape and Dimensions of the Handle

To design a tool handle, it is necessary to know the dimensions of the human hand. The handle will be created for the 50th percentile of the European male population, so the key dimensions related to the human hand will be guided by this data. The dimensions needed to know are: hand length (A) and hand width (B). Based on the size of the hand, we determine the size of the diameter of the handle (D) and the size of the grip (G) according to Eq. (1) and (2) [11]:

$$D = 0,2\,A \tag{1}$$

$$G = D\pi \tag{2}$$

The size of the grip is determined by the circumference around the edge of the handle, which is measured in the middle of the handle. Taking hand measurements for the 50th percentile male European population [11], it is obtained that D = 36.4 mm and G = 114 mm. Based on the calculated dimensions of the handle and the defined guidelines, the design of the ergonomic handle follows. The handle will have an elliptical cross-section for a grip size of 114 mm; the handle will be curved to follow the contours of the human hand, and the length of the handle will be 120 mm. The picture (Fig. 4.) shows the shape and dimensions of the handle.

Protection against saw blades is placed at a distance of 40 mm from the handle. Such a distance represents the optimal distance so that the protection does not interfere with the fingers when handling the saw. Figure 5 shows the final appearance of the hand saw handle.

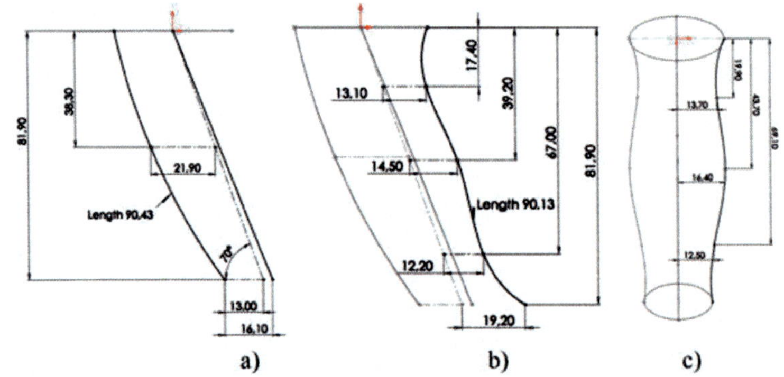

Fig. 4 Shape and dimensions of the hand saw handle

Fig. 5 The final design of the handle of the hand saw with the associated protection

4 Design Evaluation

In order to evaluate the new design of the ergonomic hand saw handle, the 3D print prototype will be tested in the grip of a human hand (Fig. 6). The testing will be carried out on a sample of 10 subjects whose anthropometric measurements do not deviate significantly from the starting value of palm length and width.

Respondents will have the opportunity to try the designed prototype and evaluate the design based on defined criteria. Table 1 shows the measured values of the length and width of the subject hands.

Given that for the purpose of this analysis a 3D print prototype was created from the plastic material ABS, criteria based on the type of handle material (handle quality and surface texture) will not be taken into consideration. Therefore, the design analysis will be based on the following criteria:

- effect on the wrist,
- grip size,
- comfort.

The respondents tested and evaluated two existing handles (Fig. 3) and a new 3D prototype (Fig. 6) of a hand saw handle based on the set criteria. They assigned a score from 1 to 5 to each criterion in the following way: 1- the product does not meet the requirements at all; 2- the product slightly meets the requirements; 3- the product somewhat meets the requirements; 4- the product mostly meets the requirements; 5- the product meets the requirements completely. Figure 7 presents the average scores of respondents for each hand saw handle.

Fig. 6 Testing of 3D printed prototype of hand saw handle

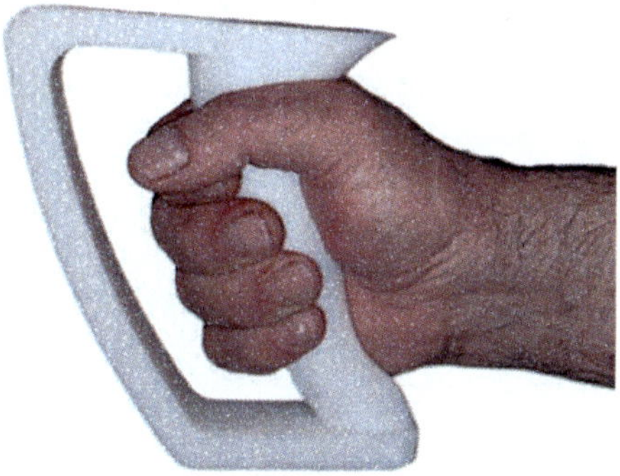

Table 1 Hand measures of the subjects

	A- Hand length (mm)	B – Hand width (mm)
Average value	183,1	82,9
Minimum value	179,7	79,6
Maximum value	186,5	86,2

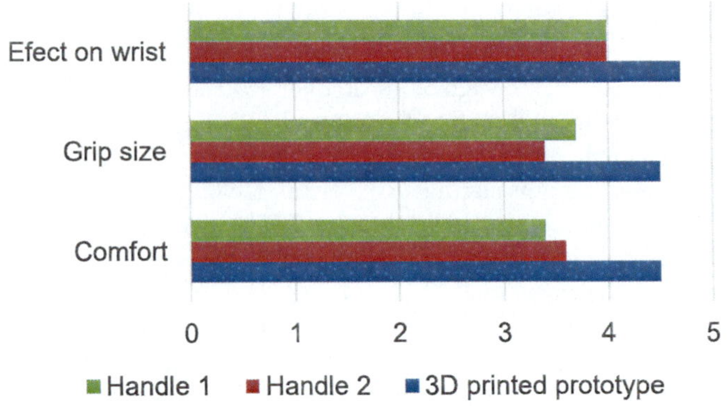

Fig. 7 The results of the evaluation

5 Conclusion

The implementation of ergonomic guidelines in the design of the hand saw handle contributes to the improvement of the design regarding the comfort and safety. The design principles defined in this paper can serve as a basis for improving the redesign process of existing hand saw handles and designing new ones. By testing the 3D print handle prototype based on criteria that can be set at this design stage, the ergonomic design of the hand saw handle is confirmed. Evaluation of ergonomic criteria such as handle quality and surface texture could not be taken into consideration at this design stage. Testing of these aspects, which also contribute significantly to ergonomic design, would be carried out on the hand saw handle produced from material that is planned for the final production. It would be advisable to direct future research towards finding an adequate surface material for the handle. In addition, due to the impossibility of creating a universal handle design that could fit all percentile groups, it would be significant to direct research into creating different sizes of hand saw handles to fit different sizes of human hands.

References

1. Putz-Anderson, V.: Cumulative Trauma Disorders: A Manual for Musculoskeletal Diseases of the Upper Limbs. Taylor & Francis, London (1992)
2. Wang, C., Cai, D.: Hand tool handle design based on hand measurements. MATEC Web Conf. **119**, 01044 (2017). https://doi.org/10.1051/matecconf/201711901044
3. Chandra, A, Chandna, P., Deswal, S.: Analysis of hand anthropometric dimensions of male industrial workers of Haryana State. Int. J. Eng. **5**(3), 242–256 (2011)
4. https://cf-t.com/blog/ultimate-hand-saw-guide. Accessed 21 Sep 2022
5. Dolsak, B., Kaljun, J.: Ergonomic Aspects of Product Design, 2nd ed. University of Maribor (2021)

6. Young, J.G., Lin, J.-H., Chang, C.-C., McGorry, R.W.: The natural angle between the hand and handle and the effect of handle orientation on wrist radial/ulnar deviation during maximal push exertions. Ergonomics **56**(4), 682–691 (2013). https://doi.org/10.1080/00140139.2013.765602
7. Wang, C., Cai, D.: Hand tool handle design based on hand measurements. MATEC Web Conf. **119**(1), 01044 (2017)
8. Putz-Anderson, V.: Cumulative Trauma Disorders: A Manual for Musculoskeletal Diseases of the Upper Limb. Taylor and Francis, London (1988)
9. Radwin Robert, Haney Jonathan: An Ergonomics Guide to Hand Tools, American Industrial Hygiene Association Fairfax, Virginia, (1996).
10. Rasovic, N., Vucina, A., Milcic, D.: Hammer Design Using Ergonomic Knowledge-based Guidelines. pp 179–187, Proceedings of the 8th International Ergonomics Conference, Springer, (2021).
11. Jurgens, H.W., Matzdorff, I., Windberg, J.: International Anthropometric Dana for Work-Place and Machinary Design, Dortmund (1998)

Ergonomics for Employees' Satisfaction in Lean Manufacturing Systems

Nicoleta Paula Neag, Anca Draghici, and Maria-Elena Boatca

Abstract The human side of lean manufacturing systems is concerned with the adequate behaviour for continuous improvements, high rates of productivity and waste reduction, but also, with supporting low stress and high satisfaction of the employees; ergonomic interventions could facilitate workplace wellbeing. The article aims to present a framework for assessing the ergonomic interventions using the usability concept to get feedback on the satisfaction of employees related to their working conditions. The research approach uses a mix of methods and tools and has been validated in the case of an automotive wiring harness assembly line (from a multinational company): the tasks in the shop floor, the control, packaging and warehouse logistic tasks, and the managerial tasks. Results have supported the idea of ergonomics continuous improvement, using the participative approach which implies employees' deep involvement and contribution to the optimization of workflow processes (from the perspective of effective and efficient use of all categories of resources) and that could reduce waste and improve development of processes and product quality.

Keywords Ergonomics · Intervention · Lean management · Lean methods · Overall Equipment Effectiveness (OEE) · Work task · SUS test

N. P. Neag (✉) · A. Draghici · M.-E. Boatca
Politehnica University of Timisoara, Timisoara, Romania
e-mail: nicoleta.neag@student.upt.ro

A. Draghici
e-mail: anca.draghici@upt.ro

M.-E. Boatca
e-mail: maria.boatca@student.upt.ro

I. Salopek Čubrić et al. (eds.), *Proceedings of the 9th International Ergonomics Conference*, Lecture Notes in Networks and Systems 701,
https://doi.org/10.1007/978-3-031-33986-8_26

1 Introduction

Ergonomics benefits has been recognized an extensive, interdisciplinary knowledge base, and their practical usefulness is undeniable; ergonomic approach has changed started from the basic "identification - analysis - elimination of the risks on the workplaces" to the new innovative solutions based on the exploitation of new emerging technologies [1]. Modern approaches using mobile application, Internet of Things, data gathering and their real time evaluation have extended the traditional ergonomics interdisciplinarity and, the sharing knowledge in the filed [1, 2] using innovative and advanced ergonomic tools based on (in the context of) Industry 4.0 concept. From the perspective of the finality of the ergonomic approach, the main objective is to ensure the wellbeing of the employee at work [3] considering human interface optimization [4–6] and the creation of sustainable products and processes, also [7, 8].

In this context, the article aims describing a proposed approach of assessing the ergonomics intervention in the case of a lean production systems (the validation context is defined by a shop floor, the manufacturing warehouse area and the offices belonging to multinational company operating in Romania in the field of automotive industry) together with employees' satisfaction regarding their new working conditions (considered as a pre-condition for ergonomics continuous improvement using the participative approach). Thus, the research refers not only to the technical aspects, but also, the human side of lean manufacturing system which is less present in the literature. Thus, a knowledge gap is fulfilled.

2 Experimental Research

The research approach is based on previous developments of [9, 10] and the normative context is given by the ISO 9241–11 and ISO 13407 standards [11, 12] within the continuous improvement approach. Based on this basic knowledge of the concept of usability, this study will describe a way of customizing the proposed framework model by specific methods and means (scientific or regulatory) for global evaluations of ergonomic interventions in the case of a lean manufacturing system, to identify continuous improvement solutions; to operationalize the way to achieve the improvements, a questionnaire was developed, which was then applied to human operators. Thus, the efficiency of ergonomic interventions was evaluated, in the terms imposed by usability concept and the new performances achieved by the lean system.

The study was carried out between March 2022 and June 2022, in a medium-sized enterprise in the automotive field, which is a significant employer on the local market in the Western Region of Romania. The manufacturing plant has 300 employees who work in three shifts and who perform their tasks on 76 different workstations related to the product assembly process. However, production line management found some deficiencies in the organization and concluded that there was room for further improvement; ergonomics could play an important role in this case due to worker complaints about stress related to musculoskeletal strengths and disorders.

2.1 Ergonomics Intervention

The neglected aspect in the adopted-implementation phase of lean management was related to the mitigation of the numerous problems invoked by human resources related the lean management implementation, but also for the effective exploitation of the new working conditions. Furthermore, an ergonomic intervention has been designed; the research approach is shown in Fig. 1. The research methods used in this phase of the experimental research were: Gemba (observation of the real production line and the workplaces); individual interviews with supervisors and human operators of the manufacturing line and the application of the Systems Usability Scale (SUS) questionnaire for the collection of feedback (as defined by [13]).

The purpose of this study was to measure the usability of the ergonomic intervention which was focused on improving eight tasks of the manufacturing line by reorienting the control and test working panels, redesign, and reorganization of the assembly working places for eliminating/diminishing the musculoskeletal disorders (MSD) (see Table 1). To do this, we compared the results from nine usability studies conducted as part of the preliminary man–machine system evaluation. Each of these studies had human operator complete tasks that were related to the usability goals of the study. Further, each human operator (user) completed the SUS questionnaire

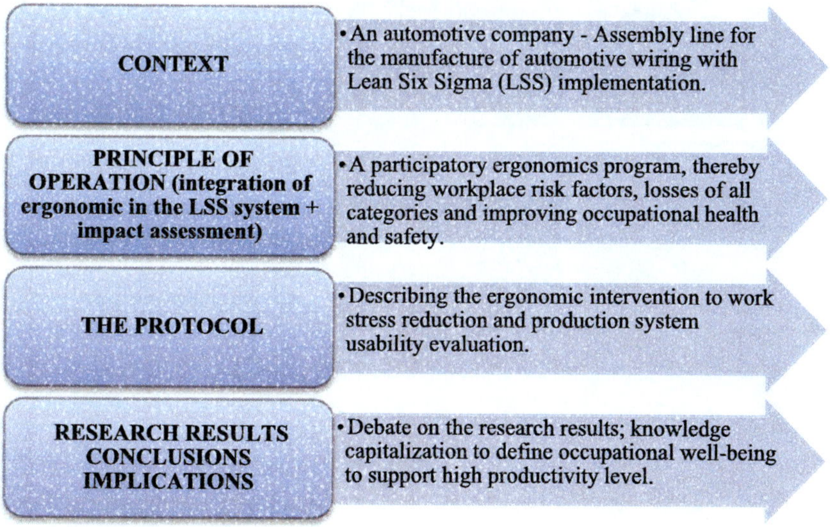

Fig. 1 Research scenario

after completing the tasks. We then compared the percent of tasks users completed without errors with their corresponding SUS scores to examine if the effectiveness of the systems.

As seen in Table 2, for each of the usability studies, the SUS test was conducted individually; different respondents perform specific tasks related to the specific activity in the manufacturing assembly line. Upon completion of the usability tests, the participant would complete the SUS. Table 3 provides the research results (mean, standard error and skew values for ErrorFree and SUS scores) related to all case studies.

For the SUS questionnaire data process and the results interpretation, the correlation method with one independent variable as SUS and one dependent variable as the percent completion error free (error free) descriptive statistics has been used. The SUS scores were calculated using the method developed by [13]; for the ErrorFree scores, we calculated the proportion of tasks completed by each participant without committing any errors. As seen in Table 3 both scores were relatively high, with ErrorFree scores having a mean of 0.86 (std. error = 0.02) and the SUS with a mean score of 73.58 (std. error = 1.6). We determined the connection between the two variables using Pearson's r coefficient to determine whether SUS scores are predictive of usability test results. The result underlines that there is a positive relation between the SUS scores of the participants and their error-free scores r (77) = 0.218,

Table 1 Details of ergonomic interventions

#	Details of the tasks and workplaces	Interventions / Ergonomics improvements
a	  Manufacturing line tasks	Horizontal reorientation of control and test panels (adjusting the height of assembly panels)
b	  Control and logistics tasks	Ergonomic organization for assemblies in orthostatic position with the implementation of visual management elements
c	 Training corner	Establish a space for training programs and coaching

Table 2 Correlations and sample size for each study

Tasks and human operators involved in the study	N (research sample)	Correlation
1. Manufacturing line tasks' evaluation - Ergonomists and other researchers - Human operators (users)	5 8	0.73 0.55
2. Design and management tasks' evaluation - Engineering Design - Implementation (prototype) - Supervisors	7 7 9	−0.82 0.29 0.07
3. Control and logistics tasks' evaluation - Line Operators - Quality control operators - Logistic operators Total users involved in the research	14 11 24 85 respondents	0.16 0.64* −0.08

Table 3 Descriptive statistics

	Mean Std	Std. error	Skewness
SUS	73.58	1.6	-0.58
ErrorFree	0.86	0.02	-2.94

$p = 0.045$, $r2 = 0.047$, indicating a reliable but small relationship between the two variables.

Furthermore, in Table 2 is presented the results of the correlation analysis between the SUS and the performance of each of the studies separately. Only for the Quality control operators' study had a significant correlation between SUS and performance ($r = 0.63$, $p = 0.039$). However, there were no reliable effects of the sample size ($p = 0.63$), number of tasks ($p = 0.55$), or the type of study ($p = 0.27$) on the correlation between SUS and performance. A correlation has been observed between users' performance on usability task completion and the associated SUS score. This finding strengthens the SUS approach (based on the SUS questionnaire) for quick usability studies. The SUS approach is effective and efficient, which truly supports the fundamental idea of usability. One of the primary research results is that although the relation between the SUS and the usability testing is reliable, with a $r^2 = 0.047$, it is very small.

2.2 *Characterizing the Lean System*

To monitor the daily activity of the machines, the frequency of breakdowns and the quality of the final products, the overall equipment effectiveness indicator (OEE) was used [14]. The value of the indicator and its significance is determined by the management of the time of use and maintenance of machinery or equipment. The calculation methodology of the OEE indicator follows the relations (1)–(4).

$$Availability\ Rate = \frac{Available\ time}{Total\ time} = \frac{Total\ time - Availability\ losses}{Total\ time} \quad (1)$$

$$Quality\ Rate = \frac{Effective\ time}{Productive\ time} = \frac{Productive\ time - Quality\ losses}{Available\ time} \quad (2)$$

$$Performace\ Rate = \frac{Productive\ time}{Available\ time} = \frac{Available\ time - Performance\ losses}{Available\ time} \quad (3)$$

$$\text{OEE}(\%) = \text{Availability Rate}(\%) \times \text{Performace Rate}(\%) \times \text{Quality Rate}(\%) \quad (4)$$

As a result (through the OEE study and practicing Go-to-Gemba in the automotive wiring factory), the losses that occurred in the production process were investigated and analyzed, thus creating a knowledge base to develop continuous improvement measures, together with ergonomic intervention. The results are presented in Fig. 2. Before the ergonomic intervention, Availability rate is at a very high level, but nevertheless the Quality and Performance rates require improvements; the OEE value (58%) is much lower than the expected one (85%). The effectiveness of the ergonomics intervention is demonstrated by the increase value of OEE(%).

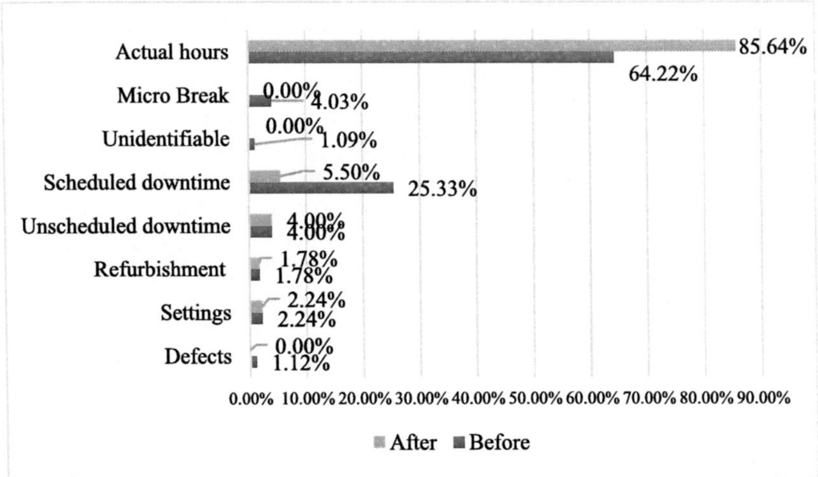

Ergonomic intervention	*Availability rate*	*Performance rate*	*Quality rate*	*OEE*
Before	*90%*	*74%*	*87%*	*58%*
After	*95%*	*94%*	*95%*	*85%*

Work: actual work, micro-interruptions and unidentified time

Indexes = sum % for machines / to their number

Fig. 2 OEE values before and after the ergonomic intervention

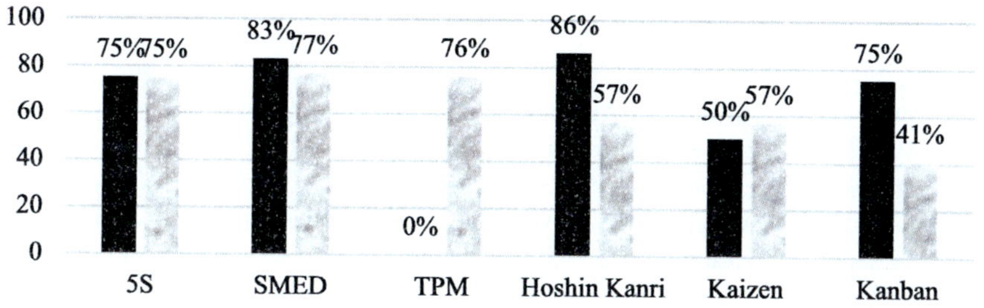

Fig. 3 Comparison of the level of implementation of lean means (marbled bars) with the effects obtained by eliminating losses (black bars) after ergonomic intervention

After determining the OEE rate, a series of informal discussions (to collect primary data) with human operators and supervisors on the wiring assembly line have completed the "big picture" of the manufacturing line. As a result of the informal discussions, employees and supervisors indicated that the losses can be eliminated 100%, not eliminated (0%) or only partially eliminated (50%), this scale leading to the simplification of the way of collecting respondents' opinions. After calculating the cumulative response percentages on seven established types of investigated losses, the results and their graphic representation were obtained (Fig. 3). The representation considered only the six lean methods and means already implemented in the company's practice.

The analysis demonstrated that the implemented lean methods do not eliminate all losses as assumed initially; deviations from the implementation plan of lean methods due to existing constraints (of time, space, or resources) in the practical organization of the manufacturing process were discovered. The comparison of the results obtained regarding the effects of loss elimination and the percentage level of implementation of certain lean methods and means (Fig. 3) allowed highlighting the level of implementation determined individually for the six methods, and which does not influence the elimination of specific losses. For example, despite a high level of implementation of Total Productive Maintenance (TPM, 76%), the effects of eliminating losses were not achieved at the minimum rates imposed (their value is 0%). A similar situation was identified for Kaizen, but the discrepancy is not so significant (57% implementation and 50% waste elimination).

3 Research Results Discussions and Final Conclusions

The approach and research described for the study of the automotive wiring harness assembly line emphasized the strong connection of usability with ergonomics to assess the satisfaction of employees when performing their tasks. Improvements made using ergonomics knowledge also have a direct positive impact on system productivity. The perspective of using such an approach is closely aligned with business objectives; efficiency, effectiveness, and satisfaction have a direct impact on labor productivity and ultimately the profitability of business processes.

Applying the SUS approach, there have been demonstrated that human operators (users) which were involved in the research have confirmed their satisfaction with the ergonomics improvements. Additionally, human operators (users) positively appreciate the solutions for ergonomics improvement that have been implemented for most of the tasks of the manufacturing line. Human operators, users value *usability*, and new developments and implementations in the field of continuous improvement must address to *learning* and *satisfaction* attributes. The most appreciated improvement was the establishment of a space for training programs and coaching.

The research limitation is related to the specific of the research context (assembly production line), the methods and tools used (SUS approach and ergonomic intervention methods within a lean manufacturing system) and the research sample. In the future more experiments are needed to test and validate the proposed approach in other research contexts. From the researcher perspective, using the SUS approach, an ergonomist may be able to say that one system is better than another but an ergonomist could not say to how much better the system was.

Finally, the research results have supported the idea of ergonomics continuous improvement (defined as a new practical approach for lean manufacturing and as supported by the studies of [15, 16]), using the participative approach which implies employees' commitment to the workflow processes optimization, and that could reduce waste and improve quality. The effectiveness of ergonomic intervention has been demonstrated by improving the performance of the lean system.

Acknowledgements This article was financially supported by the project 'Network of Excellence in applied research and innovation for doctoral and postdoctoral programs / InoHubDoc', project cofunded by the European Social Fund financing agreement no. POCU/993/6/13/153437.

References

1. Gašová, M., Gašo, M., Štefánik, A.: Advanced industrial tools of ergonomics based on Industry 4.0 concept. Proc. Eng. **192**, 219–224 (2017). https://doi.org/10.1016/j.proeng.2017.06.038
2. Horváthová, B.: Data collection for ergonomic evaluation at logistics workplaces using a sensor system. Transp. Res. Proc. **40**, 1067–1072 (2019)
3. Kata, A.: A Study on Ergonomics for Employee Wellness. International Journal of Innovative Research in Computer Science & Technology (IJIRCST) ISSN, 2347–5552 (2020)
4. Montoya-Reyes, M.: Novel ergonomic triad model to calculate a sustainable work index for the manufacturing industry. Sustainability **12**(20), 8316 (2020)
5. Szynkiewicz, W.: Utilization of embodied agents in the design of smart human-computer interfaces - a case study in cyberspace event visualization control. Electronics **9**(6), 976 (2020)
6. Gualtieri, L.: Design of human-centered collaborative assembly workstations for the improvement of operators' physical ergonomics and production efficiency: a case study. Sustainability **12**(9), 3606 (2020)
7. Lin, C.J.: Ergonomics-based factors or criteria for the evaluation of sustainable product manufacturing. Sustainability **11**(18), 4955 (2019)

8. Sarbat, I.: A structural framework for sustainable processes in ergonomics. Ergonomics **63**(3), 346–366 (2020)
9. Cirjaliu, B.: A propose approach for continuous improvement using ergonomics and quality management knowledge and methodologies. Qual. Access Success **20**(S1), 135 (2019)
10. Cirjaliu, B.: Ergonomic issues in lean manufacturing. Proc. Soc. Behav. Sci. **221**, 105–110 (2016)
11. ISO 9241–11 Ergonomic requirements for office work with visual display terminals (VDTs) - Part 11 Guidance on usability (1999)
12. ISO/IEC. 13407 Human-Centred Design Processes for Interactive Systems (1999)
13. Brooke, J.: SUS: a quick and dirty usability scale. In: Jordan, P.W., Thomas, B., Weerdmeester, B.A., McClelland, I.L. (eds.) Usability Evaluation in Industry, pp. 189–194. Taylor & Francis, London (1996)
14. OEE (Overall Equipment Effectiveness) (2011–2019). https://www.leanproduction.com/oee.html
15. Colim, A., et al.: Lean manufacturing and ergonomics integration: defining productivity and wellbeing indicators in a human–robot workstation. Sustainability **13**(4), 1931 (2021)
16. Peixoto, A.C., DinisCarvalho, J., AnaColim, N., Sousa, L.A., Rocha, J.O.: WIDEA: waste identification diagram with ergonomic assessment—towards the integration of lean and ergonomics. In: Arezes, P.M., et al. (eds.) Occupational and Environmental Safety and Health III, pp. 443–453. Springer International Publishing, Cham (2022). https://doi.org/10.1007/978-3-030-89617-1_40

Occupational Safety with Artificial Intelligence Application for Ergonomic Risk Assessment

Nicoleta Paula Neag, Maria-Elena Boatca, and Anca Draghici

Abstract The main interest is to develop effective and efficient prevention programs based on occupational health and safety (OHS) management, risk assessment (continuous monitoring and control) and thus create an organizational climate work based on employee wellbeing and productivity. Workplace risk assessment is a mandatory practice in Occupational Health and Safety (OHS) throughout Europe. However, shifting from minimum legal requirements in OHS to ergonomics involves a comprehensive approach to workplace risks and all their determinants. Due to the fast pace of technological advancements, ergonomists can leverage new software solutions based on Artificial Intelligence (AI) for workplace risks management and conception of ergonomic interventions. The article includes a review of the literature on the topic of Artificial Intelligence (AI) used in ergonomics, along with a case study on the use of AI-based software for ergonomic risks assessment in a furniture manufacturing company. The research results have shown the maturity, effectiveness, and efficiency of the proposed research approach, which could quickly be transferred into the practice of organizations.

Keywords Ergonomics · Workplace · Risk assessment · Postural analysis · Artificial Intelligence · Occupational health and safety · Prevention

N. P. Neag · M.-E. Boatca · A. Draghici (✉)
Politehnica University of Timisoara, Pta Victoriei 2, 300006 Timișoara, Romania
e-mail: anca.draghici@upt.ro

N. P. Neag
e-mail: nicoleta.neag@student.upt.ro

M.-E. Boatca
e-mail: maria.boatca@student.upt.ro

I. Salopek Čubrić et al. (eds.), *Proceedings of the 9th International Ergonomics Conference*, Lecture Notes in Networks and Systems 701,
https://doi.org/10.1007/978-3-031-33986-8_27

1 Introduction

Ergonomics is a major contributor to workplace wellbeing, providing the context and the tools necessary for prevention of injuries and occupational disorders at the same time with a continuous work system optimization. The benefits of implementing ergonomics principles in the workplace are no longer debatable, going well beyond monetary benefits: higher employee retention rate, improved organizational performance, healthier and happier employees, achievement of sustainability targets, etc. Industry 4.0 generated significant changes in the work environment, as technological advancements and innovations led to the necessity of redesigning the workplace and the work itself. Therefore, ergonomists have the right to research how these changes can impact the worker from both occupational health and safety (OHS) and wellness perspectives.

Nowadays, the latest technological advancements and increased affordability of software tools for occupational risks assessments created the context for conception of software solutions in the field of ergonomics based on artificial intelligence (AI). However, the use of such solutions at a large scale can be debatable: after all, the ergonomist is the one performing the assessment and conceiving ergonomic interventions based on the results.

The article demonstrates the use and impact of artificial intelligence in ergonomics, especially for the purpose of risk assessments, and has the following structure: a review of the literature, a case study on AI-based software used for ergonomic risks assessment in a furniture company, discussion, and conclusions.

2 Literature Review

2.1 AI Used in Ergonomics

Traditionally, workplace risk assessments and general workplace evaluations are performed using a wide range of methods and tools [1]. Depending on the particularities of the analyzed workplace and the proposed objectives, one or more methods are suitable. An important note here is that the majority of these methods rely on observations and measurements performed by an ergonomist. Therefore, concerns about the accuracy and reliability of the data can arise, considering that the observational method has a subjective character [1].

Furthermore, observations and task/workplace manual analysis are time-consuming and require an ergonomist or another type of specialist specifically trained to use the selected analysis and assessment method [2]. Innovative software tools were developed to minimize these drawbacks and to improve accuracy of assessments. In addition, software tools are an affordable option for analyzing workplaces and identifying key areas of improvement and the main risks [2].

Workplace-related musculoskeletal disorders (WMSDs) are recognized as the main category of occupational disorders, being the result of exposure to ergonomic risks during work [1, 3]. Understanding the key causes of WMSDs contributed to their prevention, representing a foundation stone in the proactive approach to ergonomic risks [3]. Prominent methods such as the RULA, REBA, OWAS, and NIOSH lifting equations, among others, have been used in ergonomics practice for a long time now, with reliable and satisfactory results [1]. Still, the drawbacks mentioned earlier are valid for these methods, too, creating the context for development of software solutions aimed at solving ergonomists' problems and improving the quality of the ergonomic assessments and interventions. Nevertheless, Industry 4.0 and intensive robotization and automation involve the necessity of new assessment instruments and an improved paradigm for ergonomic interventions [1, 4]. From pen-and-paper methods to the current technology-based approach, ergonomic risks assessments witnessed a variety of innovations that enabled adaptation to the technological advancements and improved approach to workplace wellbeing [4].

There is a variety of software tools with diverse applications. However, for the assessment of ergonomic risks, majority of the software solutions involve the use of sensors for data gathering. The main disadvantage of the sensors is the potential interference with worker's movements and manner of accomplishing the tasks. One cannot deny that wearable devices and sensors are valuable instruments for identifying ergonomic risks (e.g., awkward postures during lifting, high-risk postures, frequency of postures, and other unsafe biomechanical strains during work) [5]. Wearable devices and AI-based software tools ensure higher data accuracy and provide ready-to-use data inputs for risk assessments. In the case of many software tools, an option is included to perform risk assessments based on data collected using sensors [6, 7]. Nonetheless, the human-centered approach specific to ergonomics should always remain at the heart of AI-based software tools, as usability of such interactive systems is highly dependent on the needs and requirements of the human body and cognition [8].

Nevertheless, the introduction of AI in ergonomics and, particularly, in the management of ergonomic risks, requires an initial training of ergonomists and involves a learning curve [9]. Human-technology interaction should not be done abruptly, solely for the sake of rapid transition to technological innovations. However, with proper training and knowledge sharing, ergonomists develop the ability to use AI-based software tools efficiently: the ergonomist should not rely solely on software, but create a combination of knowledge, experience, and technology for the greater interest of achieving workplace wellbeing and optimal work systems [10]. Application of human–robot collaboration generated new occupational risks that have not been sufficiently researched so far [10]. Although artificial intelligence allows the execution of tasks without human interference or with minimal interaction (such as supervision), human knowledge remains cardinal in the conception of ergonomic interventions and ergonomic design in the workplace [11].

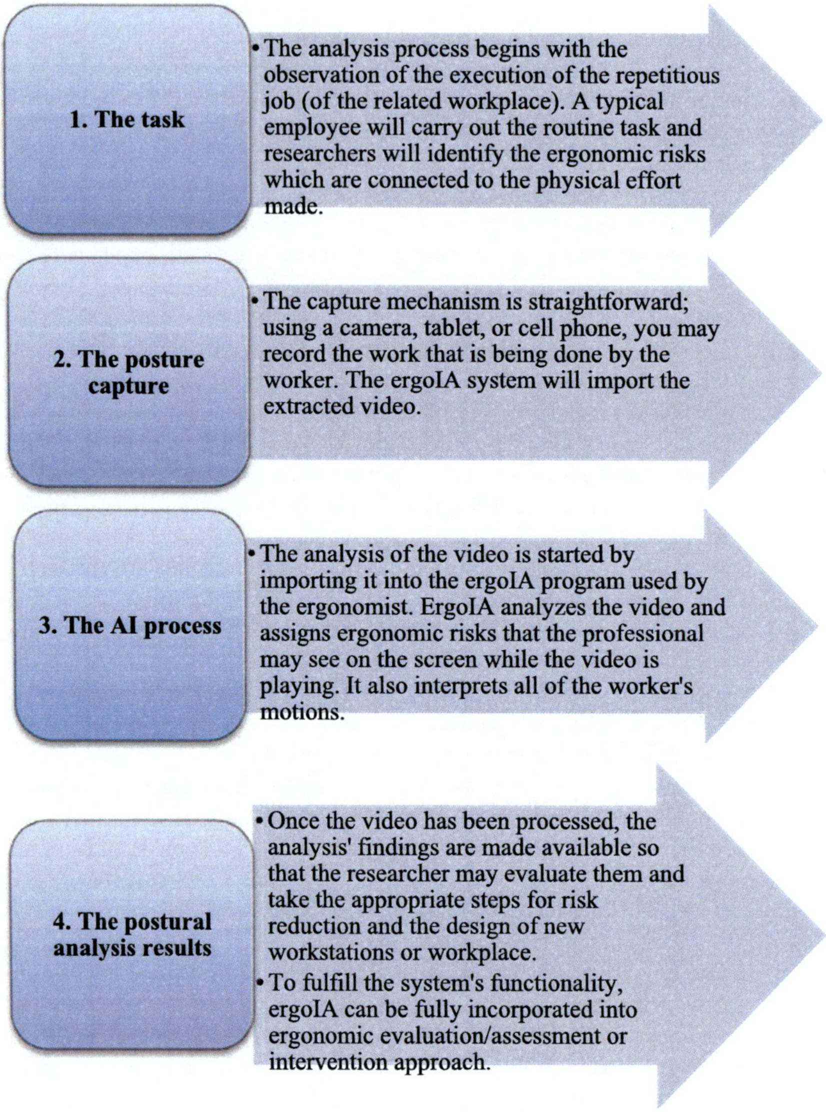

Fig. 1 Description of the ergoIA process. Source: Authors own developement

2.2 *ErgoIA for Ergonomic AI-Based Risks Assessment*

Among the software solutions dedicated to assessment of ergonomic risks, the authors identified ergoIA, a software tool enabled by AI, which was developed by Instituto de Biomecanica de Valencia (IBV) [12].

The software performs the evaluation of ergonomic risks based on the methodology of the REBA and OWAS methods, which are known for their relevance in identifying ergonomic risks [2]. The software requires minimal technological knowledge, as the analyst can record the video of the analyzed task using any device, including a smartphone. The video recording is then uploaded and ready for the analysis step. The analyst splits the video into a chosen number of frames, and the software performs a review of each frame based on the selected method (REBA or OWAS).

The evaluation output is a report containing the scores of each frame and a graphic representation of the distribution of the scores by various criteria. Therefore, the analyst can obtain a detailed analysis of the frequency and severity of ergonomic risks in general and by body regions in a short time. The process of using ergoIA software in ergonomics evaluation, assessment or intervention programs is depicted in Fig. 1. The main benefits of ergoIA are ease of use, short execution time, accuracy of results, and low technical requirements. These benefits allow application of ergoIA in a variety of industries and instances.

3 Assessment of Ergonomic Risks Based on AI - A Case Study in a Furniture Company

This case study reflects the implications of using an AI-based software tool to assess ergonomic risks for an ongoing professional activity. Particularly, ergoIA was used to assess risks incurred in a furniture manufacturing plant in Romania. The analyzed workplace (Fig. 2) is characterized by the predominance of manual operations that are performed in standing position. Workers cannot alternate standing with other postures, except for three breaks of 30 min in total (5–20–5 min). Table 1 presents the main postural strains identified during the first and second stages of the assessment process.

Fig. 2 The analysis of workplace - assembly operation (different task executions cases). Source: Copyright of the authors

Table 1 Postural strains identified for the assembly operation on the basis of the observations of the researchers

Postural strain	Observations
Work standing	There is no possibility of alternating postures during the execution of tasks
Repetitive hand and arm movements	Specific to the task of sorting wood piece accessories (nuts, bolts, hinges)
Awkward postures	Bending and twisting during sorting accessories and assembly operations
Lifting heavy loads	Wood pieces to be assembled are manipulated; load weight varies between 0.4 and ~42 kg

Source: Authors own development

The risk assessment process consisted of several stages, as follows.

1) Initial visit to the company and discussion with the operational manager, OHS representative, and quality manager.
2) Observation of the workplace and understating the specifics of the tasks;
3) Video recording of the tasks executed for the assembly operation and performing the ergonomic risks assessments using ergoIA software (as recommended by the provider [12] with the support of consultants);
4) Conception of ergonomic interventions based on identified risks, in connection with the information gathered in previous three stages.

The ergoIA software tool was used to assess ergonomic risks (in the context of postural analysis and to prevent and/or eliminate the musculoskeletal disorders). Taking into account the specifics of the analyzed workplace, the REBA method was used within the ergoIA assessment. The video recording was split into 20 frames, reflecting the workplace risks involved in the task of assembling a piece of furniture. The number of frames was decided according to the variety and complexity of the movements.

The assessment revealed a low general risk level, only 25% of the postures generating a mid-level risk (level 4). Figures 3 and 4 indicate that arms and the back are the main areas of the body negatively affected by ergonomic risks (according to the A score associated with risks in the areas of the torso, neck, and leg areas) of the assembly operation. For the rest of the body regions, ergonomic risks were found to be negligible.

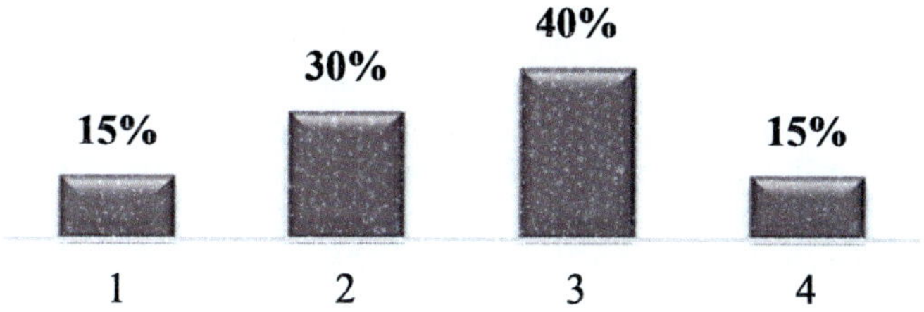

Fig. 3 Distribution of the score for the back region, REBA method. Source: Authors own development

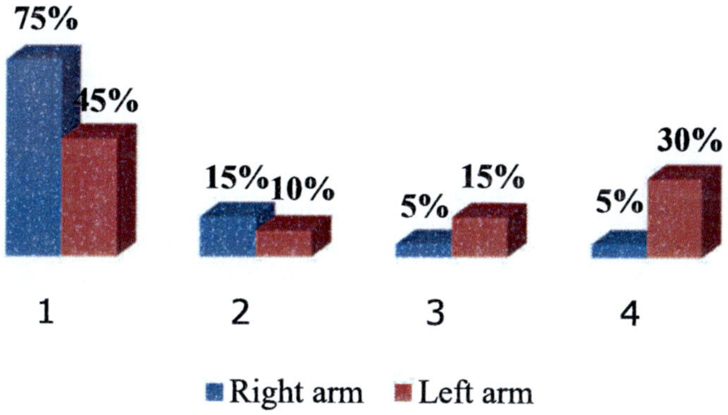

Fig. 4 Distribution of score for left and right arm, REBA method. Source: Authors own development

4 Conclusions

AI has become part of our lives in a variety of ways; Industry 4.0 and intensive automation and robotization determined the introduction of AI in ergonomics practice as a natural next step. The paper reflects key perspectives from the literature on this subject, aiming to demonstrate key benefits such as the reduction of the time required to perform risk assessments, ergonomic design in the workplace, or the conception of ergonomic interventions aimed at improving working conditions and optimizing workplaces.

An example of AI use in the field of ergonomics in ergoIA, a software tool designed for assessment of ergonomic risks using OWAS and REBA methodologies. The paper also presents how ergoIA was used to assess ergonomic risks for the furniture assembly operation, in a furniture company in Romania. The evaluation results revealed the key body regions affected by ergonomic risks. To ensure the precision of the assessment, it was combined with observation of the workplace and discussions with relevant employees and managers.

The proposed AI-based ergonomic assessment has been proved to be a valuable contribution to a general program of ergonomic intervention in the case of the furniture company (where different workplaces have been analyzed and improved). Due to its operability and efficiency, a permanent monitor and control of ergonomic risks have been designed in the factory. The next year global evaluations (done with employees and managers, assisted with occupational medical staff) will provide new insights on MSD and other stress that occurred in the company process.

Acknowledgements This article was financially supported by the project 'Network of Excellence in applied research and innovation for doctoral and postdoctoral programs/InoHubDoc', the project cofunded by the European Social Fund financing agreement no. POCU/993/6/13/153437.

References

1. Boatca, M.E.: Evaluarea interventiei ergonomiei in sisteme de productie. Ph.D. thesis, Politehnica University of Timisoara, Timisoara, Romania (2022)
2. Gajšek, B., Draghici, A., Boatca, M.E., Gaureanu, A., Robescu, D.: Linking the use of ergonomics methods to workplace social sustainability: the Ovako working posture assessment system and rapid entire body assessment method. Sustainability **14**(7), 4301 (2022)
3. Lee, Y.C., Lee, C.H.: SEE: a proactive strategy-centric and deep learning-based ergonomic risk assessment system for risky posture recognition. Adv. Eng. Inform. **53**, 101717 (2022)
4. Mohamad Rawan, M.R., Mohd Daril, M.A., Abdul Wahab, M.I., Subari, K., Manan, Q., Parveen, S.: The evolution of ergonomics risk assessment method to prevent work-related musculoskeletal disorders (WMSDS). Int. J. Online Biomed. Eng. **18**(8), 87–97 (2022)
5. Patel, V., Chesmore, A., Legner, C.M., Pandey, S.: Trends in workplace wearable technologies and connected-worker solutions for next-generation occupational safety, health, and productivity. Adv. Intell. Syst. **4**(1), 2100099 (2022)
6. Low, J.X., Wei, Y., Chow, J., Ali, I.F.: ActSen-AI-enabled real-time IoT-based ergonomic risk assessment system. In: 2019 IEEE International Congress on Internet of Things (ICIOT), pp. 76–78. IEEE, July 2019
7. Vijayakumar, K., Robert, P.: Human factor index measurement using an ISM-SEM-fuzzy approach. Sustainability **14**(13), 7635 (2022)
8. Ispăşoiu, A., Moraru, R.I., Băbuţ, G.B., Popescu-Stelea, M.: Study on the potential of artificial intelligence application in industrial ergonomy performance improvement. ACTA Technica Napocensis-Ser. Appl. Math. Mech. Eng. **64**(1-S1) (2021)
9. Sujan, M., Pool, R., Salmon, P.: Eight human factors and ergonomics principles for healthcare artificial intelligence. BMJ Health Care Inform. **29**(1), e100516 (2022)
10. Dimitropoulos, N., Togias, T., Zacharaki, N., Michalos, G., Makris, S.: Seamless human–robot collaborative assembly using artificial intelligence and wearable devices. Appl. Sci. **11**(12), 5699 (2021)
11. Kalaiyarasi, N., Gopal, T.V.: Ergonomics of human machine integration in variable autonomy. In: 2021 IEEE Conference on Norbert Wiener in the 21st Century (21CW), pp. 1–3. IEEE, July 2021
12. Instituto de Biomecánica (IBV) de Valencia (Spain), ergoIA Homepage. https://ergoia.net/?lang=en. Accessed 10 Jan 2022

The Advantages of Diary Study as a Product Features Research Method

Jesenka Pibernik, Alen Mojić, Jurica Dolić, and Lidija Mandić

Abstract This paper describes the advantages of the diary studies research method used to collect qualitative data about the user's long-term behaviors, activities, and experiences in comparison with other user research methods like interviews and surveys. The importance of choosing a particular research method in the process of developing a mobile application for diabetes is discussed. While diary studies require more time and effort to conduct than other user-research methods, they yield invaluable information about customers' real-time real-life behaviors and experiences. The paper compares data obtained from interviews with 5 respondents, a survey conducted with 427 respondents, and a diary study in which 5 respondents from a previously defined target group of users participated. The results of the log study are presented through ideas for future mobile application functions. Further analysis of qualitative and quantitative data related to frustrations, routines, and long-term behavior yielded results. Based on the confirmed hypotheses, it was concluded that the study of the diary provides a better insight into the user's behavior at the time, which has a positive impact on recognizing and defining future functions of the mobile application.

Keywords long-term behavior · diary study · user research methods · mobile application

1 Introduction

This paper investigates and compares the best practices of researching user experience through interviews and surveys with the study of diaries. User research is one of the initial steps in product development, before introducing new features or entering a different stage of the product lifecycle. It enhances the designer's understanding of what are the user's goals and pain points so he can build the solutions

J. Pibernik (✉) · A. Mojić · J. Dolić · L. Mandić
Faculty of Graphic Arts, University of Zagreb, Zagreb, Croatia
e-mail: jpiberi@grf.unizg.hr

I. Salopek Čubrić et al. (eds.), *Proceedings of the 9th International Ergonomics Conference*, Lecture Notes in Networks and Systems 701,
https://doi.org/10.1007/978-3-031-33986-8_28

they need. A better understanding of users' needs and how they align with a product and business goals helps plan a product roadmap that'll serve both the users and the company. It ensures higher scope of innovation, accuracy, and understanding of how to differentiate the product from the competitors in the market. However, conducting user research demands substantial financial and manpower investment. Therefore, the selection of the appropriate methods for each design stage can be crucial for the project budget and duration.

In the experimental part, the time it takes to collect data with each method is measured and compared so that qualitative and quantitative insights and the usefulness of methods could be explained. These results will provide insight into at what stage of the mobile application development process a particular method gives the best results. The primary goal of the research hypothesis is to prove the quality of use of the diary study in the research of user experience [1].

1.1 User Research Methods

User experience research is an essential process of getting a better understanding of the user's problems. In the generative phase of product development, the research begins with qualitative methods, to detect the user's needs and motivations. It is followed by the application of quantitative methods to test the assumptions obtained. When using a mixture of quantitative and qualitative research methods, the most objective view of the problem of user experience and design is obtained [2].

Before conducting research, it is essential to think about which method gives the best results for a certain product development phase. User research can be divided into two groups: in the first, qualitative group, research methods such as interviews and diary studies provide better insights about the pain points in certain actions of users (e.g., why they failed to measure glucose, how they feel while using certain devices). Furthermore, interviews using open-ended questions with a smaller number of users can provide findings about the user's habits. Along with the information related to the common themes and patterns, qualitative research includes the collection of opinions, motivations, etc. that can be used in highlighting the important aspects of users' stories.

In the second, quantitative group, surveys fall within more structured methods that collect measurable data on what users do. After the data was collected, assumptions drawn from qualitative research are analyzed and evaluated. A survey can be cited as a method that answers exact questions about the user's habits (e.g., "How many times do you measure sugar per day?"). The results are more reliable if the sample of the targeted users is large enough, but without qualitative data, deeper insights into user experiences cannot be detected.

1.2 The Role of the Diary Study in Exploring Long-Term User Behavior

Diaries, as a tool for writing down everyday events and observations, have been used for years, but over the last decade, it has become one of the primary qualitative methods of user research [3]. The diary study originates in the fields of psychology and anthropology, while, more recently, it has been adopted in the field of information technology (IT) as one of the methods of researching the user's needs for making products [4]. Due to the longer follow-up period, diaries are often used in medical studies of chronic diseases. The study of diaries as a research method allows the collection of data from multiple participants simultaneously, in chronological order, in the domestic environment. It gives contextual insight into the habits, patterns of behavior, and decisions of the subjects over time [5]. The duration of the diary study can vary from several days to a couple of weeks.

Research as part of the diary study is often designed to focus on one of the following areas:

- Habits - At what time of day do users use the product?
- Use scenarios - In what quantity do users use the product? What are their primary tasks? What are their workflows for completing long-term tasks?
- Attitudes and motivation - What motivates people to do certain tasks? How do users feel and think?
- Changes in behavior and perception - How easy is the system to use? How faithful are users over time? How do they perceive a brand after using a particular product?
- User journeys - how does the user interact with the product? What is the cumulative effect of multi-service touchpoints? [6]

The analysis of the data of the diary study depends on how the diary is structured. The study diary creates a large amount of attitudinal and behavioral data, which can be challenging. These difficulties can be summed up by contrasting "what people say" versus "what people do" (very often the two are quite different). To better understand the end-user experience from the user's perspective and to analyze a large amount of collected data, an affinity map and a user journey map can be created [6]. The opportunity in this phase of product development is to gather data from different types of users and find patterns in their activities and habits. Specific details of everyday rituals enable designers to propose ways of making these rituals less burdensome.

2 Research Goals

The primary goal of the paper is to prove the quality of the use of diary studies to research user experience and understand long-term behavior concerning survey research and interview. The research examines how individual methods affect the development of a product's features. For a feature to deliver a benefit, the user's needs must be met positively. Based on the defined goals, the following hypotheses are set [7]:

1. The diary study will give a better insight into the user's habits, behaviors, and needs than surveys and interviews to define an application's benefits and opportunities.
2. The diary study will generate more ideas about the possible features of the application through insight into the user's problems at the time they occur.

3 Experimental

3.1 The Survey

The survey was designed to gain insights into the experiences of the subjects that are immeasurable (qualitatively) and to obtain basic demographic and quantitative data. It consisted of general quantitative questions, related to age, gender, and medical history, and qualitative ones to gain a more detailed insight into the rituals and frustrations (pain points). The survey is formatted in Google Forms, and the collected data is quantitatively analyzed by descriptive statistical methods and qualitatively processed by a thematic analysis in Figma.

The online survey was conducted on a representative sample of 427 subjects suffering from some form of diabetes. For conducting the survey and collecting respondents, a Facebook group of diabetics was used. Of the 427 respondents, 65% were women, while 35% were men. 1.6% of respondents were aged 65+ years, 48.5% were between 30 and 65 years old, 26.9% were 18–29 years old, and 23% were aged 1–17. 88.5% of respondents live with type 1 diabetes, 7.7% with type 2 diabetes, 3.7% live with type 2 diabetes, and 3.7% with diabetes type LADA. 52.9% have been living with diabetes for 10 years or more, 18% have lived for the last 5 to 10 years and 42.1% have been diagnosed with diabetes in the last year to four years.

The obtained qualitative and quantitative data are divided by analysis and described across two groups: frustrations (pain points) - qualitative and recordings of the sugar levels/self-control – quantitative.

Frustration (Pain Points)

Using the open-ended question "How does diabetes affect your everyday life the most? Describe to us briefly what frustrates you the most.", the main problems and understanding of daily frustrations that life with diabetes brings have been obtained. Frustrations are divided into four groups that influence the long-term behavior and lifestyle of diabetics: organization of the day, glucose measurement, nutrition, and physical frustrations.

Organization of the Day

The review and analysis of the responses reveal the main frustrations related to the organization of the day. Respondents are frustrated by their inability for spontaneity and the need for constant routine planning. An example of such a response is the following statement:

Subject B: "*Frustration comes from the fact that sometimes the same routine and behavioral patterns result in completely different sugar values. Also, the need for constant planning and sometimes it is impossible to be simply nonchalant.*"

Glucose Measurement

Measurement is a major problem in the daily life of diabetics and affects the planning of the day, according to the following statements:

Subject D: "*It completely affects everyday life, dictates 24 h a day, seven days a week. The need to know the sugar level at all times to be able to respond to hypo and hyperglycemia* promptly."

Nutrition

Subjects are frustrated by the measurement of carbohydrates before each meal and the necessary diet planning, which in case of non-follow-up can lead to hyperglycemia or hypoglycemia. An example of such answers is the following statement:

Subject E: "*The necessity of feeding planning, the fact that most of my food is now my enemy, and I have to think about food 24/7*".

Physical Frustrations

The most common physical frustrations of the subjects were the inability to engage in physical activity and increase strength. An example of such answers is the following statement:

Subject G: "*High sugars put me in a state of complete physical impotence and restless need for sleep, and it happens that I sleep most of the day. Low sugars disrupt my day in terms of the time that passes until the sugar stabilizes.*"

Cross-tabulating of data gave us the context to the numbers: how different groups of people react to the same experiences. The differences can be combined into 6 different variables: depending on whether they need support (solo vs community), the way they measure sugar (individual or family member), and whether they are an advanced user of technology (manual vs automatic data entry).

3.2 The Interview

For conducting user interviews, five subjects with type 1 diabetes and type 2 diabetes were engaged. The subjects were from Croatia, from 18 to 36 years of age. All interviews were conducted via video call (Google meet). The average duration of the interview was 25 min. During the conversation, open-ended questions were used as a tool to get as useful insight as possible that was prepared within the Google docs document to navigate between topics. The interview with the participants was recorded with the consent of the participants for easier analysis. The transcripts allow the researchers to provide perspectives and quote contributors.

The interview was divided into two question sets regarding the insight into frustrations:

Insight into frustration (pain points).

Insight into goals and rituals.

All 5 subjects have a dynamic and unpredictable course of sugar as the main source of frustration, which causes problems in the balance of everyday life.

Subject A: "*Every person living with diabetes is heavily exhausted for the first couple of years (5–6 years), experiments a lot, and has no idea where to turn to get results that work well.*"

Most participants regularly monitor and measure sugar, are organized, and watch their diet.

Subject B: "*I give myself insulin for lunch, and after lunch, I do some physical activity (walking, training, cycling). After that, I learn and give myself another dose of insulin before dinner, and dinner is usually light so I can control the sugar. Before bed, I regularly check my sugar levels and react as needed.*"

The thematic analysis via affinity diagrams of the data obtained through the interview provided insight into the context of the problems and needs of the target group. Problems and needs include the long-term behavior and lifestyle of the subjects, which were addressed with the information obtained about their frustrations, motivations, environment, and behavior. By analyzing the answers about the goals that the subjects set themselves during the day we build categories; connections between them, and key themes essential for further research.

3.3 The Diary Study

For conducting the diary study, 5 subjects from the target group suffering from type 1 and type 2 diabetes were selected. The age range was from 18 to 36 years of age, and all come from Croatia. The diary study was carried out using the Google sheet tool and the In-Situ technique. It was carried out over 4 days with each participant recording certain conditions according to the instructions that they received before the start of the study. While keeping a log, participants entered the data at the time it occurred by writing down the time and date of entry. The diaries were divided into two groups: those that are filled out regularly every day (before and after eating surveys) and those that are filled out by circumstance (check-in or not feel-good polling).

The problems and needs of the participants were addressed through two approaches: problems and needs that we are sure will happen, and problems and needs that we have speculated will happen during the day. This approach covers all cases in which the participant measures sugar. The obtained qualitative and quantitative data are divided by analysis and described through two of the above approaches:

Approach 1: Pre-meal Survey and Post-meal Survey

By conducting a diary study in the form of a survey before and after eating, the main habits and behavior of participants are learned. The survey shows that 71.4% of participants are measuring sugar in their own house before and after meals (10 times), 14.3% at work (4 times), 7.1% in a restaurant (once), and 7.1% on holiday (once). In 71.4% (11 times) of cases, participants measured 15 to 30 min before meals while the other 28.6% measured an hour or more before meals. Participants measured sugar in 20% of cases 30 min after eating. In all other cases, measurements were made after an hour or more.

Approach 2: A Survey in Case Users Do Not Feel Well and Check-In

By conducting a diary study in the form of a survey in a situation where participants are not feeling well, we have gained insight into the user's habits and rituals in severe cases. With the survey, we got an insight into the user's habits before the activity, which also affects sugar and the decision to measure. The results of the check in the survey show that, out of 5 measurements, 66.7% of the measurements were made in the house before the activity was performed, while the rest of the measurement was done outside. All measurements before performing the activity are made via the Dexcom Continuous Glucose Monitoring (CGM) device.

Using an open type of question gave us an insight into the user's habits before performing physical activity. With questions about how he feels before performing a particular activity assuming sugar height and deciding on the next step, we have received specific details about everyday rituals the participant's steps, and mindset. Using those details, we saw the opportunity to make those rituals less bothersome for all the participants and develop predictive models.

4 Research Results and Conclusions

Data obtained by survey and interview were sufficient to determine the target group of users and to conceive main categories and themes. As a disease of self-discipline, diabetes revolves around instability, unpredictability, and stress. The goal of the application development is to support the independence of the user. By monitoring participants with the help of a diary study, detailed insights into nutrition and the way glucose measurements were obtained were gained. Each participant identified "special moments" they saw as milestones. Those moments need to be celebrated and encouraged via the application's features. Also, the study of diaries opened the possibility of creating an application's features with a wider knowledge of future benefits. The hypotheses posed are confirmed through analyses of the data obtained (Table 1).

Of the data obtained from the diary study, the main score points and opportunities in the creation of the application were defined. Opportunities are divided into issues that can be solved through the app, e.g., how to make people with diabetes spend

Table 1 Different research methods findings and related product features

research method	findings	influence on understanding and development of product features ideas
interview	*user pain points, frustrations, and goals*	thematic content, narratives, patterns in the users' activities, and ritual's main themes
survey	*frustration (pain points), organization of the day, glucose measurement, nutrition, physical frustrations*	how different user types react to the same experiences; differences turned into variables that can influence the product's design and/or further research
diary study	*nutrition and physical activities details, "special moments", habits, and rituals in severe cases*	exact product's features, benefits, score points, opportunities

less time tracking, measuring, and understanding their sugar levels, and those that we cannot solve but can provide ongoing support for the user.

References

1. User-Centered Design. https://www.interaction-design.org/literature/topics/user-center. https://www.interaction-design.org/literature/topics/ux-research. Accessed ed-design. Accessed 17 June 2021
2. UX Research, 17 June 2021
3. Hyers, L.L.: Diary Methods (2018). https://www.researchgate.net/publication/322886536_Diary_Methods. Accessed 17 June 2021
4. Diary studies, Wikipedia. https://en.wikipedia.org/wiki/Diary_studies#cite_note-PalenSalzman-8. Accessed 17 June 2021
5. Jarrahi, M.H., Goray, C.: Using Digital Diaries as a Research Method for Capturing Practices in Situ (2021). https://www.researchgate.net/publication/347396375_Using_Digital_Diaries_as_a_Research_Method_for_Capturing_Practices_in_Situ. Accessed 17 June 2021
6. Salazar, K.: Diary Studies: Understanding Long-Term User Behavior and Experiences (2016). https://www.nngroup.com/articles/diary-studies/. Accessed 17 June 2021
7. Mojić, A.: Understanding the long-term behavior and experience of mobile app users through a diary study. Master theses, University of Zagreb, Faculty of Graphic Arts (2021)

Evacuation Planning for Hospital Departments - Time Estimation

Judit Rauscher

Abstract Hospital evacuation is a very complex system, as patients with different conditions typically have different degrees of ability to escape. The condition of patients in hospital wards is constantly changing and patients often change function, so their ability to escape is in constant flux. When an assisted escape or rescue is required, the time needed is significantly influenced by the number of assistants involved in the rescue. However, planning and practice are affected by a number of ethical and operational issues, for which the use of evacuation simulation programs can be a good solution. In this article, as a continuation of my previous checks, I sought to answer the question to what extent the condition of the patients influences the time required for evacuation from a given hospital unit. The test series showed that the evacuation time could double for 19 patients, reaching 11–12 min. The increase in time was to be expected, but the magnitude of the increase is significant for a relatively small number of patients. This demonstrates that the impact of the patients' condition is significant and that the correct design of the rates has a major impact on the results and thus on the level of safety in reality. The present series of tests has also demonstrated that simulation methods can be used to correctly assess a situation or condition in terms of evacuation with relative ease and, with adequate care. Similar studies can be carried out during the design phase or, in the case of existing facilities, within the decision support framework.

Keywords hospital evacuation · emergency ergonomic · evacuation simulation · pathfinder · case study

J. Rauscher (✉)
Doctoral School on Safety and Security Sciences, Óbuda University, Budapest, Hungary
e-mail: judit@flamella.hu

I. Salopek Čubrić et al. (eds.), *Proceedings of the 9th International Ergonomics Conference*, Lecture Notes in Networks and Systems 701,
https://doi.org/10.1007/978-3-031-33986-8_29

1 Introduction

In general, some patients are able to move on their own with guidance, but at slower than average speeds. Some can be moved by some form of rescue equipment: assistance may be needed to transfer and move them. And there may also be those who can only be rescued with the patient's bed because they cannot be moved due to their condition.

Therefore, each hospital ward, at each point in time, will present a different evacuation process, as the condition of patients is constantly changing and patients are regularly rotated. During the planning process, whether it is during the design of the building or during the planning of the evacuation in an existing building, it is still necessary to provide some guidance for the development of evacuation strategies. When designing evacuation procedures, work organization and emergency work organization, an important aspect is to create an ergonomic and safe environment for workers. A comprehensive analysis of hospital evacuation planning and its research background was published in 2021 [1], which found that relatively little research has addressed modelling approaches.

Real-life exercises and testing in a hospital ward can be problematic in several ways. In most cases, there are ethical problems, so they cannot be practiced with real patients. It is possible to practice with simulated patients, but it is difficult to manage the practice environment in working wards. The above problems can be avoided by using evacuation simulation programs, which can be used to test an infinite number of variables by designing the right model.

Stewe Gwynne's lectures regularly show that the secret of modelling is to ask the right questions, and to build the right theoretical models and scenarios based on the right data. In his 2020 presentation, he demonstrated the importance of site assessment and real patient flows and movement mapping for a real hospital project, as this is the basis for the correctness of the theoretical model [2]. Fire simulations will only be technically valid with input data of sufficient quality and accuracy. A significant area of this is anthropometric data and the resulting ergonomic requirements that determine the dimensions needed for safe use. [3].

Determining actual number and time needed to move the different rescue devices is an important starting point for modelling. The problem was summarized by Aoife Hunt in her doctoral thesis [4]. At the Fire and evacuation Modelling Technical Conference, in several different years, several speakers have presented the potential of hospital modelling. For example, Virginia Alonso presented a comparative analysis of the use of two different modelling programs in 2016 [5]. Alberto Tinaburri presented a modelling experiment in 2018, where he demonstrated the interactions between people [6]. And I presented a case study in 2020 [7], in which I used a real hospital floor plan to investigate the difference in discharge time between day and night shift for patients with mixed patient profiles. The results of the experiment showed an 35% increase in time, which is a fairly significant time value compared to the expected 650–684 s at night instead of 474–499 s during the day.

In the present study, I have reused the same floor plan, controlling for the effect of additional parameters on the time of evacuation. Given two average patient distributions, I looked for two extreme scenarios, which are probably the most complicated (only patients who need assistance) and which can be the easiest from an evacuation point of view (only patients who can escape independently). In this study, I will investigate the influence of patient characteristics on the time to evacuations and the extent to which this parameter can be investigated simply by simulation methods. It is expected that patients in poorer condition will experience an increase in evacuation time, but the extent of this is difficult to estimate using other methods.

2 Methods

2.1 *Creating my Case Study*

In this case study I model a surgery unit, as in such units typically need to assist patients during evacuation, often lying down bedside rescue, but depending on the recovery phase they may be able to leave unaided. In line with the national minimum professional standards [8], I assume 19 beds, which requires a total of 11 nurses, so taking into account the shifts, I used a total of 12 staff for 2 wards in the model. In this paper, I'm only considering two types of persons: healthcare staff and patients.

In the set layout (Fig. 1.), there are 2 identical hospital units mirrored, and a corridor block, divided into 3 fire compartments [9]. The unit hosts single bed rooms, except for the 3-bed ward (for patients in the worst condition). The structure of the building entails a 2-corridor system (2.00 m wide), with the patient rooms to the facade, and with the supply rooms in the middle (storage, reporting room, nurse standby room and nurse working room). The doors of the patient rooms are asymmetric and two-leaved, opening outwards, with a net width of 130 cm. The fire doors in the corridor are net 180 cm wide, two-leaved and self-closing.

In this test I assume that the fire does not jeopardize the evacuation and both emergency doors are available. The goal of the evacuation is to relocate persons from unit A to the other unit B, and to establish the expected duration thereof.

2.2 *The Model Area*

I prepared the floor plan in the model area with Pathfinder program (Fig. 2.) developed by Thunderhead Engineering Ltd.. Elements of the model area characteristically have the default settings: rooms, doors and larger pieces of furniture are depicted as holes. The unit to be evacuated is on the right side of the figure, and on the left side is the safe area defined as temporary protected areas: the lounge (blue), the ward (red), and the ends of the corridors (green).

Fig. 1 Floor plan (The green title indicates the safe unit; the red title indicates the unit to be evacuated.)

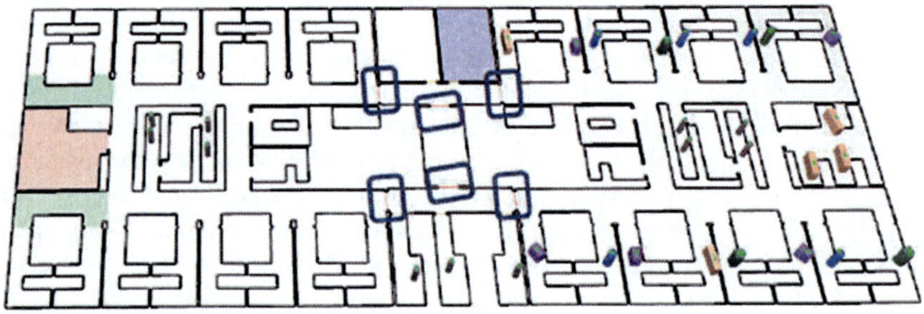

Fig. 2 Model area

For the self-closing fire doors on the corridor (blue rectangle), I set a delay to consider navigation on the door based on Boyce's 1999 measurement [4].

In the model, only the persons necessary for the sessions were placed. In unit A, patients are staying in the rooms, and the staff in the nurse counters, doctors are staying in the doctor's offices. In unit B, no patients are set in the model, only the nurses in the nurse counters.

2.3 *Characteristics and Pre-movement Time*

Healthcare Personnel

Healthcare personnel (nurses and doctors) assist the evacuation. Their speed varies throughout the process: alone with own speed, while during evacuation, with the speed defined by the evacuation equipment. I based the exact values on data in the references, entering a normal distribution in the software.

Every evacuation process is preceded by a pre-evacuation time, during which persons recognize and understand the fire alarm, interrupt their actual activities and prepare for evacuation [10]. For hospitals, the following more time periods are added: triage to determine the evacuation schedule [10], staff must arrive, medical preparation may be required, transfer into rescue device, and after rescue return back time.

Before the assisted evacuation actually starts, everyone must go to the nurse's desk in Unit A and coordination takes place, assuming a 15 s wait.

Patients

I classified patients in four groups, each with different characteristics [9].

- Patient type A is able to egress unaided, however, they use a cane, are able to move independently, but slower than healthy individuals. During evacuation, they don't require assistants.
- Patient type B is able to egress unaided, however, they use a walking frame, are able to move independently, but slower than healthy individuals. During evacuation, they don't require assistants.
- Patient type C is not able to egress unaided, they need wheelchairs and assistants. To help them in the wheelchair and to push it, presumably 1 helper is required.
- Patient type D is not able to egress unaided, they need a bed-type evacuation equipment. They are rescued by their own hospital bed, to which I allocated 2 helpers.

The basic data used to model the different individuals are presented in Table 1.

In this study, I only used the time required to use the device, not medical preparation, as pre-movement time. For patients who are self-evacuating, the time includes getting out of bed and other preparations. The pre-movement time defined for staff refers to the completion of previous operations on their part. The pre-movement values used in the model are presented in detail in Table 2.

Table 1 Person's settings in the model

Persons	Staff	A - Cane	B - walking frame	C - wheelchair	D - bed
shape/color	grey roll	blue roll	green cuboid	violet cuboid	orange cuboid
size (m)	46,5 [12]	66 [13]	70/80 [13]	110/70 [13]	215/100 [13]
speed (m/s)					
minimum	0,65	0,15	0,11	0,84	–
average	1,35 [5]	0,56 [14]	0,34 [14]	1,3 [15]	0,5 [16]
maximum	2,05	1,18	1,04	1,98	–
SD	0,25	0,22	0,20	0,34	–

Table 2 The applied time periods before evacuation are shown in the following table:

Applied delay (s)	min	average	max	SD
staff [5]	30	71	246	60
patient A [17]	30	60	90	20
patient B [17]	30	60	90	20
patient C [6]	32	41	52	8
patient D [6]	60	78	120	19

2.4 Scenarios

In this case study, persons are located in identical positions in all variations and for repeated runs, only the settings of the individuals change.

. I used 12 staff and 19 patients, with 4 different distributions:

- Version 1 – 8–8 type A and B, and 3 type D patients (only in ward)
- Version 2 - 4 type A patients, 5–5 type B, C, and D patients
- Version 3 – 2 type A patients, 5 type B, 6–6 type C and D patients
- Version 4 – 10 type C and 9 type D

As the most data settings contained normal distribution, to produce useable results, several sessions are required. In this paper, I did 20–20 sessions, with random person settings (but their position and type remain the same in every session!). This is a strong simplification of reality, but I would like to exclude other variables from the result.

3 Results of Simulations

In all scenarios, the 4 nursing staff expected during the day are present in the units, and the staff members assisting in the evacuation are from the same floor. A review of the run results showed that staff generally had to return to the Unit A area several times. Usually, the last patient is one of the bed user patients of the ward, therefore, based on the distances, the theoretical evacuation sequence was also established.

Figure 4 show the evacuation process in the first version, when patients are typically self-evacuating. It can be observed that in 2/3 of the elapsed time, the self-evacuating patients have already reached the temporary protected space (blue room) and in the subsequent period only bed rescue takes place.

Figure 5 show the evacuation process in the fourth version, when all the patients are rescued by staff. It can be observed that after 2 min, essentially no one has been able to leave the area, arrival and preparation processes are taking place. It can also be seen that at ~ 360 s, when in the previous version essentially everyone had already left the study area, 12 of the 19 patients were still in Unit A.

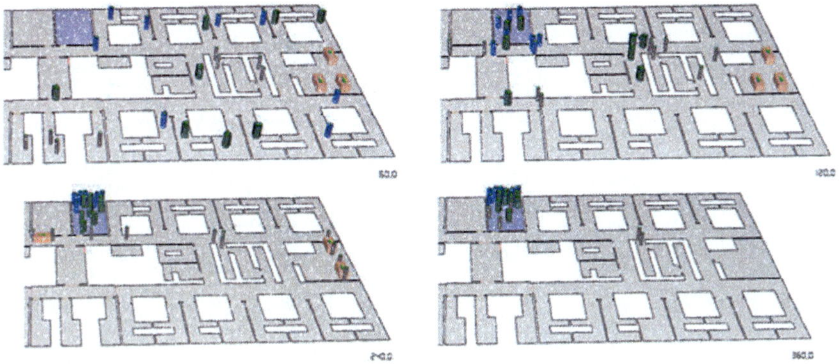

Fig. 4 Evacuation process in Version 1 at the time of 60, 120, 240, 360 s

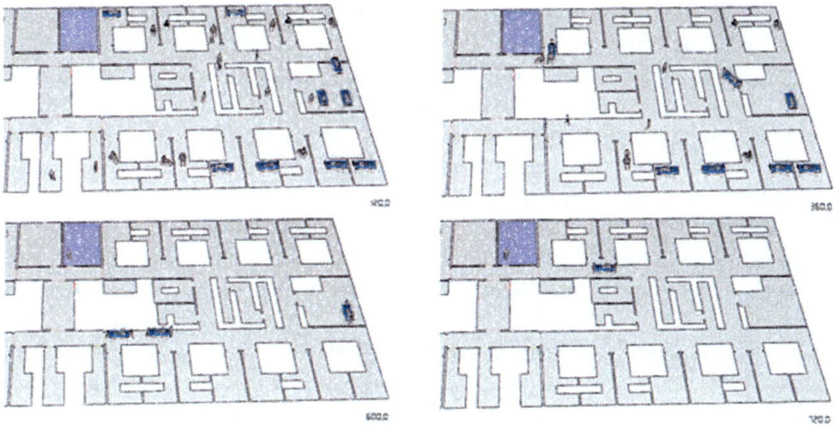

Fig. 5 Evacuation process in Version 4 at the time of 120, 360, 600, 720 s

Table 3 shows the results of the different runs, with their statistical parameters. I investigated the time at which the last patient left the unit A affected by the fire during the escape and rescue.

In the first version, when there were almost only patients escaping on their own, the average evacuation time was almost 6 min. This was essentially doubled when in version 4 there were only patients to be rescued in the unit, which in numerical terms is between 11–12 min, which in the case of an actual fire already represents a significant risk in terms of life safety.

Table 3 Simulation results and interval estimations

		version 1	version 2	version 3	version 4
simulation value (s)	min	286	459	494	605
	average	341	498	554	682
	max	384	582	607	818
	SD	30	35	34	60
interval estimation (s)	lower	322	476	532	645
	upper	359	520	575	719

4 Conclusion

In this paper, I present the results of a series of simulation tests in which I have tested a hospital surgical department from a new perspective. In my first study [9], I tested and compiled how to build such a model, from a scientific and technical point of view. In the previous case study [7], I examined whether discharge time - with the same patient population - is significantly influenced by the time of day or night, as this determines the number of support staff. Now I wanted to find out to what extent patient characteristics influence the time of discharge with the same number of assistants.

In the case study I examined 4 different patient rates, showing the two "extremes". Based on the simulations performed, I demonstrated that the evacuation time could be doubled, reaching 11–12 min with 19 patients rescued. This can be a real risk in the event of a fire.

Using simulation capabilities, I have demonstrated that relatively simple methods can be used to test newer and newer conditions in a way that does not cause ethical or operational safety problems. But to get good results, one needs to work with appropriate input parameters and a well-constructed theoretical model. Even with this safe and relatively simple tool, it would still take a relatively long time to determine the result of all the variants - in this simple case, where only the patients' condition changed, that would be 19^4, more than 130,000 variants. Therefore, further research and testing is needed to create some design scenarios that can be applied during the design of buildings, in the creation and review of hospital evacuation strategies.

A further research and development direction could be the creation of an even simpler tool to estimate the expected evacuation time, even on a daily basis, taking into account the number of staff and the number of patients.

References

1. Yazdani, M., Mojtahedi, M., Loosemore, M., Sanderson, D., Dixit, V.: Hospital evacuation modelling: a critical literature review on current knowledge and research gaps. Int. J. Disaster Risk Reduction **66**(3), 102627 (2021)

2. Gwynne, S., Hunt, A., Weigh, S., Kabalan, B.: Modelling and mapping people movement in hospitals. In: Fire and Evacuation Modelling Technical Conference Proceedings (2020)
3. Szabó, G.: Testméretek mozgástartományok, 35 p. (2017)
4. Hunt, A.: Simulation hospital evacuation, Ph.D. thesis, University of Greenwich (2016)
5. Alonso, V.: Egress Modelling in Health Care Occupancies. In: Fire Protection Research Foundation report (2014)
6. Tinaburri, A., Ponziani, F.A., Ricci, V.: Agent based modelling of meta-communication with assisted people during emergency egress. In: Fire and Evacuation Modelling Technical Conference Proceedings (2018)
7. Rauscher J.: Evacuation Strategy Planning – Special Requirements for Hospital Departments. In: Fire and Evacuation Modelling Technical Conference Proceedings (2020)
8. 60/2003. (X.20.) ESZCSM regulation on Minimum Professional Requirements for the Provision of Health Services
9. Rauscher, J., Kovács, T.: Kórház kiürítési stratégia és a kiüríthetőség ellenőrzése. Védelem Tudomány **4**(2), 23–44 (2019)
10. Society of fire protection engineers: SFPE guide to human behavior in fire. 2nd edn. Springer, Cham (2019)
11. Pan American health organization: hospitals don't burn! Hospital Fire Prevention and Evacuation Guide. Washington D.C. PAHO (2018)
12. Pheasant, S.: Bodyspace, 2nd edn. Taylor&Francis Ltd., London (2003)
13. Rauscher, J., Nagy, R.: A betegszállító eszközök és közlekedő területek tervezésének összefüggései I. Védelem Katasztrófavédelmi Szemle **26**(1), 5–10 (2019)
14. Veres, G..: PhD Thesis. Óbuda University, Doctoral School of Security Sciences (2018)
15. Society of Fire Protection Engineers: SFPE Handbook of Fire Protection Engineering, Table 64.25. 5ft edn., Springer Science+Business Media, New York (2016)
16. Heizler, G..: Mozgáskorlátozott személyek menekítése (1988)
17. Gwynne, S., Galea, E.R., Parke, J., Hickson, J.: The collection of pre-evacuation times from evacuation trials involving a hospital Outpatient area and a University Library facility. In: Proceedings of the seventh international symposium fire safety science (2002)

The Ergonomics of Blood Pressure Measurment

Uwe Reischl and Conrad Colby

Abstract The most commonly used method to measure blood pressure in adults is middle upper arm sphygmomanomtery. However, different cuff-sizes can result in different pressure readings. Pressure readings will be higher for larger arm circumferences and lower for smaller arm circumferences when the same cuff is used. The goal of this study was to determine an adjustment factor that will allow measurements obtained for any combination of arm circumference and cuff size to be compared to each other. The results of this study showed that the percentage (%) value of upper arm circumference coverge by the cuff can serve as such an adjustment factor. Data obtained from nineteen test subjects showed that a 1% change in upper arm coverage results in a 1 mmHg change in both systolic and diastolic pressure readings. Using the proposed % adjustment factor allows a patient's blood pressure to be compared to blood pressure values obtained using any combination of cuff bladder size and arm circumference. This will also allow pressure readings to be matched to the American Heart Association's recommended upper arm coverage of 80%.

Keywords Sphygmomanometer · Cuff Size · Upper Arm Circumference · Adjustment Factor

1 Background

Sphygmomanometry is the most commonly used method to measure blood pressure in adults. Identifying a "correct" cuff size has been a challenge for many years [1–5]. Differences in upper arm circumference and the use of different sphygmomanometer cuff sizes are known to impact pressure readings [6–10]. The arm circumference and cuff-bladder size each have opposite effects on sphygmomanometer pressure

U. Reischl (✉) · C. Colby
Boise State University, Boise, ID, USA
e-mail: ureischl@boisestate.edu

C. Colby
e-mail: ccolby@boisestate.edu

I. Salopek Čubrić et al. (eds.), *Proceedings of the 9th International Ergonomics Conference*, Lecture Notes in Networks and Systems 701,
https://doi.org/10.1007/978-3-031-33986-8_30

readings. Sphygmomanometer manufacturers suggest limits on arm dimensions. However, these differ among suppliers. The goal of this study was to link pressure readings, cuff bladder length, and upper arm circumference with an "adjustment factor" allowing any sphygmomanometer cuff bladder to be used for adults with varying upper arm dimensions. Such an adjustment factor will also permit blood pressure values to be matched to the recommendations made by the American Heart Association [11–15].

2 Sphygmomanometer Use

Sphygmomanometers are used in determining a person's blood pressure indirectly. To perform such a measurement, a sphygmomanometer cuff is wrapped around the upper arm of a person and inflated until the blood flow in the brachial artery in the upper arm is occluded. A stethoscope is then placed on the skin to listen to the sounds generated by the turbulence of the arterial blood when the artery is being compressed by the sphygmomanometer cuff. As the heart beats, blood is forced through the arteries at an increased pressure. This is called systolic pressure. This is then followed by a decrease in pressure called diastolic pressure.

The sphygmomanometer cuff is initially inflated above the systolic pressure to block the blood flow in brachial artery completely. The manometer valve is then opened gradually to relieve the cuff pressure on the upper arm. When the cuff pressure is slightly below the systolic pressure, blood begins to flow again creating turbulence sounds. These sounds are called Korotkoff sounds and continue until the pressure applied to the artery falls below the diastolic pressure after which no sound can be heard. For healthy adults, the systolic pressure is 120 mmHg and the diastolic pressure is 80 mmHg. The Korotkoff sounds can be detected using the stethoscope when placed on the arm on top of the arterial bifurcation located on the proximal side of the elbow. The placement of the cuff and the stethoscope is illustrated in Fig. 1.

Automated digital sphygmomanometers are also available. These systems do not require a manual inflation of the sphygmomanometer cuff or the use of Korotkoff

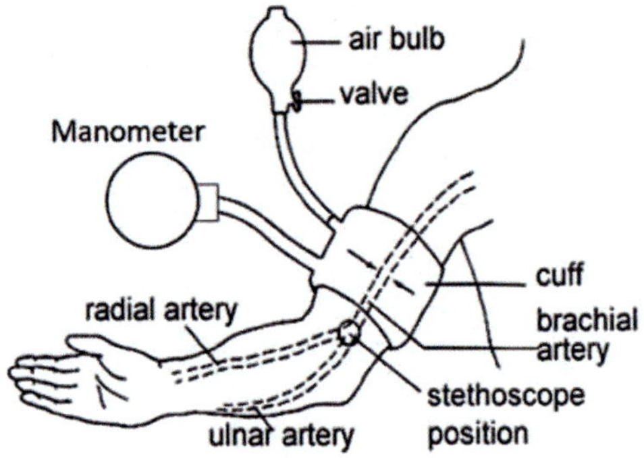

Fig. 1 Sphygmomanometer cuff and stethoscope placement on the upper arm of a patient

Table 1 Summary of cuff bladder dimensions

Sphygmomanometer Cuff No.	Bladder Length (cm)	Bladder Width (cm)
# 1	17	8
# 2	21	15
# 3	23	13
# 4	21	10

sounds. These systems are based on the detection of arterial pulse patterns assoicated with the arterial occlusion.

3 Methods and Procedures

3.1 Simulations

Four cuff bladder sizes were tested to identify the relationship between arm circumference and resulting sphygmomanometer pressure readings. The dimensions of the four bladders used are listed in Table 1. Initial tests were conducted using a simulator which wqas able to document the impact of changes in cuff bladder size and changes in arm circumference on pressure readings. This was done using cuff bladders with different dimensions and changing the simulator cylinder circumference incrementally. Four bladder circumference coverages were evaluated based a 10.9 kP simulator chamber pressure. The simulator used in this study is illustrated in Fig. 2. The simulations were followed by measurements on nineteen adult subjects. Bladders #1 and #2 were used for the human subjects. Figure 3 illustrates the location of the cuff bladder inside a sphygmoanometer cuff.

3.2 Human Subjects

Fourteen famales and five males participated in this study. The mean age of the subjects was 28 years (range 19–72 years), mean height was 166.6 cm, and mean weight was 64.4 kg. Upper arm circumference was determined using a standard anthropometric tape measure. Baseline blood pressure readings were obtained using the #2-sphygmomanometer cuff. Each subject rested 5–10 min prior to a measurement. Subjects were seated in a chair with back support. The left arm rested on a table at heart level. Legs were uncrossed and feet were flat on the floor. No talking was allowed during the measurement period.

After completing the baseline measurements, ten subjects were evaluated again using the smaller #1 sphygmomanometer bladder. Blood pressure values were

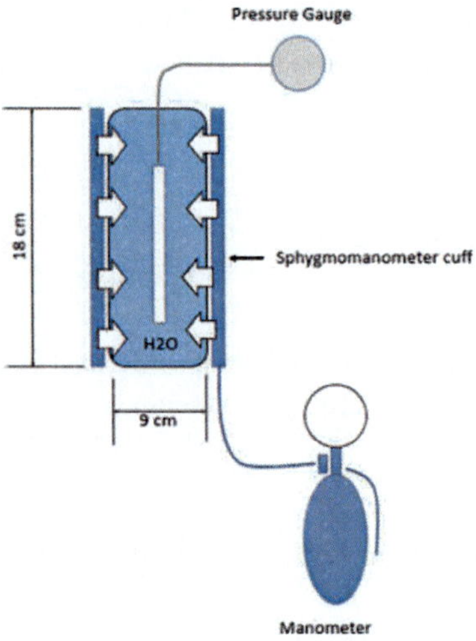

Fig. 2 Simulator used to assess the changes in sphygmomanometer pressure readings resulting from changes in cuff bladder dimensions

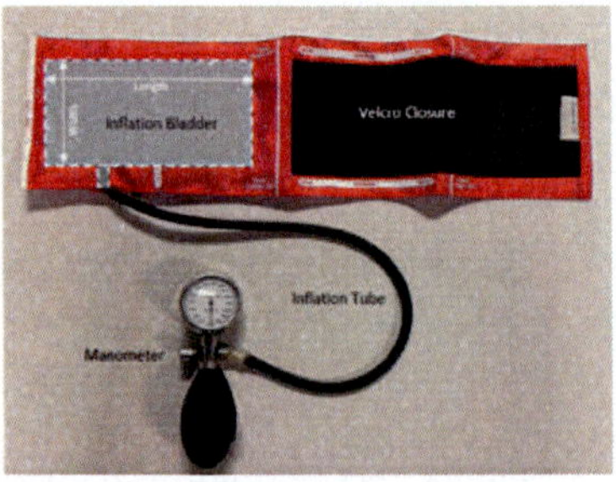

Fig. 3 Illustration of a sphygmomanometer showing the location of the inflatable bladder inside the cuff

obtained for the remaining nine subjects using the #2 sphygmomanometer cuff again, however, this time their upper arm was wrapped with a 15 cm-wide, 2 mm-thick felt cloth. This was done to increase their upper arm circumferences. All blood pressure measurements were made using the WGNBPA-730 automatic oscillometric sphygmomanometer in order to reduce potential investigator bias.

4 Results

The tests showed that the sphygmomanometer inflation pressure that was required using bladder #2 to achieve the internal 10.9 kP simulator reference pressure increased linearly as the circumference of the simulator chamber was increased from 29 to 38 cm. Segments of 15 cm-wide, 2 mm-thick felt cloth were wrapped around the chamber to increase the circumference. The pressure changes were proportional to the changes in chamber circumference.

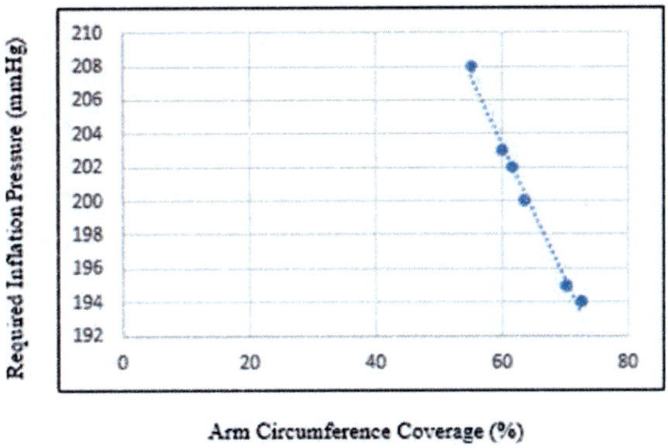

Fig. 4 Decrease in the required sphygmomanometer pressure when cuff bladder circumference coverage was increased

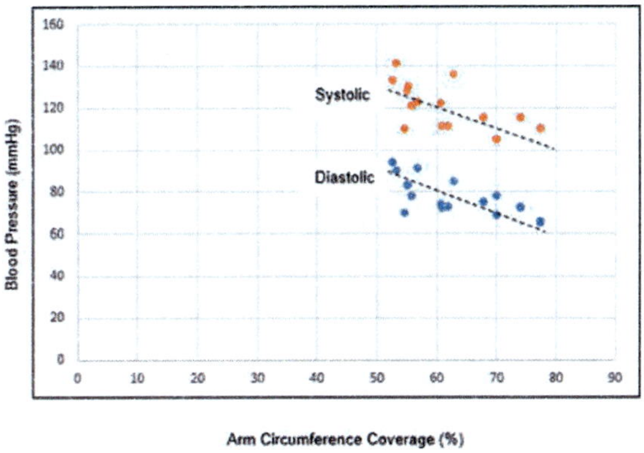

Fig. 5 Decrease in test subject systolic and diastolic pressure readings when cuff bladder circumference coverage was increased

Figure 4 illustrates the linear decrease in sphygmomanometer bladder inflation pressure readings when the cuff bladder circumference coverage was increased. Figure 5 illustrates the relationship between arm circumference coverage and the blood pressure values for the nineteen volunteer subjects. It can bee seen that when arm circumference coverage was increased, the measured blood pressure values decreased.

5 Analysis

The results obtained in this study are consistent with the laws of fluid dynamics, i.e., P = F/A. An increase in the test subject upper circumference led to an increase in required sphygmomanometer pressure. However, an increase in the length of the cuff bladder, i.e., an increase in arm circumfrence coverage, required a lower inflation pressure. This shows that using the percentage (%) of arm circumference coverage provided by a cuff bladder provides the basis for the proposed "adjustment" factor.

Regression analysis of test subject data showed that a 1% change in circumference coverage results in a 1 mmHg change in the blood pressure. This implies that when the arm circumference coverage is reduced by 1%, the measured systolic and diastolic pressure values will increase by 1 mmHg. Conversely, when the bladder coverage is increased by 1%, the pressure value that results will decrease by 1 mmHg.

6 Discussion

The American Heart Association (AHA) recommends a standard sphygmomanometer bladder upper arm coverage of 80%. [11]. Given the wide range of arm sizes found in the general adult population, and a limited number of sphygmomanometer sizes that are commercially available, it would be difficult to comply with such a recommendation. However, using the proposed adjustment factor allows the use of a single sphygmomanometer size to measure blood pressure in adults having different arm circumference dimensions.

Example:

A patient's upper arm circumference is 32 cm.

- The sphygmomanometer bladder length being used is 23 cm.
- The resulting blood pressure values are:
  - 140 mmHg systolic
  - 95 mmHg diastolic.

Adjusting to the AHA recommended 80% coverage, the following corrections can be calculated:

1. Patient's upper arm circumference: 32 cm
2. Sphygmomanometer cuff bladder length: 23 cm
   a. Arm circumference coverage = 23/32 = 72%
   b. AHA recommended coverage = 80%
3. Anthropometric adjustment factor: 72% - 80% = - 12
4. 1% change in circumference coverage = 1 mmHg change in pressure reading
5. Pressure readings adjusted to the AHA recommendation will be:
   a. Systolic: 140 mmHg – 12 mmHg = 128 mmHg
   b. Diastolic: 95 mmHg – 12 mmHg = 83 mmHg

The above example illustrates the steps used to correct for differences in upper arm circumference coverage. Adjusting sphygmomanometer reading to the 80% criteria can help reduce the potential mis-diagnosis of hypertension in muscular and obese patients.

7 Limitations

The findings are consistent with the basic laws of fluid dynamics. It shuld be recognized that the simulator used in this study represented a "closed" system while a human arm would represents an "open" system which can respond to external pressure changes in a non-linear fashion. Nevertheless, the study demonstrates that pressure readings obtained when using different combinations of cuff bladder sizes and arm circumferences can be unified by using the percentage (%) value in upper arm circumference coverage.

The proposed adjustment factor of 1 mmHg for a 1% change in arm circumference coverage is based on data obtained from nineteen healthy subjects only. Testing additional subjects may results in new data suggesting the need to modify the adjustment factor in the future.

Compliance with Ethical Standards This project was reviewed and approved for use of human subjects by the Boise State University Institutional Review Board (186-MED19-019).

References

1. O'Brien, E.: Review: a century of confusion: which bladder for accurate blood pressure measurement? J. Hum. Hypertens **10**, 565–572 (1996)
2. Fonseca-Reyes, S., Fajardo-flores, I., Montes-Casillas, M., Forsyth-Macquarrie, A.: Differences and effects of medium and large adult cuffs on blood pressure readings in individuals with muscular arms. Blood Press. Monit. **14**, 166–171 (2009)
3. Bakx, C., Oerlemans, G., van den Hoogen, H., van Weel, C., Thien, T.: The influence of cuff size on blood pressure measurement. J. Hum. Hpertens. **11**, 439–445 (1997)
4. Sprafka, J.M., Strickland, D., Gomez-Marin, O., Prineas, R.J.: The effect of cuff size on blood pressure measurement in adults. Epidemiology **2**, 214–217 (1991)
5. Pickering, T.G., Hall, J.E., Appel, L.J., et al.: Recommendations for blood pressure measurement in humans: an AHA scientific statement from the council on high blood pressure research professional and public education subcommittee. J. Clin. Hypertens. **7**, 102–109 (2005)
6. Manning, D.M., Kuchirka, C., Kaminski, J.: Miscuffing: inappropriate blood pressure cuff application. Hypertension **68**(4), 763–766 (1983)
7. Jones, D.W., Appel, L.J., Sheps, S.G.: Measuring blood pressure accurately: new and persistent challenges. JAMA **289**(8), 1027–1030 (2003)
8. Marks, L.A., Groch, A.: Optimizing cuff width for noninvasive measurement of blood pressure. Blood Press. Monit. **5**(3), 153–158 (2000)
9. Stolt, M., Sjonell, G., Astrom, H., Hansson, L.: Factors affecting the validity of the standard pressure cuff. Clin. Physiol. **13**(6), 611–620 (1993)
10. Russell, A.E., Wing, L.M., Smith, S.A., Aylward, P.E., McRitchie, R.J., Hassam, R.M., West, M.J., Chalmers, J.P.: Optimal size of cuff bladder for indirect measurement of arterial pressure in adults. J. Hypertens. **7**(8), 607–613 (1989)
11. Prineas, R.J.: Measurement of blood pressure in the obese. Ann. Epidemiol. **1**(4), 321–326 (1991)
12. Bovet, P., Hungerbuhler, P., Quilindo, J., Grettve, M.L., Waeber, B., Burnand, B.: Systematic difference between blood pressure readings caused by cuff type. Hypertension **24**(6), 786–792 (1994)

13. Croft, P.R., Cruickshank, J.K.: Blood pressure measurement in adults: large cuffs for all? Epidemiol. Community Health **44**(2), 170–173 (1990)
14. Mattoo, T.K.: Arm cuff in the measurement of blood pressure. Am J. Hypertens. **15**(2), 675–685 (2002)
15. Pickering, T.G., Hall, J.E., Appel, L.J., et al.: Recommendations for blood pressure measurement in humans and experimental animals. Hypertension **45**, 142–161 (2005)

Ergonomic Aspects of the Daughters of Mary, Help of Christians', Pastoral Habit and Suggestions for Its Improvement

Irena Šabarić, Franka Karin, and Mihaela Bodrožić

Abstract Nuns' clothing is a type of uniform easily recognisable as such. Each order has its own rules and a habit that indicates belonging to it. As with any type of uniform, there are certain disadvantages that make it difficult to perform daily tasks. The aim of this paper is to propose improvements for habits based on the survey conducted among the Daughters of Mary Help of Christians—Salesian Sisters of Don Bosco order. As the result, adjustments were proposed stemming from the analysed questionnaires with the aim of adapting the ergonomic aspects of the habit to the order's daily activities.

Keywords Nuns' habit · Salesian Sisters · Uniform adaptation · Ergonomics

1 Introduction

A uniform represents a person's profession or indicates their role in society. It communicates visually. Ergonomic requirements for uniforms in certain professions have been neglected in the past, e.g., nuns' clothing. It consists of items a person wears as an external sign of dedication to God, poverty, and belonging to a certain religious family [1]. When religious orders were being founded, no thought was given to the ergonomics of clothing. Nuns had struggled with inadequate uniforms for centuries until the Second Vatican Council (1962 to 1965) when the proposal to adapt the uniform to their needs was finally accepted.

I. Šabarić (✉) · F. Karin · M. Bodrožić
University of Zagreb Faculty of Textile Technology, Zagreb, Croatia
e-mail: irena.sabaric@ttf.unizg.hr

F. Karin
e-mail: franka.karin@ttf.unizg.hr

M. Bodrožić
e-mail: mihaela.bodrozic@gmail.com

I. Salopek Čubrić et al. (eds.), *Proceedings of the 9th International Ergonomics Conference*, Lecture Notes in Networks and Systems 701,
https://doi.org/10.1007/978-3-031-33986-8_31

2 Uniform and Work Clothes

The word uniform means a type of clothes that is connected with a particular group of people and is characterized by the fact that it is the same; not changing or different in any way [2]. A uniform is a set of clothing items strictly prescribed by the rules, the role of which is to show belonging to a certain profession of an official. It is defined by rules that depend on the activity. According to EN 240:2003, "Protective clothing, general requirements", work clothing must fulfill the aspect of ergonomics in accordance with anthropometric conditions and satisfy functionality. It is a prescribed rule that work clothes are worn over private clothes or are completely replaced with them and must meet standards that are in accordance with the activity performed and the work environment. Basic health and ergonomic requirements derive from the European Directive 89/686/EEC and their importance cannot be denied, namely design, comfort, harmlessness of materials and aging [3].

2.1 Nun' Clothing

When the first monasteries started forming throughout Europe, groups of women consecrated their lives as well. In the beginning, they wore items of monks' ceremonial clothing. Religious groups grew in numbers and religious clothing developed accordingly. Numerous female religious communities had different habits as a sign of belonging to a certain order. Nun's clothing covered everything but the face and palms. The body, completely covered by heavy and impermeable layers of clothing, had no contact with air and light. As a result, nuns often fell ill. The clothes were mainly woollen, worn in many layers, heavy, and difficult to move in. In the nineteenth and twentieth centuries, nuns started actively participating in life outside the convents helping in hospitals and schools. Difficulty they had moving due to the ill-adjusted habit became a significant shortcoming of their clothing. Consequently, a change in the clothing of religious communities was approved at the Second Vatican Council. As heavy dresses, complex headdresses, and long veils made the life of nuns difficult and dangerous, guidelines were accepted for the adaptation of their clothing to the requirements of the modern world [4, 5]. The head covering with a stiff band tied around the face was replaced by a white hoop to which a shorter veil was attached. The length of the skirt was shortened to the calf, the sleeves allowed more mobility, and the fabric itself was lighter to wear and maintain [6].

3 Daughters of Mary Help of Christians - Salesian Sisters of Don Bosco

One of the communities in which both the old and the new habits can be seen is the Daughters of Mary Help of Christians, popularly known as Salesian Sisters of Don Bosco. They are members of the women's religious congregation within the Roman Catholic Church who actively participate in the life of communities. They support the culture of life and are inspired by the Gospel [7]. The congregation was founded on August 5, 1872 in the small village of Mornese in the Italian province of Liguria by St. John Bosco. Maria Domenica Mazzarello was the first Mother superior. As the congregation grew, it received its own set of the rules of life, work, and dress, which were adjusted at the Second Vatican Council to the new decisions and guidelines that are still in force today. The Daughters of Mary Help of Christians are one of the most recognizable nuns in the world known for their charisma and work with the youth, as well as for their clothes.

3.1 *Daughters of Mary Help of Christians' Clothing*

Salesian sisters in the first years since their foundation when they wore the unsuitable habits are shown at Fig. 1. Uniform consisted of a long veil attached to a white cap that fit tightly to the head and a stiff white collar to cover their neck and chest. The most important part of the outfit was the habit or the upper dress. It was ankle-length, cinched at the waist, had two sets of long sleeves, and was made of heavy wool. A petticoat made of lighter material such as cotton was worn under the habit. Maintaining such clothing was demanding and caused problems with hygiene and health. As a result, sisters fell ill and many died. In addition, the clothing restricted the movement and nurses' work with children. Nowadays, Daughters of Mary Help of Christians' habit is simplified and consists of a formal dress for ceremonies, a short veil, a skirt and an undershirt called a *modestino* worn under the upper shirt [8].

3.2 *Daughters of Mary Help of Christians' Official Religious Clothing*

Daughters of Mary Help of Christians' clothing consists of an upper dress with a pocket on the left side, a pleat at the center front and a separate waistband with buckles sewn on to adjust the width. A veil is placed around the head with a metal ring sewn into it keeping the shape of the veil. On its right side, there are two clips and a button which, with the help of a rubber band on the left side, regulate the fit of the veil. In addition, a skirt that reaches below the knee is worn under the upper

Fig. 1 Salesian nuns in old habits [9]

dress as well as a short-sleeved undershirt (*modestino*) with a sewn-on collar 2 to 3 cm wide that sits close to the neck and fastens with one button at the back.

4 Experimental Part

Despite the fact that today's religious clothing is adapted to the needs and work of nuns, it was found in conversations with them there is still room for improvement. For this research, aimed at improving the ergonomic aspects of religious clothing, the Daughters of Mary Help of Christians were chosen, It was conducted via an online Google forms questionnaire. This method was applied due the remoteness of the convents located in Italy, Central, and Eastern Europe. According to data from 31 December 2020, the FMA are present in 97 countries: in Africa, 628 in 26 countries, in America, 3,366 in 23 countries, in Asia, 2,886 in 22 countries, in Europe, 4,597 in 22 countries, in Oceania, 58 in 4 countries [10]. Europe was chosen for research because of its proximity to the sisters from Croatia, but also because their uniforms consist of the same items of clothing. There are certain variations within the garments that make up the uniform of the FMA sisters depending on the part of the world in which they operate. In India, for example, the sari is an integral part of the uniform and is not included in this study because a different region was chosen. Sisters in Croatia forwarded the survey to monasteries in Europe so that as many respondents as possible could fill out the survey in a short time. Data was gathered from September 14 to September 20, 2021. The questionnaire was completed by 65 nuns aged 25 to over 70. 19 nuns were from Poland, 12 from Slovakia, 10 from

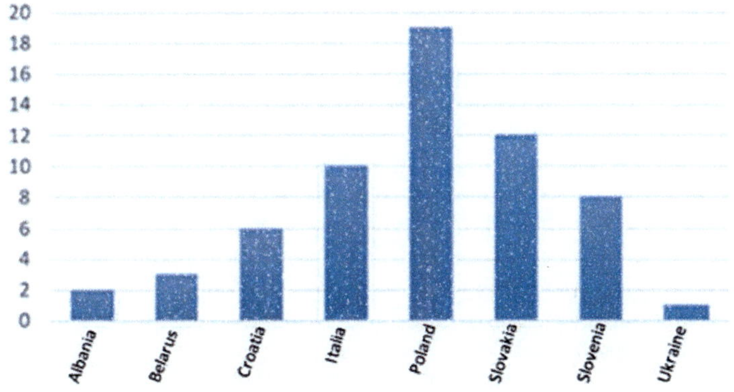

Fig. 2 Completeness of the survey by country. [11] *Source* Taken from Author, F. et al. (2021)

Italy, 8 from Slovenia, and 6 from Croatia. The number of other nurses from other countries is shown in Fig. 2. The number of respondents from each country depended on the size of the queue in the mentioned countries and the division of the survey within the queue. Three questionnaires were invalid. The questionnaire consisted of 37 items divided into two sections. The first section covered general information such as age, the number of years since taking vows, the country of work, while the second related to the advantages and disadvantages of religious clothing. The questions were formulated in such a way that allows multiple answers due to the complexity of the research topic, in form of check boxes to simplify the analysis.

5 Results

Pastoral activities significantly affect the choice of clothing. Therefore, it is useful to know what percentage of the respondents need adjustments in order to perform them as easily as possible. In the survey, 31% of respondents declared that they are physically active in pastoral activities on daily basis, 38% of respondents are active several times a week, 8% of respondents declared that they are active several times a month, and 8% that they are active once a week. Only 15% of respondents stated they are not active. One of the questions in the survey dealt with the number of years since taking vows. The years are counted from the simple vows taken during the first year of novitiate at Daughters of Mary Help of Christians at which point the girl wears religious clothes for the first time. After nine or ten years, the nuns take solemn vows. The survey showed that younger nuns who have not yet taken solemn vows, as many as 21 of them, reported the most shortcomings of habits. Having put on nun's clothing relatively recently, it can be assumed they have the most problems adapting to them. Other nuns have taken solemn vows 10 to more than 40 years ago.

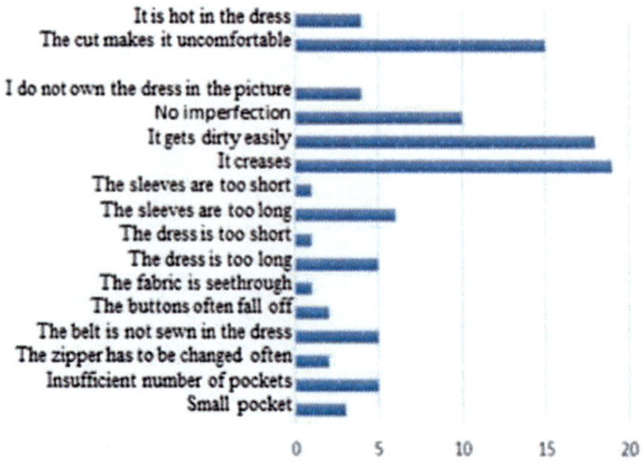

Fig. 3 Disadvantages of the nun's dress. [11] *Source* Taken from Author, F. et al. (2021)

The main disadvantages of religious clothing, or habit, reported by most nuns are how hot it is to wear, the stiffness of tthe material which restricts movement, and the veil that is fitted too tightly around the head which makes it uncomfortable. The biggest disadvantages are the creasing of the habit and how quickly it gets dirty. As minor flaws, the nuns cite the sleeves and the dress as too long, the lack of pockets, and a belt that is not sewn into the upper dress. A suggestion to improve the habit is made to measure tailoring in order to fit different body types (Fig. 3).

The veil is worn by 95% of the respondents every day and its biggest drawback is the hoop that fits too tightly around the head. It easily creases and gets dirty as well. The suggestion for improvement is a longer band that regulate the fit around the head. *Modestino* is worn by 35% of the nurses every day and by 19% several times a week. Its biggest shortcomings are the tightening of the collar around the neck, how easily it gets dirty, and the material that wears out quickly. The suggestion for improvement is adjusting the collar to prevent tightening. The survey found that the type of material is one of the main disadvantages of the entire habit. Namely, 72% of the respondents stated the type of material is very important to them because the biggest problems of religious clothing can be prevented by adjusting the material [11].

5.1 Suggestions for Improving Religious Clothing

Following the survey, the proposal for improving the religious dress is a combination of the original form of the upper dress with modifications determined necessary in order to allow nuns free movement in pastoral activities and better functionality of the garment. The pleats on both front and back of the dress, sewn in at the shoulders, are kept. The waistband is attached only at the front while the rest is loose to enable

Fig. 4 A proposal for short- and long-sleeved polo shirt. [11] *Source* Taken from Author, F. et al. (2021)

width regulation with two or three buckles at the back. The length of the skirt reaches below the knee and has two deep pockets which are a shortcoming of the original. According to the results of the survey, suggestions for improving the veil should be aimed at increasing comfort. The section of the veil that goes around the head is fixed with a sewn-in metal hoop. On the right side, two clips and a button are added which enable a better fit together with a longer rubber band on the left side. A suggestion to improve the *modestino* is to adjust the collar allowing a comfortable fit, the tightness regulated by the three-button fastening at the back of the neck. In addition, a fitted, knitted polo shirt is suggested in order to suit different body types, with longer sleeves that can be turned back if necessary. On the back panel, a pleat is recommended allowing the expansion of the panel and consequently more comfort in wearing. A logo on the upper left side of the front panel is proposed supporting a sense of belonging to the Sisterhood (Fig. 4).

After reviewing the survey results and proposing improvements for the official religious clothing, an interview was conducted with a nun from the Croatian-Slovenian province in order to further adapt the garments to the nuns' needs. She gave a positive opinion on the proposal of the polo shirt with an additional suggestion that the polo shirt be available in a 3/4 sleeve for older nuns who do not like short sleeves and for active nuns whose work is hindered by long sleeves. Regarding the proposal to fasten the waistband at the back, she expressed the opinion that it is not the best solution for older nuns still active in pastoral work as their arms and shoulders hurt and the current design is easier to put on. She considered the proposals for the dress and skirt innovative and interesting. However, their production requires skills in tailoring and sewing which all nuns do not possess. As many nuns make their own religious clothing, the new proposals could be difficult to implement [11].

6 Conclusion

As a type of uniform, nuns' clothing had its default form and was not adapted to the activities of nuns for centuries. This changed only after the Second Vatican Council held from 1962 to 1965 when a new, more functional form was confirmed. The nuns' clothing is increasingly adapted to the needs of their day-to-day life and work keeping in mind ergonomic aspects as well. Salesian Sisters of Don Bosco were chosen for the purpose of this research, but it can be assumed other orders have similar needs for adaptation. Although a great variety of available materials exists nowadays and clothing is constantly adapted to wearers' needs, this paper shows a need for further improvements and even new items of clothing such as the polo shirt. Improvements in the clothing are mainly adaptations to increase comfort and enabling easier performance of daily duties. Shortcomings in the suggested improvements are evident in the fact that nuns often make their own clothes but do not possess necessary skill to implement some of the suggestions. Uniforms are sewn according to already set parameters and rules where the uniform defined for their order must be respected and cannot be changed during the sewing process. The nurses are not skilled enough in the process of cutting and sewing, and there could be problems adjusting the cut to a certain body type. While in the case of clothing purchased in a store, everything still depends on the manufacturer and the clothing sizes used.

References

1. Paulus, J.II: Posvećeni život – posinodalna apostolska pobudnica. Družina, Ljubljana (1996)
2. Hrvatska enciklopedija Leksikografskog zavoda Miroslav Krleža; "Odjeća". http://www.enciklopedija.hr/natuknica.aspx?id=63206. Accessed 30 Oct 2022
3. Cambrige Dictionary: Uniform. https://dictionary.cambridge.org/dictionary/english/uniform. Accessed 30 Oct 2022
4. Duchenne, I.V.: The development of the religious habit of the faithful companions of jesus. Costume **6**, 86–90 (2016)
5. Michelman, S.O.: Changing old habits: dress of women religious and its relationship to personal and social identity. Sociol. Inqury **67**, 353–361 (1997)
6. Trzebiatowska, M.: Habit does not a nun make? Religious dress in the everyday lives of polish catholic nuns. J. Contemp. Relig. **25**, 51–65 (2010)
7. Župa Svetog Duha.https://zupa-duhasvetoga-jarun.hr/salezijanci/kceri-marije-pomocnice/. Accessed 10 July 2022
8. Giovanni, B..: Constituzioni dell'Istituto delle Figlie di Maria Ausiliatrice. Instituto Figlie Maria Ausiliatrice, Torino (1878)
9. Salesian Sisters of St. John Bosco Daughters of Mary Help of Christians. https://www.google.com/search?q=Salesian+nuns+in+habits&source=lnms&tbm=isch&sa=X&ved=2ahUKEwiEm6L8hpz7AhVmRPEDHSL6D_MQ_AUoAXoECAIQAw&biw=1920&bih=969&dpr=1#imgrc=NZCuHxJeut8yDM&imgdii=y7uT5HyCethfBM. Accessed 7 Nov 2022
10. Italy – Daughters of Mary Help of Christians towards their XXIV General Chapter. https://www.infoans.org/en/sections/news/item/13608-italy-daughters-of-mary-help-of-christians-towards-their-xxiv-general-chapter. Accessed 3 Nov 2022
11. Bodrožić, M.: Projektiranje kolekcije odjeće za aktivni dio pastorala sestara Kćeri Marije Pomoćnice prema anketnom istraživanju. Sveučilište u Zagrebu Tekstilno-tehnološki fakultet, Zagreb (2022)

"Neurotafl" Software and Hardware System for Inclusive Education

Artem Smolin, Juri Didevich, Alena Dzhumagulova, Andrei Balkanskii, Anna Spiridonova, Larisa Sopronenko, Yuliya Magina, and Aleksandr Volosiuk

Abstract This article is devoted to research in the field of using information technology for individual and group work of blind and visually impaired children, including those with intellectual disabilities. As a part of this research a unique hardware-software complex "Neirotafl" was developed. The correctional-pedagogical experiment was carried out, as well as the study of anxiety in children 7–12 years old.

Keywords Information technology · Neurointerface · Blind · Visually impaired · Intellectual disability · Anxiety · Haptics · Accessibility · Accessible interface · Multimodal perception

1 Introduction

Recent developments in information and computer technology are increasing the opportunities for educational and cultural institutions to use them in inclusion. The Salamanca Statement of Principles, Policies and Practices in Special Needs

---

J. Didevich—Self-employed

A. Smolin · A. Dzhumagulova · A. Balkanskii · A. Spiridonova · L. Sopronenko · A. Volosiuk (✉)
ITMO University, Saint-Petersburg, Russia
e-mail: aavolosiuk@gmail.com

A. Smolin
e-mail: smolin@itmo.ru

J. Didevich
Saint-Petersburg, Russia

Y. Magina
K. K. Grot Boarding School №1, Saint-Petersburg, Russia

A. Volosiuk
St.Petersburg Electrotechnical University, Saint-Petersburg, Russia

I. Salopek Čubrić et al. (eds.), *Proceedings of the 9th International Ergonomics Conference*, Lecture Notes in Networks and Systems 701,
https://doi.org/10.1007/978-3-031-33986-8_32

Education was concluded in 1994. Since then the term "inclusive education" has appeared.

Russia signed the Convention on the Rights of Persons with Disabilities in 2008, thereby opening up opportunities to create an accessible environment and digital inclusion. At the moment, accessibility principles are used in software and hardware products for learning, changing emotional backgrounds and creating cultural works.

There are innovative inclusive technologies using artificial intelligence, machine learning and implemented with the help of neuro-interfaces and vibrotactile feedback [1]. This work is devoted to one of such technologies.

Employees of the "Neuroculture" laboratory together with ITMO University have created a unique hardware-software complex Neurotafl, the purpose of which is to conduct research with the aim of socio-cultural adaptation of people with disabilities.

The innovation of this project is in the combination of technical means combining the tactile-kinesthetic interface, the system of data analysis and interpretation by algorithms of the physiological and emotional calculations received during the audio-motor associations of the users.

The purpose of such interaction is gradual, controlled and purposeful development of the stimulated modalities in congenital or early acquired blindness, problems with sensory and emotional deprivation, or correction of various shifts in behavior, somatic state, and individual neuropsychiatric disorders [2].The software allows individual goal setting, control and analysis of the results of the interaction in a clear graphic form of computer interface, as well as the purposeful implementation of activities by volunteers under the general supervision of specialists, teachers and psychologists, to record the results obtained with the software tools of the system.

2 The Use of Information Technology in Inclusive Education

The age of information technology development also offers its own approaches to teaching visually impaired children, but due to the high cost of equipment it is not popular in Russian schools. Children are taught using a computer or a tablet with specially designed software.

Thus, an article [3] tells about teaching and learning graphics to visually impaired children. The authors investigate the method of voicing to convey information, and also use it to teach children through a game in a special application on the Android platform.

The app was tested on 6 visually impaired children: 3 children 10–12 years old and 3 children 13–16 years old.

In order to keep children interested, interactive games were developed: in the first one a child must repeat an image (the game "Hear the invisible"); in the second one a child must follow an object by drawing a figure (the game "Follow the rabbit"); in the third one they must find extrema.

Since the sample is very small, it is impossible to draw any significant conclusions from them, but the article itself attracted exactly by the use of this non-standard method of transmitting information. Some ideas, like adding a sound and changing its tone on certain zones, can be useful in adapting visual learning content not only for children, but also for adults, which can help them more clearly perceive the images. Presenting knowledge in a game form allows making the learning process self-motivating and fast.

Also, a study [4] examines the user-oriented design and analysis of a special system with a touch screen to display graphic visual information to people with visual impairments. A study consisting of locating tactile points was conducted to test the effectiveness and efficiency of the system and to determine user interaction patterns with the touch screen. Results showed that participants correctly located tactile dots with an accuracy rate of 69.83% and an average time of 15.34 s. Of the 116 total trials, accuracy increased over the course of the trials.

In the study [5] a hardware-software complex consisting of a vibro-suit and an Android tablet is presented. The software developed by the authors converts the entered text or recognized speech of a sighted and speaking interlocutor into a sequence of vibration signals (a vibration pattern that carries a certain verbal meaning) transmitted to the suit worn by the deaf-blind interlocutor. The deaf-blind interlocutor, in turn, draws a sequence of vibrations on the tablet (similar to the way we enter a picture key to unlock a smartphone), which is converted into text or automatically voiced text. To ensure that the deaf-blind interlocutor could not make a mistake when entering a drawing on the tablet, the authors of the study developed a transparent cover, the topography of which, when applied to the screen, exactly repeats the location of reference points for the drawing. A sighted interlocutor can also use "graphic key" input if he or she knows the alphabet of the deaf-blind. The aim was to facilitate communication for the deafblind community, empowering individuals to initiate and engage in social interactions with other people without the direct need of an intervener.

Studies like this one show that teaching children requires a special approach in everything from the interface to the presentation of the material.

3 Hardware-Software Complex Neurotafl

The hardware-software complex Neurotafl is an instrument of the new generation, which allows users to express themselves through individual artistic experience. Motivated by music and various movement sounds, it has a beneficial effect on the physiological processes in the body. It improves self-perception, increases the ability to coordinate and develop motor skills.

Table 1 Table of modalities

Sensory perception	Sensory organs	Modality
Sight	Eyes	Visual
Hearing	Ears	Auditory
Touch	Skin	Tactile-kinesthetic
Smell	Nose	Olfactory
Taste	Tongue	Flavor
Balance	Balance organs	Vestibular

3.1 Creation Concept

The authors of the project consider the body with its perceptual and motor abilities as a whole within the framework of the theory of "embodied cognition," or embodied cognition. The essence of this entirely new approach is that there is a complex relationship between sensory-motor experience, behavior, emotion, and decision-making. The constructor of our emotions is the brain, which works in conjunction with physical factors, and sensation is connected to both unconscious and conscious thinking-that is, our motor skills and our perceptions are incorporated into the workings of consciousness. Our perception is multimodal in nature. All of our senses and modalities (as presented in Table 1) work in perceiving the world, and they mutually influence each other.

Multimodal perception and cross-modal behavior are functional capabilities of the human cognitive system, for example, playing a musical instrument depends very much not only on hearing, but also on tactile and visual feedback to the instrument.

3.2 Technological Structure of the Neurotafl Complex

The hardware-computer complex Neurotafl is an interactive real-time system that records and analyzes the user's actions at the moment of interaction. The complex recognizes and converts movements on the surface of the tactile-kinesthetic interface, as well as actions with interactive physical objects, into music or sound.

The main purpose of this interaction is to stimulate motor skills through the projection of auditory and tactile modes of sensory perception, as cognitive abilities are stimulated and developed through the encouragement of motor activity.

The interaction process takes place in the form of a game, using audio-motor sensation transfer reactions - when a stimulus from one modality, such as motor skills, tactile sensations, texture and shape is associated with a stimulus from another modality - sound and music. This type of interaction can develop thinking processes and functions in the participants involved in the interaction.

The basis of the hardware-computer complex Neurotafl is the use of artificial neural networks. Neural networks are used to solve complex problems that require analytical calculations similar to those performed by the human brain.

Some control and adaptation functions are performed using statistical analysis data obtained in the course of interaction, with the help of machine learning systems that allow the interaction to identify hidden and problematic patterns of user behavior, such as the lack of responsiveness to various types of sound and tactile stimuli.

Neurotafl software allows controlling statistic and qualitative parameters of interaction and includes the following functional program modules:

- Computer vision module - provides physical interaction with the system using physical objects;
- Acoustic reflectometry module - provides tactile interaction with the surface;
- A classifier of recognizable actions and interaction elements based on machine learning;
- Narrative system containing fixed sets of audio and music sequences and converting incoming data into audio, music, and tactile sequences;
- statistical module for analyzing user actions;
- graphic interface of management, control and program settings;
- statistical analysis graphical interface.

The software, except for the control and statistical interfaces, uses fast Fourier transforms, digital signal processing, real-time audio and audio synthesis.

Figure 1 shows the appearance of the Neurotafl complex.

The base of the complex is made of steel pipes. The design of the base allows you to adjust the height of the table from 600 to 900 mm. There are holes in the frame racks for fixing the table at the required height by means of special clamps.

The tabletop is a birch plywood construction. The upper removable part of the product is made of a set wooden board, a panel of translucent monolithic polycarbonate and a light filter. In the design of the tabletop there are necessary technical holes for installing equipment, as well as for fixing the product on the frame.

The body is made of aluminum composite panels 4 mm thick. Necessary technical holes are provided on the product to install the equipment, as well as protective and decorative elements of the product.

The protective panel of monolithic polycarbonate is installed on the composite panel enclosure through the distance holders. It serves to protect equipment and switching elements from mechanical damages during exploitation of the product.

It is important that all structural elements are made of hypoallergenic and fireproof materials and are reliably fastened to each other. On the table-top there are explanatory texts with tactile Braille relief tactile font.

The JVC TK-C9300E universal camera from JVC is used for the computer vision system.

The camera works in infrared mode in the range of 850 nm. For this purpose, an appropriate infrared filter is applied to the lens of the camera through an adapter.

In order to illuminate interior surface of the table and for stable work of the camera the infrared illuminator was made, which works in the range of 850 nm.

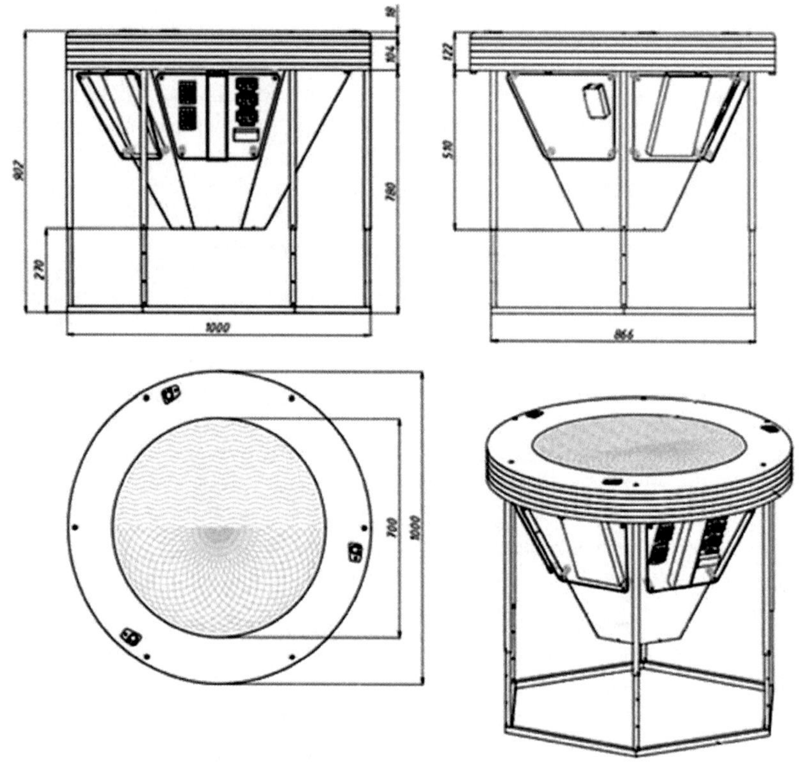

Fig. 1 Hardware-software complex Neurotafl

3.3 Interactive Elements of Physical Interaction

Eight cubes (70 × 70 × 70 mm) made of oak lamellas were made as interactive elements (Fig. 2).

Each element is equipped with the following nodes and modules:

- MPU9250 digital gyroscope sensor module;
- MPU-6500 accelerometer;
- Triaxial compass AK8963.

The integrated Digital Motion Processor (DMP) is responsible for data processing.

Processing, reception and transmission module board is equipped with a built-in wi-fi module. The board has a micro usb connector as well as eleven digital GPIOs with an analog I/O (3.2 V max). The advantage of the board is the miniature size and the Wi-fi module.

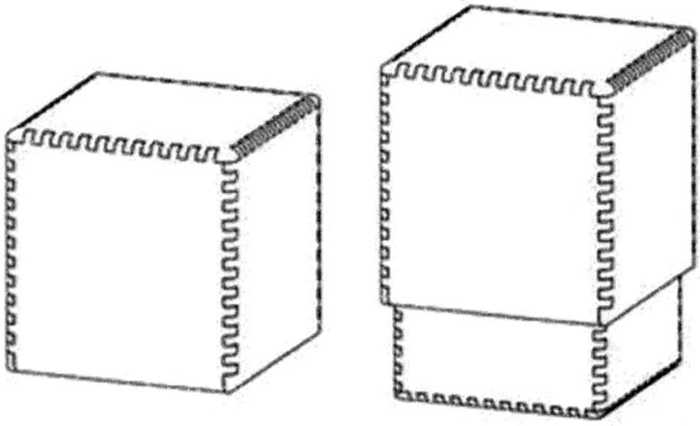

Fig. 2 Cube Interactive Element

3.4 Technical Features of the System

The system of machine analysis of user actions monitors the actions performed by the user on the surface of the "table" and with the objects of physical manipulation "cubes".

1. Basic analysis parameters:
   - kinesthetic (tactile) interaction
   - recognition of objects on the surface;
   - trajectory recognition;
   - automatic numbering of objects;
   - object coordinates;
   - object dimensions;
   - velocity of the object;
   - the acceleration of the object;
   - velocity and amplitude of the impact on the surface.
2. objects of physical interaction
   - recognition of the object's faces (6 faces of a cube);
   - velocity and acceleration of the object;
   - rotation of the object relative to any of the faces.
3. parametric analysis of interaction-related sound events in the course of each session
   - Auditory model spectrum;
   - Loudness (Spectral Energy);
   - Brightness (Spectral Centroid);
   - Noisiness (Spectral Flatness);
   - Pitch (F0) Estimation;

– Beats per minute.

To control sound, music and tactile elements we use algorithmic transformation of data received from the system of computer analysis of user's actions. In direct work it is supposed to use three moments of training: geometrical figures, complex elements and the coordinate system. For the visually impaired we use specially developed methods and forms of work, designed for auditory and tactile perception.

Special attention is paid to the development of:

1. muscular sense;
2. residual vision;
3. concretization of ideas about kinds and forms of objects;
4. surrounding objects;
5. types of implements;
6. development of visual-practical thinking;
7. formation of spatial orientation skills.

4 Study of the Effectiveness of Lessons Conducted with the NEUROTAFL Hardware and Computer Complex at the School No.1 Named After K.K. Grot for Blind and Visually Impaired Children

Twelve students between the ages of 7 and 12 took part in the activities of the project "Neurotafl", conducted at boarding school № 1 named after K.K. Grot for the blind and visually impaired children.

The Sample and Research Procedure. The sample consisted of blind, visually impaired children, as well as visually impaired children with mental disorders from 7 to 12 years old.

Number of sessions per person: 2 sessions per week for 1 h. A total of 16 h per person in three months (March–May). The total number of hours: 192. The main form of activity in the course of conducted classes is educational and playful activity of students.

Interactive sound and music programs with the use of hardware and computer complex NEIROTAFL were created for the games. The priority activities of the lessons are: individual work, work in pairs with the use of the NEUROTAFL hardware-computer complex; participation in a conversation and discussion; participation in exercises and games; creative work; reflection and feedback.

During the lessons children were interested in interactive fairy tales that were conducted with the use of the NEUROTAFL computer complex. Children were interested in perceiving fairy tales aurally and tactilely using interactive capabilities of the system: they listen to the sounds of composing the tales and listening to the results, some were interested in tactile interaction: they felt the texture of the interactive elements (cubes), touched the table surface and felt vibration. The emotional background of the children changed with each session (Fig. 3).

Fig. 3 Students at the school No.1 named after K.K. Grot uses Neurotafl

The Purpose and Objectives of the Classes. The purpose of the lessons is socio-cultural development, development of emotional intelligence and revealing the creative potential of blind and visually impaired children with disabilities (disabilities). At carrying out lessons the following tasks have been put:

1. Improvement of the psycho-emotional state of the students.
2. Decrease in the level of school, self-esteem, interpersonal anxiety.
3. Development of the students' creative potential.
4. Development of communication skills and self-control skills.
5. Development of thinking processes and the ability to reflect and empathize.
6. Improvement of students' quality of interaction with the environment.

Formative Experiment. Elements of storytelling therapy were used as exercises. Fairy tale therapy is a correctional method, which uses composing fairy tales and stories for realization of correctional influences. Fairy tale therapy is used as one of the techniques for changing dysfunctional settings and forming alternative behavioral scenarios. Fairy tale therapy is used by psychologists in work with children with emotional disorders - first of all, with fears, anxiety, hypersensitivity (oversensitivity) [6].

We have developed and conducted sessions with the use of interactive audio-musical fairy tales with the use of the hardware and computer system NEUROTAFL.

The hardware monitored the expression of emotions and the dynamics of their experiences in the course of the lessons. Also by means of the complex NEUROTAFL during the interaction with sound-music and semantic narratives of interactive fairy tales the pupils were taught to communicate as well as to perform self-regulation of behavior.

The technique of studying socio-situational anxiety. D. Kondash's "Socio-situational Anxiety Scales" method was used to study anxiety. With the help of this technique we can find out anxiety level according to three main parameters: school anxiety that manifests itself in educational activity, interpersonal anxiety that develops in interaction with peers and significant adults, and also self-evaluation anxiety which expresses itself in ideas we have about ourselves.

Testing Results. Initial testing revealed increased anxiety in the children. The results are reflected in the table of changes in the indicators after repeated testing (Table 2).

Statistical analysis using the Mann–Whitney U-criterion showed significant differences in the index of general and school anxiety. The use of questionnaires is difficult for visually impaired children with intellectual disabilities, so the observation method was used. Scores for visually impaired children with intellectual disabilities using the observation method were as follows: 9 (Ll.) medium/low, 10 (Dn.) high/raised, 11 (Ol.) high/medium, 12 (As.) high/medium levels.

Results of the forming experiment.

- Growth of interaction parameters: expression of positive emotions, decrease of school, self-esteem and interpersonal anxiety level.
- Positive dynamics of thinking abilities and abilities to empathy.
- Correction and compensation of secondary deviations in development of blind and visually impaired children.

Table 2 Indicators of social-situational anxiety and blind and visually impaired children without intellectual disabilities

Participant No	General Anxiety		School Anxiety		Self-evaluation Anxiety		Interpersonal Anxiety	
	Before	After	Before	After	Before	After	Before	After
1	51	45	16	14	17	15	13	12
2	46	43	14	13	13	12	12	12
3	76	73	30	23	25	22	24	18
4	60	59	13	10	22	20	21	21
5	66	58	25	20	18	17	23	21
6	61	55	24	23	13	11	24	21
7	55	50	20	17	21	19	20	16
8	59	53	32	28	26	24	25	20

5 Conclusions

The use of hardware-computer complex "Neurotafl" in the formative experiment with the participation of blind and visually impaired children 7–12 years old, including those with intellectual disabilities, allowed to achieve a number of psychological and pedagogical, corrective and developmental results for blind and visually impaired children, namely:

1. Growth of parameters of interaction: expression of positive emotions decrease in level of school, self-esteem and interpersonal anxiety;
2. Positive dynamics of thinking abilities and empathy abilities;
3. Correction and compensation of secondary deviations in development of blind and visually impaired children;
4. As a result of anxiety reduction the quality of interaction with other subjects of education and with the environment has improved.
5. Advantages of the method are non-invasiveness, non-medical, absence of age restrictions, absence of side effects. It promotes normalization of the psycho-physiological condition of the child.
6. During sessions, improvement of mood, reduction of hyper-activity, emotional tension, fatigue and aggressiveness are also observed.
7. Corrective-developmental activities contribute to restoration of speech and cognitive functions in children with psychological and speech delays. The children show a decrease in the severity of stereotypes, and their communication skills and understanding of spoken language gradually improve. Repeated observations have shown improvement in motor functions.

Thus, this is a highly effective technique that allows correction of developmental deviations and serves the purpose of adaptation and rehabilitation of children with disabilities who experience difficulties in learning and social adaptation.

References

1. Clynes, M.: Music, Mind and Brain: The Neuropychology of Music. Plenum Press, New York (1982)
2. Mamaichuk I.I.: Psycho-correctional technology for children with developmental problems. Rech, St. Petersburg (2006)
3. Radecki, A., Bujacz, M., Skulimowski, P., Strumillo, P.: Interactive sonification of images in serious games as an educational aid for visually impaired children. Br. J. Educ. Technol. **51**, 1–25 (2019)
4. Bateman, A., Zhao, O.K., Bajcsy, A.V., Jennings, M.C., Toth, B.N., Cohen, A.J., Oliveira, M.A.: A user-centered design and analysis of an electrostatic haptic touchscreen system for students with visual impairments. Int. J. Hum. Comput. Stud. **109**, 102–111 (2018)

5. Theil, A., ET AL.: Tactile board: a multimodal augmentative and alternative communication device for individuals with deafblindness. In: 19th International Conference on Mobile and Ubiquitous Multimedia (MUM 2020), pp. 223–228. Association for Computing Machinery, New York (2020)
6. Volkova, I.P.: Modern paradigms of theory and practice of socio-psychological adaptation and integration of visually impaired. Bull. St. Petersburg Univ. Psychol. Health **3**(12), 21–29 (2008)

Ergonomic Design of the PET Bottle for Maximum Usability

M. Soldo, N. Rašović, and A. Vučina

Abstract The paper's primary goal is to analyse and define guidelines and methodology for the ergonomic design of PET bottles. In designing any product, in addition to production requirements (design for manufacturing), ergonomic and aesthetic requirements must be considered to provide the consumer with practicality, usability and aesthetic appeal. However, design for aesthetics is not considered, which constrained the research activities. The conclusion summarises the ergonomic design of PET bottles according to ergonomic guidelines defined in the paper.

Keywords Design · PET bottle · Ergonomics · Surface modelling

1 Introduction

PET packaging for drinking water offers diverse possibilities to designers. They must be innovative in new product design that satisfies all consumer needs while respecting the manufacturing, ergonomic requirements, legal regulations, standards, availability of products and ultimately imposed time and material limitations. Due to the demands of the global market, where the competition between companies is more ruthless than ever, the designer position is unenviable. Hence, design flexibility concerning the packaging shape results in various possible solutions. An important characteristic of water packaging is its practicality and usability in everyday use, where shape and dimensions play a prominent role. Most PET bottle design research focused on designing bottles through gripping force, for example, through two main

M. Soldo (✉) · N. Rašović · A. Vučina
Faculty of Mechanical Engineering, Computing and Electrical Engineering, University of Mostar, Mostar, Bosnia and Herzegovina
e-mail: mario.soldo@fsre.sum.ba

N. Rašović
e-mail: nebojsa.rasovic@fsre.sum.ba

A. Vučina
e-mail: adisa.vucina@fsre.sum.ba

I. Salopek Čubrić et al. (eds.), *Proceedings of the 9th International Ergonomics Conference*, Lecture Notes in Networks and Systems 701,
https://doi.org/10.1007/978-3-031-33986-8_33

tasks: transporting and pouring [1] or the difference between the human hand and adaptive prothesis gripping force while holding a bottle [2]. Many types of research include visual aspects before ergonomics in bottle design to detect consumer positive or negative subjective judgmental [3] or attract emotional attachment to the customer through the bottle design [4]. Some research analyses the aesthetic design of PET bottles regarding the moment of force and grasping [5]. The paper aims to define a method for decision-making in the ergonomic design process that would affect the final design of the bottle to maximise usability. The research seeks to design a bottle that would be adapted to people regardless of age and gender. Therefore ergonomic and manufacturing knowledge-based guidelines for PET bottle design are defined for maximisation of usability. After defining ergonomic and production requirements, SolidWorks has been used to design PET bottles. The conclusion contains future research activities.

2 Ergonomics Design and Methodology

2.1 *Requirements and Goals*

The paper aims to describe the ergonomic design process for a water bottle volume of 0.5 l, which will be practical and usable for most adults and teenagers. The bottle's shape and dimensions must be practical with the intention of a simple function performing. These functions include balancing when placed on a level surface and are suitable for human factors and ergonomics, i.e., hands and mouth, fit into bottle holders since they have increased use for this product. The ideal material for a bottle is glass, but due to the cost and processing time nowadays, the most common 0.5 l water bottles are made of plastic. PET was chosen for the bottle's material since of the frequency of use for this kind of product (Table 1), and the main goals of PET material are acceptable for human health and recycling. The bottle must be as lightweight as possible, use minimal material through production, and meet all strength requirements. Water retention, packaging, distribution, transport, storage, reuse and recycling must be examined for strength.

2.2 *Current Design*

The main factor influencing the bottled water brand's recognition and position on the market is the design of the bottles, which must be designed in such a way that they are aesthetically and ergonomically attractive to the user and thus satisfy all the ergonomic needs of the user. In order to start the process of ergonomic design of the bottle, the main characteristics (Table 1) of several bottled flasks of water of 0.5 l currently on the market in Bosnia and Herzegovina were firstly investigated (Fig. 1).

2.3 Production and Ergonomic Guidelines

Guidelines for the PET bottle design from the aspect of the production process and use are well-considered. The selected materials and production process are decisive factors in the bottle design process. Since all analysed bottles are made of PET material (Table 1.) using the stretch blow moulding process [6], designers must ensure consistency and reliability throughout the entire manufacturing process. For the stretch blow moulding process, manufacturing experience suggests a number of constraints the designer must achieve. Given the requirements in mind, the design guidelines from the production aspect were defined:

- Square surfaces with sharp edges are undesirable due to errors that will occur due to inconsistent production.

Table 1 Main characteristics of several brands of bottled water in Bosnia and Herzegovina

Brand	Bottle volume	Bottle mass	Bottle height	Cross-section shape	Max. length of cross-section	Plastic cap diameter	Bottle material
Brand_1	0.5 l	20 g	210 mm	circular	60 mm	25 mm	PET
Brand_2	0.5 l	24 g	210 mm	circular	62 mm	21 mm	PET
Brand_3	0.5 l	28 g	220 mm	square	56 mm	32 mm	PET
Brand_4	0.5 l	16 g	220 mm	circular	60 mm	25 mm	PET
Brand_5	0.5 l	22 g	230 mm	circular	63 mm	22 mm	PET

Fig. 1 Several half-litre bottled water markets in Bosnia and Herzegovina

- Complications may occur due to the disproportional expansion of the cross-section concerning the overall design.
- Rounding on all edges, ribs, and corners would give a uniform wall thickness while reducing stress during stretching.
- PET material will give the best properties if impact energy is absorbed elastically through the process.

An essential factor in the design of bottles concerning the phase of use is the diameter of bottle holders in cars, bicycles and gym equipment, usually from 74 to 86 mm [7]. So now guidelines can be defined for the use phase as:

- The bottle should be placed in a holder with a diameter of 75 mm.

Ergonomic guidelines must be defined and used in the design process since the product's primary function, and shape are usually strongly connected [8]. In order to define guidelines for the ergonomic design of the bottle, anthropometric measurements and biomechanics of the human hand must be taken into account:

- The height of the grip on the bottle should be equal to the hand's width.
- The grip length should be at least 51% of the bottle's circumference.
- The grip surface on the bottle must not be slippery.

2.4 *Ergonomic Shape and Dimensions of PET Bottle*

According to anthropometry, over 95% of women and 99% of men have a hand length greater than 165 mm, measured from wrist crease up to the most extended finger. Measurements from the tip of the thumb to the tip of the most extended finger are roughly equal to the hand's length (Fig. 2). When using ergonomic guidelines, it is crucial to remember that the grip length should be at least 51% of the bottle[7].

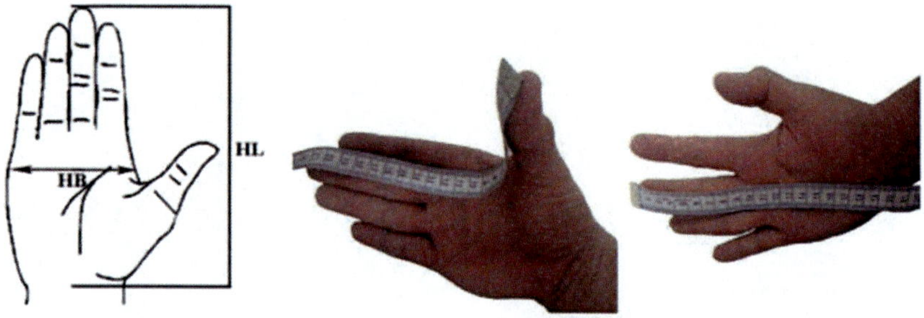

Fig. 2 Left: Hand length (HL) and breath (HB). Right: Hand measurement correlation. (Source of right Fig.: Taken from Lovett, J. et Engineering Design of a Disposable Water Bottle for an Australian Market (2013) [7])

The intention is to design a bottle shape that will suit as many users as possible. Since the hand's length and width vary depending on the gender and percentile group, the smaller dimension of the female hand is taken into consideration, i.e. for women of the 5th percentile groups (Fig. 2), which should correspond to the bottle's ergonomic design for both sexes and all percentile groups. Also, these smaller female hands must cover a minimum of 51% of the bottle's circumference. So every second greater hand should comfortably hold the bottle with a higher percentage of circumference and less pressure. The pressure will be highest when the grip is 51% of the bottle's circumference.

The easy experiment of holding any bottle or glass (Fig. 3) concludes that the average male hand could cover more than 75% of the bottle circumference. So it was decided that the adopted hand length of 165 mm would achieve a grip angle of 260° (72% of circumference). It means that the maximum permissible circumference (100%) is 229.22 mm (adopted dimension of 230 mm), which is equal to a diameter of 73 mm. 99th percentile group of men have a hand length of 219 mm (it can grab 95.5% of bottle circumference). Even that percentile group can grab bottles without difficulty by contacting larger fingers and thumbs.

Considering the guidelines for bottle holders, the diameter of the bottle will be 73 mm.

Table 2 and Table 3 shows how much per cent each percentile group (for female and male) would cover the maximum permissible adopted circumference of 230 mm, and is the adopted maximal circumference acceptable for each percentile group? By studying tables, it is evident that the adapted dimension for the bottle is corrected and assumed.

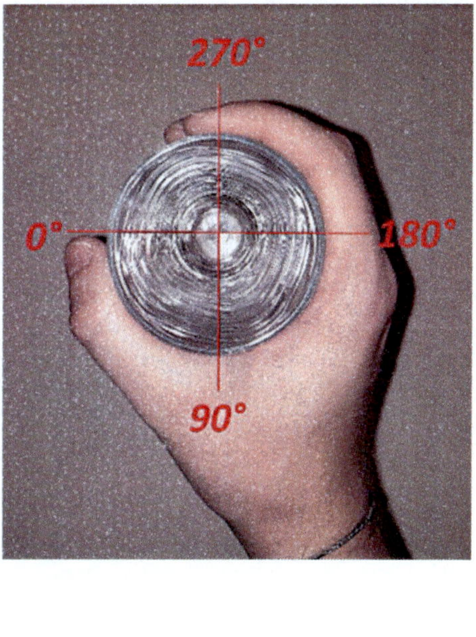

Fig. 3 Angles across holding the glass

Table 2 The percentage of coverage bottle circumference of 230 mm according to the percentiles group for females

Percentiles	Hand length (HL) [9]	Circumference bottle covered	Acceptable (Yes/No)
1st	159 mm	69.1%	Yes
5th	165 mm	71.7%	Yes
50th	180 mm	78.3%	Yes
95th	197 mm	85.6%	Yes
99th	205 mm	89.1%	Yes

Table 3 The percentage of coverage bottle circumference of 230 mm according to the percentiles group for males

Percentiles	Hand length (HL) [9]	Circumference bottle covered	Acceptable (Yes/No)
1st	173 mm	75.2%	Yes
5th	179 mm	77.8%	Yes
50th	193 mm	83.9%	Yes
95th	211 mm	91.7%	Yes
99th	219 mm	95.2%	Yes

Bottle height is essential to define the bottle's total volume of liquid. Based on the analysis of available bottles on the market (Table 1), 230 mm for the maximum height of the bottle is adopted. Later during the moulding, it is possible to change the height of the bottle to achieve a volume of 0.5 l. Hand breath (HB) which will cover a certain percentage of the bottle height, will be set at the height of 30 mm from the bottom of the bottle. The logo or label sets to 120 mm from the bottom of the bottle. The logo can be 3D permanent on the bottle so it will remain permanently visible through multiple uses.

To get a better bottle adaption to the human hand and to avoid the bottle slipping during use, the position of the fingers when holding the bottle was analysed. A cylindrical clay part with a diameter of 73 mm and a length of 230 mm was made for analysis (Fig. 5). After taking a handprint in clay (in this case, the handprint belongs to the 95th percentile of males, i.e. the group with the largest hand breath), main dimensions are taken and used for rough grip part sketches of the bottle. In this way, users with larger and smaller hand sizes can hold the bottle easily.

In order to successfully hold the bottle in hand, the contact points on hand must apply directly opposing forces on either side of the bottle (Fig. 4). The mathematical calculations of the required force (Fig. 3) to hold the bottle in hand is:

$$F_N = \frac{m \cdot g}{2 \cdot \mu} = 9,9N \qquad (1)$$

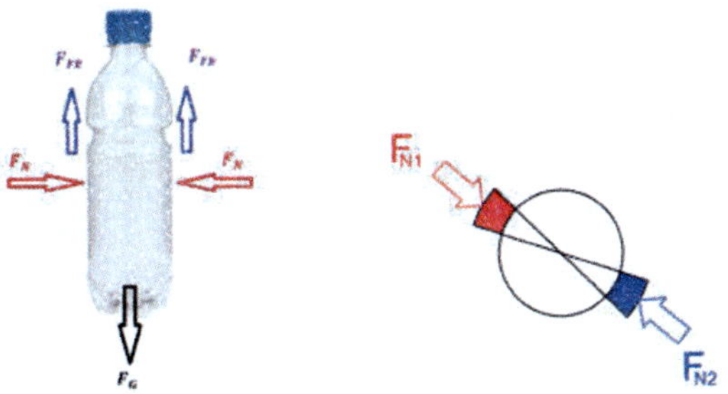

Fig. 4 Left: The active forces on holding the bottle. Right: The influence of separated normal forces on the circular section of the bottle

Fig. 5 Taking and measuring hand and fingers print in cylinder clay

where are: F_N - normal force; $\mu = 0,26$ - skin coefficient of friction [10], $m = 0,525kg$- mass of the bottle (with plastic cap) and $g = 9,81m/s^2$ – mean value of gravitational acceleration.

Although the calculated normal force will not change, the area covered by the hand will be defined by how much pressure is required. The effective contact area will increase for each degree above 51% of the bottle circumference for holding the bottle.

Figure 6 presents the PET bottle design modelled in SolidWorks, respecting previously defined ergonomic and production guidelines. The bottle is modelled by Surface modelling. Surface elements in modelling free forms, enabling the creation of faces using simple modifications, are widely used.

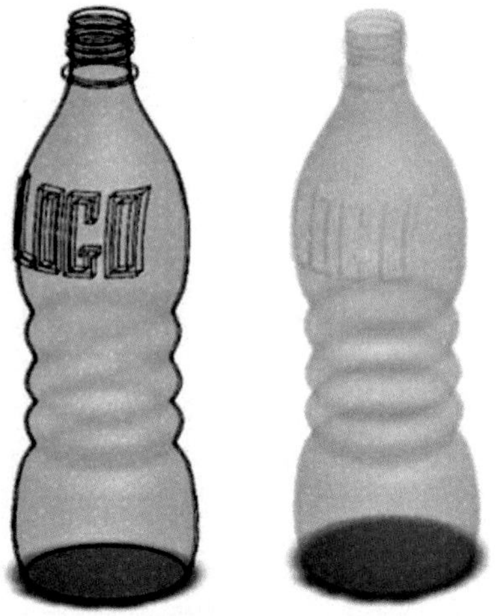

Fig. 6 PET bottle model designed through command Surface in program SolidWorks with and without contour and material applied

3 Conclusion

This paper defines ergonomic guidelines and methodology for the half-litre PET bottle design. In addition to these guidelines, guidelines for phase of use were indispensable to obtaining a usable product. The measurements of the smallest female hand were used in the design, and the optimal bottle diameter of 73 mm was obtained. With an adopted 73 mm bottle diameter, even the smallest dimensions of a woman's hand could achieve 69.1% coverage of the bottle circumference. Also, the bottle can be held by the largest male hand (95.2% coverage of the bottle circumference). This design also satisfies the ergonomic needs of a large part of the teenage population. The position of the fingers when holding the bottle was also analysed. The analysis of the position of the bottle holder when holding the bottle was used in the part design for holding the bottle, which enabled a better hold and avoided slipping. In that way, all percentile groups can hold the bottle without increasing grasping pressure.

Since aesthetics is a significant factor for this type of product, guidelines for the aesthetic design of the bottle will be defined in future research. The final product must be designed from the perspective of all phases of the PET bottle life cycle, such as packaging, distribution, and recycling.

References

1. Cepriá-Bernal, J., Pérez-González, A., Mora, M.C., Sancho-Bru, J.L.: Grip force and force sharing in two different manipulation tasks with bottles. Ergonomics **60**(7), 957–966 (2017)
2. Kargov, A., Pylatiuk, C., Martin, J., Schulz, S., Döderlein, L: A comparison of the grip force distribution in natural hands and in prosthetic hands. Disabil. Rehabil. **26**(12), 705–711 (2004)

3. Kasikovic, N.: PET bottle design, analysis of correlation between visual aesthetic impression and subjective judgments of bottle characteristics. ResearchGate (2014)
4. Widiyati, K., Aoyama, H.: A study of kansei engineering in pet bottle silhouette. In: Proceedings of ASME Design Engineering Technical Conferences 2 (PARTS A AND B), pp. 849–858 (2011)
5. Widiyati, K., Aoyama, H.: The ease of grasping to evaluate aesthetically pleasing pet bottle design. J. Adv. Mech. Des. Syst. Manuf. **7**(5), 849–861 (2013)
6. Lim, L.T., Auras, R., Rubino, M.: Processing technologies for poly(lactic acid). Prog. Polym. Sci. **33**(8), 820–852 (2008)
7. Lovett, J.: Engineering Design of a Disposable Water Bottle for an Australian Market (2013). https://www.semanticscholar.org/paper/Engineering-design-of-a-disposable-water-bottle-for-Lovett/e3a1118339d0621eac044ec678b44359b6158369
8. Harih, G., Dolšak, B.: Tool-handle design based on a digital human hand model. Int. J. Ind. Ergon. **43**(4), 288–295 (2013)
9. Department of Defense Human Factors Engineering Technical Advisory Group Washington: Human engineering design data digest. US Army Missile Command. Redstone Arsenal, AL (2000)
10. Derler, S., Gerhardt, L.C., Lenz, A., Bertaux, E., Hadad, M.: Friction of human skin against smooth and rough glass as a function of the contact pressure. Tribol. Int. **42**(11–12), 1565–1574 (2009)

The UV Protection of Cationized Cotton/ Polyester Blend

Anita Tarbuk, Ivana Čorak, Snježana Brnada, and Tihana Dekanić

Abstract Clothing provides some UV protection which depends on the physical and chemical properties of the fiber and fabric. Cationization, as a modification with ammonium compounds in alkaline medium, results in the change of the surface charge, as well as in higher fabric thickness and lower porosity. Since these parameters contribute to better UV protection, in this research, the cotton/polyester 50/50 blended fabrics were cationized with Rewin OS by exhaustion method. An anionic optical brightener Uvitex RSB was applied in concentrations of 1, 2, 5 and 10% owf by exhaustion. The whiteness and yellowing were determined on remission spectrophotometer Spectraflash SF 300 (Datacolor) according to ISO 105-J02:1997 and DIN 6167:1980 respectively. The UV protection was determined according to AS/NZS 4399:2017 on the transmission spectrophotometer, Cary 50 Solascreen (Varian). The porosity and thickness were determined. Porosity was calculated based on the fabric cover factor after analysis of microscopic images using DinoLite microscope and the fabric thickness was performed on a thickness gauge. All parameters were observed and evaluated in correlation in relation to fabric UV protection.

Keywords Cotton/polyester blend · Cationization · Optical brightener · Porosity · Thickness · UV protection

1 Introduction

Excessive exposure to the sun, special to its ultraviolet radiation (UV-R), can lead to skin and eye damage, tumors and subsequently death. For this reason, the proper UV protection is necessary. As second skin, textiles offer the highest quality and most reliable UV protection [1–4]. Transmission, reflection or absorption of UV-R changes depending on the physical and chemical properties of the fiber and fabric

A. Tarbuk (✉) · I. Čorak · S. Brnada · T. Dekanić
University of Zagreb Faculty of Textile Technology, Prilaz Baruna Filipovića 28a, 10000 Zagreb, Croatia
e-mail: anita.tarbuk@ttf.unizg.hr

I. Salopek Čubrić et al. (eds.), *Proceedings of the 9th International Ergonomics Conference*, Lecture Notes in Networks and Systems 701,
https://doi.org/10.1007/978-3-031-33986-8_34

i.e. type of fibers, moisture content, fabric thickness, porosity, density, yarn fineness, pre-treatments, modification and finishing treatments, concentration of dyes, optical brighteners, UV absorbers, if applied, etc. For instance, synthetic fibers have better protective properties than natural fibers due to the presence of aromatic rings; the optical brighteners absorb UV-R due to fluorescence phenomenon; lower porosity prevents transmission; textile care process and the detergent which is used all leads to better UV protection. Changing these properties full UV protection with textiles is possible [1–13].

Cationization is a modification with ammonium compounds or epihalohydrins in alkaline medium. Usually cellulose-based materials, i.e. cotton, are cationized. A cationic agent blocks negative -OH groups to form cellulose ethers resulting in the change of the fiber surface charge. It can be performed in after-treatment, or during mercerization process. If it is carried out during mercerization, in addition to the known properties obtained by mercerization, cationization additionally changes the charge of cotton cellulose [14–17]. Regardless of application method, the swelling occurs and leads to higher fabric thickness and lower porosity.

Since this modification has been investigated only for cellulose fibers, in this paper the effect of cationization of cotton/polyester blended fabric and the influence to its UV protection was researched through its mechanical properties and porosity.

2 Materials and Methods

The cotton/polyester 50/50 blended fabrics supplied by Čateks d.o.o. (Čakovec, Croatia) were used. Fabric in basket weave 2/2 (B 2/2) of mass per unit area 160 g/ m^2, had yarn density of warp 42 threads/cm and weft 17 threads/cm. Fabric in 5-end satin weave (5eS) of mass per unit area 165 g/m^2, had yarn density of warp 43 threads/cm and weft 20 threads/cm. Fabrics were scoured and bleached under industrial conditions.

Fabrics were cationized with cationic reactive polyammonium compound Rewin OS (CHT-Bezema) by exhaustion in Polycolor machine (Mathis) in a bath containing: 50 g/L Rewin OS, 10 g/L NaOH, with LR 1:10 at 90 °C for 45 min. Fabrics were neutralized and rinsed afterwards, then air dried.

Optical brightener, a triazine derivative of stilbene disulfonic acid, Uvitex RSB (Fig. 1) was applied in wide range of 1, 2, 5 and 10% owf by exhaustion having LR 1:30 at 100 °C for 30 min in stainless-steel bowls of Linitest (Hanau). After treatment, the fabrics were air dried.

The streaming potential was measured on Electrokinetic analyzer (EKA, Anton Paar) using stamp cell and the electrokinetic potential (zeta, ζ, ZP) was calculated according to the Helmholtz-Smoluchowsky equation [18]. Zeta potential was measured as a function of pH of the 1 mmol/L KCl and the Isoelectric point (IEP) was determined.

Spectral remission (R [%]) was measured on a remission spectrophotometer Spectraflash SF 300 (Datacolor). Fabric whiteness (W_{CIE}) according to ISO 105-J02:1997

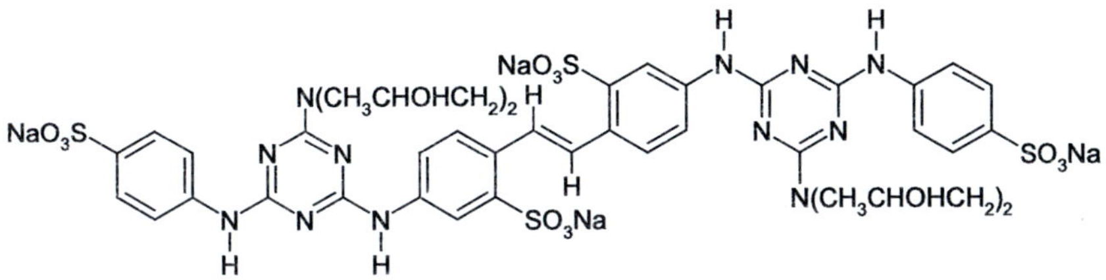

Fig. 1 Uvitex RSB, Triazine derivative of stilbene disulfonic acid (λ_{max} = 435–440 nm)

and Yellowing Index (YI) according to DIN 6167:1980 were calculated automatically. Tint Deviation (TD) and its coloristic meanings were determined according to Griesser [19].

UV-A and UV-B transmissions were measured on a transmission spectrophotometer Cary 50/Solascreen (Varian). The results were calculated and evaluated according to AS/NZS 4399:2017.

The porosity of woven fabrics is defined as the ratio of total empty area of fabric to total volume of woven fabric which contains textile material and air [20]. It was calculated based on the fabric cover factor after analysis of microscopic images using DinoLite microscope and the fabric thickness was performed on a thickness gauge.

3 Results and Discussion

In this paper, the UV protection of cationized cotton/polyester blended fabrics was researched. For that purpose, the cationization of cotton/polyester blended fabrics, woven in basket (B2/2) and 5-end satin (5eS), was performed using cationic agent Rewin OS. The cationization effects were studied through the interface phenomena, i.e. the changes in zeta potential and adsorption of anionic optical brightener. In order to compare cotton/polyester fabrics before and after cationization the yarn density and fabric thickness were determined and electrokinetic analysis was performed. The results are collected in Table 1. In Fig. 2 the zeta potential vs. pH of electrolyte 0.001 mol/L KCl is presented.

From the results it can be seen that density did not change significantly, but thickness did. The reason is in fiber swelling in alkaline medium. In neutral and alkaline medium both cotton/polyester fabrics, regardless of weave, have negative (<-42 mV) zeta potential due to dissociation of negative surface groups. Cotton cellulose has hydroxyl (-OH) groups, but also carboxyl that revels in chemical bleaching (-COOH). According to the literature [14–18] zeta potential of cotton is around -25 mV.

Polyester addition in blend lowers zeta potential because of small number of carboxyl groups, as well as its hydrophobic character because of which it does not have the ability of swelling and water adsorption (hydration) [18]. Therefore, in

Table 1 Yarn density for warp and weft, fabrics thickness, zeta potential (ZP) at pH 9 and Isoelectric point (IEP) of cotton/polyester blended fabrics: woven in basket 2/2 (B2/2) and 5-end satin (5eS) before and after cationization (C)

Sample	Warp density [No/cm]	Weft density [No/cm]	Thickness [mm]	ZP at pH 9 [mV]	IEP
B2/2	42.0	17.0	0.408	−42.3	2.5
5eS	43.0	20.0	0.456	−44.5	2.4
B2/2-C	41.0	18.0	0.522	21.9	–
5eS-C	40.0	19.0	0.592	24.0	–

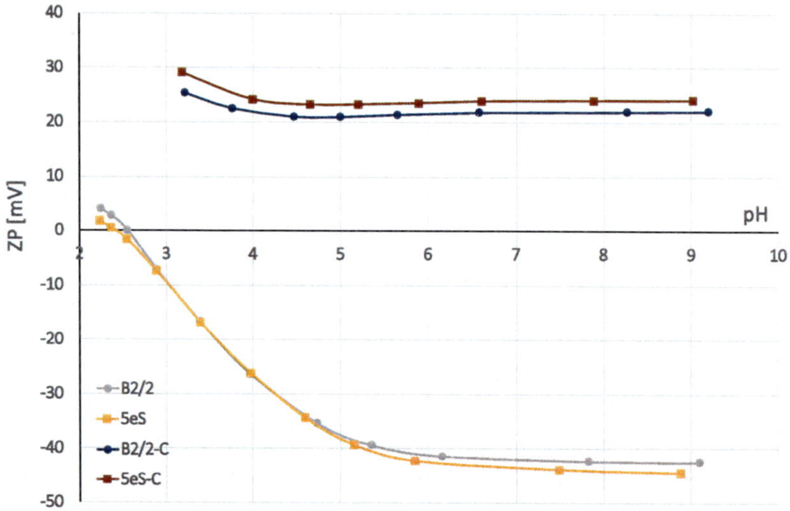

Fig. 2 The zeta potential (ZP) vs. pH of 0.001 mol/L KCl of cotton/polyester blended fabrics: woven in basket 2/2 (B2/2) and 5-end satin (5eS) before and after cationization (C)

whole pH range from 3 to 9, ZP is negative, i.e. −44 mV at pH 9 and −9 mV at pH 3. The Isoelectric point can be observed at pH lower than 2.5. Cationization with Rewin OS changes zeta potential. Ammonium groups (-NH_2) from cationic compound are present on fiber surface, resulting in positive zeta potential > 22 mV in whole pH range. Therefore, there is no IEP. There is the difference in ZP between fabrics regarding the weave. 5eS has lower number of interlacements than basket 2/2, therefore it can swell more. For that reason, cationic compound can be adsorbed more to fiber surface, and the resulting zeta potential is higher.

The fabric spectral characteristics presented by the degree of whiteness according to CIE (W_{CIE}), yellowing index (YI), maximum remission (R_{max} [%]), wavelength at which it was achieved (λ_{max} [nm]), and Tint Deviation (TD) with its coloristic meanings are shown in Tables 2 and 3. From the results it can be seen that both cotton/polyester fabrics have remission at 700 nm, and whiteness degree of 78. No appreciable deviation in tint from the white scale can be observed. Uvitex RSB, as triazine derivative of stilbene disulfonic acid has medium affinity for cotton, but some amount can be adsorbed to polyester component as well. Treatment with the lowest concentration of optical brightener leads to the significantly higher whiteness (W_{CIE} > 120) due to fluorescence emission at 440 nm. Remission from fabrics treated with Uvitex RSB corresponds to a red–purple tone and shifts to blue at higher concentrations, or to neutral at 5%. It can be seen that the highest spectral remission as well as whiteness were achieved at the concentration of 5% owf.

When using a higher concentration of optical brightener than 10% owf, the spectral remission is slightly lower. Fabrics treated in concentrations lower than 5% owf are

Table 2 Degree of whiteness (W_{CIE}), yellowing index (YI), tint deviation (TD), maximum remission (R_{max} [%]) and wavelength (λ_{max} [nm]) before and after modification and after optical brightening of cotton/polyester bled fabric in basket weave

Fabric	W_{CIE}	YI	TD	λ_{max} [nm]	R_{max} [%]
B2/2	78.4	1.43		700	83.70
B2/2-C	76.9	2.07		700	83.90
B2/2-RSB-1	120.5	−14.63	R1 – trace	440	103.72
B2/2-RSB-2	127.7	−17.53	R1 – trace	440	108.97
B2/2-RSB-5	135.8	−21.27		440	115.45
B2/2-RSB-10	134.7	−20.79		440	114.73
B2/2-C-RSB-1	120.2	−15.05	G2 – slightly	440	108.16
B2/2-C-RSB-2	119.6	−14.85	G2 – slightly	440	108.38
B2/2-C-RSB-5	110.9	-11.97	G4 – markedly	440	106.13
B2/2-C-RSB-10	107.1	-10.72	G4 – markedly	440	104.61

Table 3 Degree of whiteness (W_{CIE}), yellowing index (YI), tint deviation (TD), maximum remission (R_{max} [%]) and wavelength (λ_{max} [nm]) before and after modification and after optical brightening of cotton/polyester bled fabric in 5-end satin weave

Fabric	W_{CIE}	YI	TD	λ_{max} [nm]	R_{max} [%]
5eS	78.2	1.32		700	84.96
5eS-C	76.7	2.32		700	84.31
5eS-RSB-1	123.1	−15.56	R1 - trace	440	105.67
5eS-RSB-2	129.5	−18.07	R1 - trace	440	110.68
5eS-RSB-5	136.3	−21.36		440	116.14
5eS-RSB-10	134.9	−20.76	G1 - trace	440	116.03
5eS-C-RSB-1	120.1	−15.04	G2 - slightly	440	107.16
5eS-C-RSB-2	117.9	−14.20	G2 - slightly	440	108.38
5eS-C-RSB-5	110.3	−11.69	G4 - markedly	440	106.38
5eS-C-RSB-10	104.7	−9.70	G4 - markedly	440	104.09

red in traces (R1) than the white standard, while at other concentrations they show true neutral whiteness.

The cationization reduces the whiteness degree, and some yellowing occurs. However, cationization changes the surface charge, and the adsorption of optical brighteners is significantly higher. Analyzing the whiteness degree, it can be seen that for cationized optically brightened fabrics whiteness is lower because the quenching of fluorescence phenomenon occurs even when the lowest concentration was applied. High concentrations leads to the layering of Uvitex RSB molecules what prevents excitation of lower layers, resulting in concentration quenching of fluorescence and subsequently to the reduction of whiteness. Additionally, the molecules at high concentrations build dimers and do not have the ability to fluoresce any more [1, 6, 13, 21]. Both fabrics, regardless of weave, show similar quenching phenomenon. This was confirmed by the tint deviations, i.e. G2 "slightly greener than white standard" for 1% owf and for the highest concentration of 10% owf it is G4 "markedly greener than white standard".

The UV protection was determined according to AS/NZS 4399:2017. The results of measured UV-A and UV-B transmission and the mean value of the ultraviolet protection factor (UPF) are presented in Table 4. From the results it can be seen that all fabrics made of cotton/polyester blend provide excellent protection. The fabrics are extremely dense, and the polyester component has the ability to absorb radiation. Although they all provide excellent UV protection, differences between mean UPF values are visible.

The porosity of woven fabric is influenced by its constructional and structural characteristics, yarn, fiber and finishing processes, so it affects some properties as the permeability of air, water vapor, water and light, which is directly related to UV protection. Fabrics in the satin weave have a large number of warp and weft

Table 4 Mean UPF, UV-A and UV-B transmission, and porosity of cotton/polyester blended fabrics: woven in basket 2/2 (B2/2) and 5-end satin (5eS) before and after optical brightening

	B2/2				5eS			
Fabric	Mean UPF	τ_{UVA}	τ_{UVB}	Porosity [%]	Mean UPF	τ_{UVA}	τ_{UVB}	Porosity [%]
	171.65	0.13	11.1	1.93	252.07	0.1	11.1	0.19
.-C	1000	0.1	0.1	0.56	1000	0.1	0.1	0.06
.-RSB-1	1000	0.1	0.1	0.11	1000	0.1	0.1	0.08
.-RSB-2	1000	0.1	0.1	0.44	1000	0.1	0.1	0.06
.-RSB-5	1000	0.1	0.1	0.25	1000	0.1	0.1	0.13
.-RSB-10	1000	0.1	0.1	0.71	1000	0.1	0.1	0.61
.-C-RSB-1	1000	0.1	0.1	0.10	1000	0.1	0.1	0.16
.-C-RSB-2	1000	0.1	0.1	0.39	1000	0.1	0.1	0.06
.-C-RSB-5	1000	0.1	0.1	0.58	1000	0.1	0.1	0.54
.-C-RSB-10	1000	0.1	0.1	0.52	1000	0.1	0.1	0.09

floats, which results in a higher cover factor of the fabric, and is directly related to the lower porosity compared to basket weave fabric. Analyzing the results of porosity, a correlation between the proportion of vertical pores and UPF protection can be observed. The initial fabric in the satin weave has 10 times less proportion of vertical pores compared to the basket woven fabric and without any finishing treatment provides 32% better UV protection. In both fabrics, optical brightening leads to decrease in the proportion of vertical pores (basket weave - 71% and satin weave - 68%). The reason for this difference is fabric weave. In fabrics in which two adjacent wefts and/or warp floats equally, as in the case of fabrics in rib and basket weaves, the threads tend to come closer to each other, changing the shape of the cross section. According to Hamilton's theory their common diameter is considered as one and counted like a square root of adjacent number of threads [22]. It was proved that not only that yarns come close together, but they also enter under each other [23] increasing the distances between equal pairs of threads, and thus the proportion of vertical pores. That is why the fabric in basket 2/2 weave has a higher porosity compared to 5-end satin and, therefore has a lower UPF (171.65). In the finishing process (optical brightening and cationization), the proportion of vertical pores is initially significantly reduced due to shrinkage of the fabric. Therefore, the lowest concentration achieves the highest possible value UPF = 1000. It should be noted that the maximum UV protection does not decrease with the occurrence of fluorescence quenching.

4 Conclusion

In this paper, the UV protection of cationized cotton/polyester blended fabrics of different weave (basket and 5-end satin) was researched. The effect of cationization with a cationic agent, Rewin OS, was researched after modification and after optical brightening with stilbene derivative Uvitex RSB, in a wide concentration range. It has been shown that for unmodified fabric 5% owf is necessary for the best whiteness, whilst for cationized ones even 1% is to high concentration when quenching of fluorescence already occurs. All the fabrics showed excellent UV protection due to high density and polyester in the blend. The difference in porosity of fabric in basket 2/2 and 5-end satin weave showed difference in UV protection. However, fabric shrinkage and yarn movement and overlapping during finishing processes led to the maximum possible UV protection (UPF = 1000), which does not decrease due to quenching.

Acknowledgements The work has been supported by Croatian Science Foundation under the project UIP-2017-05-8780 **HPROTEX**.

References

1. Tarbuk, A., Grancarić, A.M., Šitum, M.: Skin cancer and UV Protection. AUTEX Res. J. **16**(1), 19–28 (2016)
2. Tarbuk, A., Grancarić, A.M., Šitum, M., Martinis, M.: UV Clothing and Skin Cancer. Coll. Antrop. **34**(S2), 179–183 (2010)
3. Reinert, G., Fuso, F., Hilfiker, R., Schmidt, E.: UV-protecting properties of textile fabrics and their improvement. Text Chem. Color **29**(12), 36–43 (1997)
4. Saravanan, D.: UV protection textile materials. AUTEX Res. J. **7**(1), 53–62 (2007)
5. Grancarić, A.M., Penava, Ž, Tarbuk, A.: UV Protection of cotton – the influence of weawing structure. Hemijska industrija **59**(9–10), 230–234 (2005)
6. Tarbuk, A., Grancarić, A.M., Šitum, M.: Discrepancy of whiteness and UV protection in wet state. Coll. Antrop. **38**(4), 1099–1105 (2014)
7. Cox Crews, P., Katchman, S., Beyer, A.: Influences on UVR transmission of undyed woven fabrics. Text Chem. Color **31**, 17–26 (1999)
8. Tarbuk, A., Grancarić, A.M., Jančijev, I., Sharma, S.: Zaštita od ultraljubičastog zračenja površinski modificiranom poliesterskom tkaninom. Tekstil **55**(8), 383–394 (2006)
9. Zhou, Y., Cox Crews, P.: Effect of OBAs and repeated launderings on UVR transmission through fabrics. Text Chem. Color **30**(11), 19–24 (1998)
10. Botteri, L., Dekanić, T., Tarbuk, A., Đorđević, D.: The influence of FWA chemical constitution to the whiteness and UV protection of cotton and cotton/polyester blend. Adv. Technol. **10**(2), 66–72 (2021)
11. Dekanić, T., Pušić, T., Soljačić, I., Vojnović, B., Volmajer Valh, J.: The influence of iron ions on optical brighteners and their application to cotton fabrics. Materials **14**(4995), 17 (2021)
12. Dekanić, T., Soljačić, I., Pušić, T.: Impact of artificial light on optical and protective effects of cotton after washing with detergent containing fluorescent compounds. Tenside, Surfactants, Deterg. **51**(5), 451–459 (2014)
13. Dekanić T., Tarbuk, A., Pušić, T., Grancarić, A.M., Soljačić, I.: Light Conversion for UV Protection by Textile Finishing and Care. In: Sunscreens: Properties, Role in Skin Cancer Prevention and Health Effects (Ed. S. H. Sharp). New York: Nova Science Publishers, pp. 143–172 (2015)
14. Grancarić, A.M., Tarbuk, A., Dekanić, T.: Elektropozitivan pamuk. Tekstil **53**(2), 47–51 (2004)
15. Tarbuk, A., Grancarić, A.M., Leskovac, M.: Novel cotton cellulose by cationisation during mercerisation - Part 2: Interface phenomena. Cellulose **21**(3), 2089–2099 (2014)
16. Marković, J., Vunderl, M., Grgić, K., Tarbuk, A.: Kationiziranje otpadne pamučne tkanine – utjecaj na mehanička svojstva. Tekstil **67**(1–2), 27–33 (2018)
17. Čorak, I., Brlek, I., Sutlović, A., Tarbuk, A.: Natural dyeing of modified cotton fabric with cochineal dye. Molecules **27**(1100), 13 (2022)
18. Grancarić, A.M., Tarbuk, A., Pušić, T.: Electrokinetic properties of textile fabrics. Color Technol. **121**(4), 221–227 (2005)
19. Griesser, R.: Assessment of whiteness and tint of fluorescent substrates with good inter-instrument correlation. Color Res. Appl. **19**(6), 446–460 (1994)
20. Ragab, A., Fouda, A., El-Deeb, H., Abou-Taleb, H.: Determination of pore size, porosity and pore size distribution of woven structures by image analysis techniques. J. Text. Sci. Eng. **7**(5), 1–9 (2017)
21. Soljačić, I.: O optičkim bjelilima. Tekstil **21**(5), 377–398 (1972)
22. Hamilton, J.B.: 7—A general system of woven-fabric geometry. J. Text. Inst. Trans. **55**(1), 66–82 (1964)
23. Malčiauskiene, E., Milašius, A., Laureckiene, G., Milašius, R.: Influence of weave into slippage of yarns in woven fabric. Medziagotyra **17**(1), 47–51 (2011)

Analysis of the Possible Impact of Cabin Distractions on the Performance of Young Female Road Vehicle Drivers

Sandro Tokić, Davor Sumpor, Mirta Zelenika Zeba, and Srećko Đuranović

Abstract In this second preliminary study, the possible impact of cabin distractions on the performance of young female road vehicle drivers was analyzed. Six female drivers under the age of 30 were filmed with GoPro cameras while driving. The potential negative impact on the performance of female drivers was analyzed using the method of analyzing the glance behavior of female drivers. Female drivers performed two driving segments: baseline and driving under the usual cabin distractions while keeping the road sections and external conditions on the road consistent for each driving segment and each type of cabin distraction. The driving segment under the cabin distractions included smoking while driving, consuming soft drinks from a PVC bottle, and using a mobile phone. A mobile phone was fixed on a rack under the central rear-view mirror and used for two types of cabin distractions: talking using a video link and solving the cognitive task set through a video link conversation. The performance of drivers exposed to cabin distractions was compared with the performance in baseline driving mode, both using the most important measures of the method of analyzing the gaze parameters of female drivers: the percentage of gaze in individual zones, the duration of each gaze by zones, the number of transitions between zones, and the number of zones in which the drivers looked. The results of this preliminary study will be used to design a final larger research blueprint on a larger sample of drivers with adequate gender - age representation.

Keywords cabin distractions · female subjects · performance · glance

S. Tokić (✉) · D. Sumpor · S. Đuranović
Faculty of Transport and Traffic Sciences, University of Zagreb, Vukelićeva 4, 10000 Zagreb, Croatia
e-mail: stokic@fpz.hr

M. Z. Zeba
Polyclinic for Rehabilitation of Listening and Speech SUVAG, Kneza Ljudevita Posavskog 10, Zagreb, Croatia

Faculty of Electrical Engineering and Computing, University of Zagreb, Zagreb, Croatia

I. Salopek Čubrić et al. (eds.), *Proceedings of the 9th International Ergonomics Conference*, Lecture Notes in Networks and Systems 701,
https://doi.org/10.1007/978-3-031-33986-8_35

1 Introduction

Research into distraction's impact on road vehicle drivers' performance still needs to be explored to fully understand its impact. Due to advancements in technology, drivers now engage in secondary distraction activities that can impair their performance while driving more than ever before [1]. Distracted driving can generally be detected by tracking head and eye movements [2]. While driving a motor vehicle, drivers actively search the traffic environment, noticing stimuli from which they extract the information they need for safe driving of the vehicle. Easiest way to detect the direction of a driver's gaze is to use a camera aimed at the face and annotate the gazes in the zones as they appear in the video. This method is cheap, fast, and relatively reliable, and has been used in some scientific studies such as the one by Birrell et. al [3].

This research is a continuation of the research whose partial results were published in the 2020 preliminary study [4], so the research design of this research described in chapter 2 will be identical and taken from a previous preliminary study. The research was carried out at the scientific university campus Borongaj as part of the support of the University of Zagreb. While driving the vehicle the subjects were recorded. Two main goals are to try to bypass one of the system main drawbacks, which is sensitivity to external condition lighting [5]. The second one is testing and improving the research design.

In this research, the so-called baseline driving means everyday driving that contains all the drivers' usual activities, without secondary distraction tasks. The average values and limit ranges of the amount of driver glance parameters for baseline driving must be determined before evaluating the factors of cabin distractions. Eye movement measures such as glance duration, glance frequency, percent of gaze on road center, and others, will be analyzed from the video data on the sample of six female drivers.

2 Research Draft

The driver's activities specified by the scenario to cause cabin's distraction are as follows:

A) Setting up navigation on a cell phone: location and driving route setting by using Google maps application (n = 1, average duration 45 s). **B3) Mobile phone conversation, video call**: talk to assistant by using Viber application (n = 2, duration in the range 55 - 109 s). **D1) Cognitive task through mobile phone conversation, video call:** (n = 2, duration in the range 102 - 114 s). Two-back task implies that the assistant reads a series of letters one by one every 3 s, and the participant needs to say "same" when he hears the same letter that is identical to the letter two positions before. **E) Complex task** (n = 1, average duration 101 s), and to serve multiple commands on the center console: Adjusting radio (turn on the radio, change few radio stations, adjust the sound volume), adjusting air conditioner (turn on air conditioning, set up temperature, turn on the ventilator for the specific zone) and setting other commands (switching internal air circulation on/off). **F) Smoking** (n = 1, average duration 199 s): take a pack of cigarettes from the glove compartment, then take one cigarette out of the box, return the pack of cigarettes to the glove compartment, take a lighter and pretend you light a cigarette, put away the lighter, hold the cigarette with the dominant hand, for the end put it in the ashtray. **G) Drinking** (n = 1, average duration 167 s): with the dominant hand, the driver must retrieve the PVC bottle from the seat behind him or the stand next to him, open it, put the cap away, take a few sips of the drink, close the bottle, and return the PVC bottle to the same place where he took it (repeated 4 to 6 times, until the end of the section in the scenario is reached).

In addition to the driver, three other persons participate in the research. The first one is a passenger next to the driver that gives participant instructions according to the scenario and turns on/off the cameras and the Endomondo platform. The second passenger in the back seat monitors the scenario implementation and gives a signal to the assistant. The assistant outside the car performs activities B3 and D1, which involve the use of video connection in the Viber application on a mobile phone.

The driver's glance behavior was analyzed based on eleven possible driver view zones: left side of the windshield, right side of the windshield, left side mirror, rear-view mirror, right side mirror, instrument panel, navigation (central part of the windshield), center console, left window, right window, others (beneath the steering wheel).

It is important to note that for driver activities A, B3 and D1 the cell phone is placed on the central part of the windshield, on the stand on the left and under the rearview mirror, within the "Navigation" zone, for the following reasons: to minimize the impact of manual distraction due to holding the mobile phone with the dominant hand during the above activities, also because it is now a common situation in real traffic (drivers use mobile phones over a wireless connection with an earpiece, using headphones without accessories fixed on a stand), and because legal regulations in the Republic of Croatia require drivers to keep both hands on to the steering wheel while driving. Also, research [6] confirms that the effect of using a mobile phone while driving on the performance of the driver is the same regardless of whether the driver holds the mobile phone with his hand during the conversation or not.

The 1120-m-long route in ZUK Borongaj is divided into seven sections according to Table 1 and Fig. 1, of which only 13.5% is the length of the change of direction (four sharp turns to the left).

On the route in ZUK Borongaj shown in Fig. 1, there is interaction with public transport buses, teachers' and students' cars, and pedestrians. The bus stops are located on Sects. 3 and 7, with pedestrian crossings just before the stop. At the four pedestrian crossings marked in Fig. 1, students simulated pedestrians in the presence of pedestrians, if necessary. The average speed of the vehicle in which the camera filmed the driver was 25 km/h, according to the data from the Endomondo platform.

The matrix of the driver's activities imposed by the scenario by sections and laps is shown in Table 2. The drivers drove seven laps according to the scenario shown

Table 1 Separation of the route at ZUK Borongaj into seven sections

Section label	Distance L_s(m)		Length L (m)	Short name	Full name of the section
	Initial	Final			
1	000	250	250	Flat part west	Fully flat one-way section
2	250	300	50	Sharp left	Shart left turn + intersection
3	300	580	280	Flat part south	Fully flat two-way and one-way section
4	580	630	50	Sharp left	Shart left turn
5	630	830	200	Flat part east	Fully flat one-way section
6	820	870	50	Shart left	Sharp left turn + intersection
7	870	1120	250	Flat part north	Fully flat two-way section

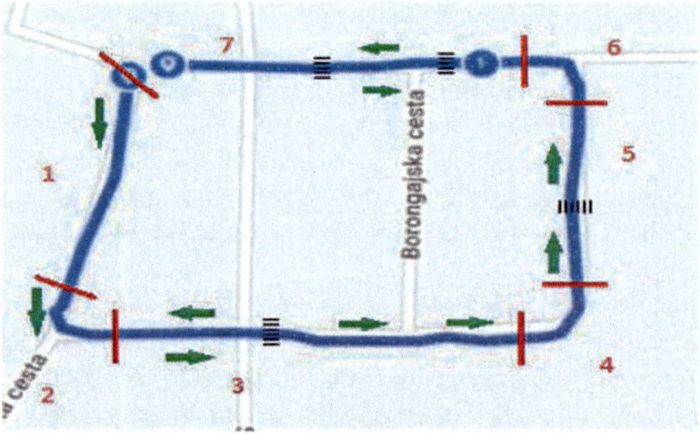

Fig. 1 Separation of the route in ZUK Borongaj into seven sections

in Table 2, including lap number zero, which represent the baseline driving without additional driver activities imposed by the scenario.

The drivers are told that they must not stop for no reason, that they must comply with traffic rules and regulations, drive at a speed at which they feel comfortable and safe, but in a way so that they can perform tasks.

The empty fields in Table 2 imply the baseline driving without any driver's activities imposed by the scenario. Route sections with a change of direction are not excluded from the analysis of the results using the driver's glance behavior method because they take over 13.5% length of the route and because all activities of the driver imposed in the scenario from Table 2 also contain a change of direction due to driving through bends, except for activity setting the navigation on the mobile phone.

Drivers were recorded by GOPRO HERO8 Black cameras, placed on the central part of the windshield, on the stand on the left and under the rearview mirror.

Table 2 Matrix of driver's activities imposed by the scenario by sections and laps

Lap number	Label and full name of the section						
	Flat part west	Sharp left	Flat part south	Sharp left	Flat part east	Shart left	Flat part north
	1	2	3	4	5	6	7
0							
1	A		D1	D1	D1	D1	D1
2	B3	B3	B3		E	E	E
3	F	F	F	F	F	F	F
4			D1	D1	D1	D1	D1
5	B3	B3	B3		E	E	E
6			G	G	G	G	G

Table 3 Percentage of gaze length shares and number of zones used depending on the driver's activities (female drivers 1–6)

		% of total gaze length*						
		Driver 1	Driver 2	Driver 3	Driver 4	Driver 5	Driver 6	All drivers Mn / min - max.
	Baseline driving	82.4	77.8	76.2	77.4	76	74.3	77.4 / 74.3 – 82.4
F	Smoking	76.8	74.3	74	76.5	72.4	67.9	73.7 /67.9 – 76.8
G	Drinking	84.3	75.5	78.9	78.2	75.2	73	77.5 /73 – 84.3
D1	Cognitive task through mobile phone conversation (video call)	89.5	87.4	91.1	87.7	73.6	78.4	84.6 /73.6 – 91.1
A	Setting up navigation on a cell phone	44.7	45	66.6	51.5	70.8	30.3	51.5 / 30.3 - 70.8
E	Complex task	69.1	52.7	62.5	69	59.8	49.2	60.4 / 49.2 – 69.1
B3	Mobile phone conversation (video call)	65.2	50.4	58.3	69.6	62	35.2	56.8 / 35.2 – 69.6

Note: % of total gaze length in the main zone "Left side of the windshield"

3 Results

In this paper, four parameters were analyzed. Number of glances in a zone, percentage share of gaze length in the main zone "Left side of the windshield", Mean (M) value of gaze length as well as minimum (min) and maximum (max) values of the same parameter. The results of the previous research justify the continuation of the research on a larger sample consisting only of female respondents. Tables 3, 4 and 5 show results that have a high level of overlap with previous results.

Driver's activities F and G causes the same behavior patterns. All analyzed drivers make hand movements at the beginning and end of the activity, to reach for a bottle or cigarettes and at the same time follow the hand movements with their eyes. After taking the mentioned object, they direct their gaze to the main zone and continue driving without any major difficulties. According to the results in the table, all analyzed parameters are very similar to the results in the baseline driving so we can conclude that based on the direction of the gaze, it is very difficult to detect such behavior. To detect such behavior, it is necessary to observe the movements of hands and objects in the vehicle cabin. Driver's activity D1 affects, in addition to increasing the percentage of glances in the main zone and reducing the number of used zones. An increase in average gaze duration was also observed in the main zone, except for Driver 2. These results are in line with the results of other studies that during

Table 4 Number of zones used depending on the driver's activities (female drivers 1–6)

		Number of zones used						
		Driver 1	Driver 2	Driver 3	Driver 4	Driver 5	Driver 6	All drivers Mn / min - max.
	Baseline driving	9	9	10	9	10	10	9.5 / 9 – 10
F	Smoking	9	10	11	10	10	9	9.8 / 9 – 11
G	Drinking	8	10	9	9	10	10	9.3 / 8 – 10
D1	Cognitive task through mobile phone conversation (video call)	5	8	8	7	9	7	7.3 / 5 – 9
A	Setting up navigation on a cell phone	3	4	4	3	5	6	4.2 / 3 – 6
E	Complex task	8	10	9	9	10	10	9.3 / 8 – 10
B3	Mobile phone conversation (video call)	8	9	9	9	6	8	8.2 / 6 - 9

Table 5 Average, min. and max. glance duration (s) depending on the driver's activities

		Glance duration (s)						
		Driver 1 M / min.- max.	Driver 2 M / min.- max.	Driver 3 M / min.- max.	Driver 4 M / min. – max.	Driver 5 M / min. – max.	Driver 6 M / min. - max	All Drivers M_M / min. - max
	Baseline driving	3.14 / 0.25 - 9.75	2.67 / 0.22 - 9.67	3.21 / 0.18 - 9.85	3.69 / 0.17 - 9.99	2.85 / 0.21 - 9.75	2.34 / 0.19 - 9.99	2.98 / 0.17 – 9.99
F	Smoking	3.32 / 0.25 - 9.75	2.61 / 0.22 - 7.85	3.28 / 0.29 - 9.62	3.2 / 0.23 - 9.75	2.63 / 0.25 - 9.50	2.45 / 0.24 - 8.99	3.36 / 0.22 – 9.75
G	Drinking	5.10 / 0.25 - 8.99	2.81 / 0.25 - 6.75	3.79 / 0.32 - 9.33	3.67 / 0.25 - 9.75	2.54 / 0.25 - 9.57	2.94 / 0.24 - 9.75	3.48 / 0.24 – 6.75
D 1	Cognitive task through mobile phone conversation (video call)	5.66 / 0.25 - 9.25	0.70 / 0.25 - 8.85	5.93 / 0.38 - 9.23	6.25 / 0.25 - 8.75	3.1 / 0.25 - 9.5	5.47 / 0.49 - 8	4.52 / 0.25 – 9.5
A	Setting up navigation on a cell phone	1.25 / 0.25 - 3.99	0.99 / 0.25 - 5.15	1.69 / 0.15 - 7.62	2.13 / 0.13 - 3.25	1.37 / 0.25 - 7.99	0.62 / 0.25 - 1.5	1.34 / 0.13 – 7.99
E	Complex task	1.83 / 0.25 - 9.25	1.68 / 0.15 - 6.75	2.18 / 0.25 - 9.46	2.02 / 0.25 - 9.75	1.75 / 0.25 - 7.49	1.67 / 0.25 - 8	1.86 / 0.15 – 9.75
B 3	Mobile phone conversation (video call)	1.79 / 0.25 - 7.99	1.03 / 0.15 - 5.17	1.91 / 0.27 - 7.62	2.3 / 0.25 - 8.5	1.78 / 0.25 - 9.99	1.33 / 0.25 - 6.5	1.69 / 0.15 – 9.99

Note: Glance duration in the main zone "Left side of the windshield"

execution of cognitive tasks, car drivers' gaze behavior changes so that they look more towards the road in front of them [7.]. Driver's activity A in relation to other secondary distraction tasks drastically reduces the percentage share of gaze length in the main zone with a two-fold reduction in the number of used zones compared to baseline driving. Based on this, the conclusion is that out of all the tasks, this one was the most difficult for drivers to perform. These results are consistent with recent results on the impact of cell phone use during vehicle automation [8]. Drivers in activities E and B3 behave very similarly. A decrease in the percentage share of gaze length in the main zone as well as the average duration of gaze length is evident.

The 3rd level of automation is the most dangerous, because during automation the driver does not control the vehicle and collect information from the traffic environment but still it must be ready to take over commands when the system request [9].

4 Conclusion

Repeated research on different sample shows that by analyzing the directionality of the driver's gaze, it is possible to predominantly detect secondary distraction tasks that have elements of visual distraction such as A, E and B3. In contrast to a task like D1 which is difficult to recognize because drivers look predominantly in the main zone. For cognitive distraction tasks, gaze direction is not a good indicator of driver distraction. F and G have elements of visual and manual distraction, so for their detection measures of gaze direction can be partially used. In this research, it was also shown that for female respondents the biggest challenges are the task A. The female respondents had difficulties when performing the task in such a way that they could not always complete the navigation setup and often typed in wrong location, which made it even more difficult for them. Not infrequently, compensatory measures such as speed reduction mitigate the influence of the secondary distraction task. Regarding the parameters observed during this activity, there was a significant reduction in the number of zones used, as well as the average time spent looking at each zone. This research is one in a series of studies with the aim of in-depth research into the influence of manual, visual and cognitive distraction on the performance of road vehicle drivers. Many questions are still open and there is plenty of room for improving research. The following is a continuation of the research according to the same research design on older respondents of both sexes (age group from 30 to 60 years and age group over 60 years). Potential future research will focus on the influence of alcohol and distraction simultaneously on the performance of road vehicle drivers.

References

1. Ismaeel, R., Hibberd, D., Carsten, O., Jamson, S.: Do drivers self-regulate their engagement in secondary tasks at intersections? Exam. Based Naturalistic Driving Data, Accid. Anal. Prev. **137**, 105464 (2020). https://doi.org/10.1016/j.aap.2020.105464
2. Lee, J.D., et al.: Distraction Detection and Mitigation: Appendices, National Highway Traffic Safety Administration, Washington, DC (2013)
3. Birrell, S.A., Fowkes, M.: Glance behaviours when using an in-vehicle smart driving aid: A real-world, on-road driving study. Transp. Res. F: Traffic Psychol. Behav. **22**, 113–125 (2014). https://doi.org/10.1016/j.trf.2013.11.003
4. Bolčević, E., Tokić, S., Sumpor, D., Horina, J.L.: A preliminary study of the possible impact of non-related driving tasks on the car driver's performance using glance-based measures. In: Sumpor, D., Jambrošić, K., Jurčević Lulić, T., Milčić, D., Salopek Čubrić, I., Šabarić, I. (eds) Proceedings of the 8th International Ergonomics Conference, ERGONOMICS 2020. Advances in Intelligent Systems and Computing, vol. 1313, pp. 33–40. Springer, Cham (2021). https://doi.org/10.1007/978-3-030-66937-9_4
5. Vora, S., Rangesh, A., Trivedi, M.M.: Driver gaze zone estimation using convolutional neural networks: a general framework and ablative analysis. IEEE Trans. Intell. Veh. **3**(2), 254–265 (2018)

6. Caird, J.K., Simmons, S.M., Wiley, K., Johnston, K.A., Horrey, W.J.: Does talking on a cell phone, with a passenger, or dialing affect driving performance? an updated systematic review and meta-analysis of experimental studies. Hum. Factors **60**(1), 101–133 (2018)
7. Wang, Y., Reimer, B., Dobres, J., Mehler, B.: The sensitivity of different methodologies for characterizing drivers' gaze concentration under increased cognitive demand. Transp. Res. F: Traffic Psychol. Behav. **26**, 227–237 (2014). https://doi.org/10.1016/j.trf.2014.08.003
8. Noble, A.M., Miles, M., Perez, M.A., Guo, F., Klauer, S.G.: Evaluating driver eye glance behavior and secondary task engagement while using driving automation systems. Accid. Anal. Prev. **151**, 105959 (2021). https://doi.org/10.1016/j.aap.2020.105959
9. Guo, H., Zhang, Y., Cai, S., Chen, X.: Effects of level 3 automated vehicle drivers' fatigue on their take-over behaviour: a literature review. J. Adv. Transp. **2021**, 1–12 (2021). https://doi.org/10.1155/2021/8632685

Influence of Simulated Outdoor Wear on the Performance of Sportswear for a Professional Football Player

Sena Toktaş, Ivana Salopek Čubrić, and Antonija Petrov

Abstract The importance of sports and active sportswear has increased worldwide. Comfort, functionality, and attractive design are the main characteristics that every sportswear should fulfill. All these characteristics significantly affect the performance and efficiency of active athletes. Since undisturbed mobility is required in sports, clothing should be fully adapted to the athlete's body. Knitwear is most often used to make sportswear due to its good permeability to water vapor, elasticity, and stretchability. As a result, in recent years the demand for knitted fabrics has risen sharply. Among the listed properties, a very important property of sportswear is the resistance of the material to abrasion. Abrasion resistance of textiles is a complex phenomenon, and the properties of the fibers are affected by the yarns, the fabric structure, and the finishing processes. Abrasion resistance has a direct effect on the usable properties of the product. The direct effect of abrasion under the action of mechanical friction reduces the quality of the product itself, damages it, encourages pilling, fades, etc. This study aims to investigate how simulated outdoor wear, which dominantly refers to the abrasion process, affects the properties of a series of sportswear materials intended for use by a professional football player. The simulation of wear is performed using the AquAbrasion tester. In the study, the tensile properties of a selected set of materials were tested before and after the simulation on the AquAbrasion tester. In order to observe differences between different actions of the forces, both strip and ball burst methods are used. The results of the material elongation and acting forces are further discussed and brought into the context of material properties and final use by a professional athlete.

Keywords Athlete · Sportswear · Wear · Tensile Properties · Abrasion

S. Toktaş
Ege Üniversitesi, Mühendislik Fakültesi, İzmir, Turkey

I. Salopek Čubrić (✉) · A. Petrov
University of Zagreb Faculty of Textile Technology, Zagreb, Croatia
e-mail: ivana.salopek@ttf.unizg.hr

A. Petrov
e-mail: antonija.petrov@ttf.unizg.hr

I. Salopek Čubrić et al. (eds.), *Proceedings of the 9th International Ergonomics Conference*, Lecture Notes in Networks and Systems 701,
https://doi.org/10.1007/978-3-031-33986-8_36

1 Introduction

The demand for knitted materials has grown rapidly in recent years. Sportswear is an important production area in our daily life, so our quest for comfort does not and will not end. Many studies have been carried out to maximize comfort features and these studies will continue [1, 2]. The properties of the fabric, such as weight, thickness, stretchability, and abrasion resistance, are dependent on a number of factors like fiber properties [3, 4]. Natural fibers have good water absorbency and this seems like a good feature, but it is not preferable for the perspiration that occurs during sports activities to be absorbed by the fabric. It will give an unpleased cold wet feeling [5]. Most sportswear manufacturers use polyester, polyester blends, polyamide, and other synthetic fibers as they are considered the most suitable for the properties sought [6–9]. After the fibers turn into yarn, the structure they will form by bringing them together is also important [10–12]. In sportswear, it is desired that the garment fit perfectly on the wearer. In long-term and intense sports activities such as football, it is expected that the garment integrates with the body in the best condition, can stretch with the body, and breathe. These expectations are met with the knitted fabric's hollow structure due to its loop structure [13–15]. With the physical tests of textile products, we can measure the comfort properties of the fabric and the sustainability of these properties. Also, many tests can be done to measure the durability of the fabric. Different fabrics have different strengths and there are several types of the durability of fabrics. Strength types can be tested separately with suitable devices. For example, tearing durability, stretchability, and abrasion resistance [16–18]. Many factors affect the wear phenomenon, the formation of which is quite complex. As a result of abrasion, a decrease in fabric strength properties, weight, thickness, changes in color, and unpleasant pilling and even melting can be seen on the fabric surface [19]. Wear and abrasion resistance is the focus of numerous studies. These two terms should not be confused. Textile materials lose their usefulness as a result of abrasion. Wear resistance is a very general definition the term that covers almost every form of material loss. Abrasion resistance is usually used as a general term meaning the result of two surfaces rubbing against each other [20–23]. The abrasion resistance of textile products is important in many areas such as furniture and technical textiles as well as in clothing. In the abrasion tests, the fibers may be displaced out of the fiber as a result of friction, and together with it, a certain felting may occur in the product. Abrasion, which is the most important problem that occurs during use, is an undesirable feature that determines the durability of textile products. In addition, mechanically damaging the fabric also affects its properties such as colorfastness and even water repellency [24]. During the movement of the person, abrasion occurs as a result of fabric rubbing against each other or other surfaces, for example, the ground in certain parts of the fabric. As a result of wear, changes in certain mechanical properties of the fabric and losses in product mass occur. This makes the textile product less unusable. Since knitted fabrics can be produced more easily at a lower cost, they will find a wider area of use over time. Due to technological advances and increasing demands, the development of new testing techniques and equipment is likely to continue.

2 Experiment

The experimental is focused on the propensity to pilling and the influence of different forces of materials used for sportswear that were abraded before the investigation.

2.1 *Materials*

For the investigation are used three representative knitted fabrics. The fabrics are for the production of football sportswear and consist of 100% polyester yarn. The material properties are given in Table 1.

2.2 *Methods*

For this study, the Aquabrasion instrument (Fig. 1) was used. It is a wet abrasion tester that has a controlled pump system that allows the application of liquid on the surface of the material, thus replicating the condition of sweating or rain. The instrument can perform tests both in wet and dry conditions. Samples have been abraded under different pressures – 9 kPa and 12 kPa, as well as without pressure. The number of defined abrasion cycles for all the samples was 500.

Assessment of pilling propensity is given by visual comparison of the sample before abrasion using the AquAbrasion instrument and the sample after the defined number of cycles with corresponding pressure. Grades are ranked from 1 to 5 with the corresponding standard ISO 12945–2:2000 [25]. If the fabric has no surface changes, i.e. there is no surface fuzzing, the corresponding grade for piling would be 5. To measure the tensile properties of materials that have previously been exposed

Table 1 Knitted fabrics

Fabric ID	KF1	KF2	KF3
Surface mass, g/m^2	155	151	145
Yarn linear density, tex	12	10	10
Fabric thickness, mm	0.618	0.491	0.556

Fig. 1 AquAbrasion instrument

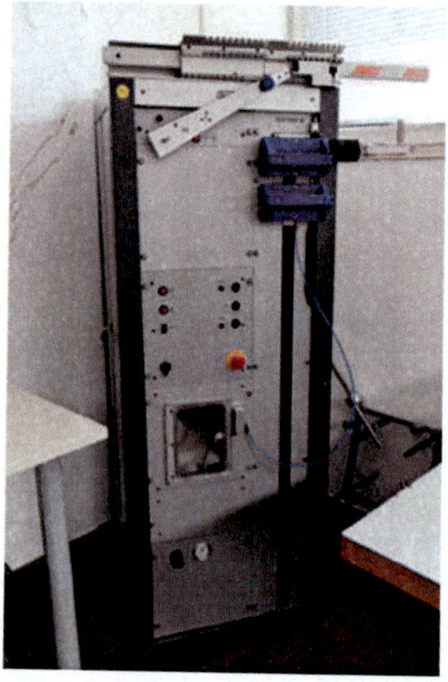

Fig. 2 Tensile tester - strip method

to abrasion on the Aquabrasion instrument, two different tensile testers have been used. The first one was a tensile tester produced by Textechno company (Fig. 2) - strip method. The second one was a tensile tester Apparechi Branca equipped with a ball bursting mechanism and all samples also have been measured by applying the burst method (Fig. 3).

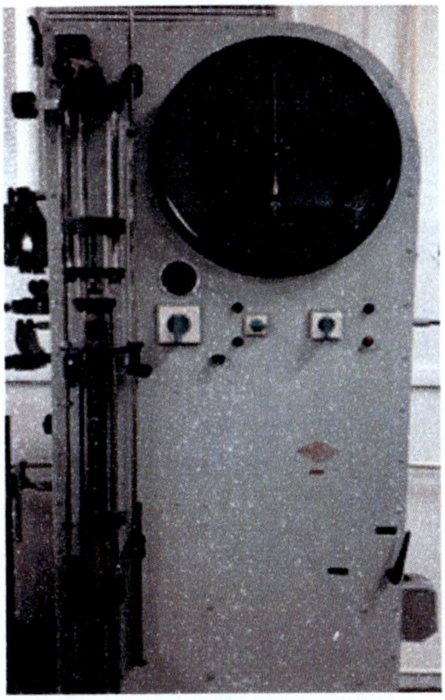

Fig. 3 Tensile tester - burst method

3 Results and Discussion

The results of the visual assessment of materials exposed to 500 cycles of abrasion are shown in Table 2. The results of materials testing on a tensile tester using a strip method, under different pressures (0, 9, and 12 kPa) are given in Table 3.

Table 2 shows the visual assessment of materials after exposure to 500 abrasion cycles. For samples KF2 and KF3, after 500 abrasion cycles without pressure and under a pressure of 9 kPa, there is no visual change on the surface of the material. Increasing the pressure to 12 kPa in both samples, moderate surface fuzzing is observed with rather low grades. Pills of varying density and size partially cover the sample surface. Sample KF1 shows very different changes. At pressures (0, 9, and 12 kPa), and after 500 abrasion cycles, the same visual assessment remains. It can be seen that slight surface fuzzing and partially formed pills occur. As far as the results for the specimen KF1 are concerned (Table 3), it could be seen that the elongation of materials exposed to the pressure of 9 kPa and 12 kPa is significantly higher than the one of material that was abraded without the pressure. The values are in the range of 79.84% - 102.51%. The coefficient of variation for the observed property is rather

Table 2 Visual assessment of materials (grades on the scale of 1–5)

Fabric ID	Without pressure	Under 9 kPa	Under 12 kPa
KF1	4	4	4
KF2	5	5	3
KF3	5	5	3

Legend: 3 - Moderate surface fuzzing and/or moderate pilling, 4 - Slight surface fuzzing and/or partially formed pills, 5 - No change

Table 3 Tensile properties of materials - strip method

Pressure	Property		AVG	S	CV	MIN	MAX
0 kPa	Elongation	**KF1**	79.84%	2.72	3.4	77.92	81.76
	Maximal Force		413.65 N	46.61	11.27	380.7	446.61
9 kPa	Elongation		112.80%	3.47	3.08	110.35	115.26
	Maximal Force		435.87 N	32.92	7.55	412.6	459.15
12 kPa	Elongation		102.51%	3.02	2.94	100.38	104.64
	Maximal Force		389.00 N	12.2	3.14	380.37	397.62
0 kPa	Elongation	**KF2**	67.62%	4.04	5.98	64.76	70.48
	Maximal Force		491.37 N	51.79	10.54	454.75	527.99
9 kPa	Elongation		82.80%	4.04	4.87	79.95	85.65
	Maximal Force		446.53 N	5.18	1.16	442.87	450.19
12 kPa	Elongation		86.64%	1.4	1.61	85.65	87.63
	Maximal Force		439.78 N	11.05	2.51	431.97	447.59
0 kPa	Elongation	**KF3**	86.83%	3.17	3.65	84.59	89.07
	Maximal Force		561.20 N	3.22	0.57	558.92	563.48
9 kPa	Elongation		83.92%	9.77	11.64	77.01	90.83
	Maximal Force		494.47 N	89.31	18.06	431.31	557.62
12 kPa	Elongation		79.55%	5.09	6.40	75.95	83.15
	Maximal Force		474.28 N	42.35	8.93	444.34	504.23

Legend: AVG – average value, S – standard deviation, CV – coefficient of variation, MIN – minimum value, MAX – maximum value

even (2.9–3.4%). The values of the maximum force applied to the specimens vary significantly – from 389 to 435 N, with relatively high variation within the results obtained for the specimen prepared for the measurement without applied pressure. The results obtained for the elongation of specimen KF2 differ even more considering the pressure applied, being the highest for the specimen exposed to 12 kPa. Contrary, the elongation of the material KF3 has a reverse proportional relationship with the pressure applied, i.e. the decrease of the elongation with the increase of the pressure is well seen (86.63% without pressure, 83.92% under 9 kPa, and 79.55% under 12 kPa). Regarding the maximum force for this specimen, the decrease of values due to applied pressure is also noticed. These results have shown that rather similar materials made of the same yarn (100% PES) still exhibit different behaviour due to abrasion and that the values of both appearance and tensile properties differ significantly.

In the test performed with the burst method, a noticeable difference is detected between the results (Table 4). A lower breaking force was applied to KF1 than other fabrics in all three different measurements. Considering the results of the bursting height (mm) of the fabrics, the values are rather uniform. Their averages are 32.6, 31.3, and 29.3, respectively (KF1-KF2-KF3). The breaking force differs significantly,

Table 4 Tensile properties of materials – bursting method

Fabric ID	Measurement	Breaking force, N	Bursting height, mm
KF1	AVG	352	32.6
	CV, %	19.35	1.77
KF2	AVG	761	31.3
	CV, %	3.24	3.68
KF3	AVG	617	29.3
	CV, %	2.62	3.94

with the 761 N for KF2, giving the highest result. There is also a high difference between the highest and lowest CV values. KF1 was the highest with 19.35%, while KF3 gave the lowest CV value with 2.62%. KF2, on the other hand, is extremely close to KF3, with a value of 3.24%.

The comparison of the breaking forces of materials before abrasion and after conducting abrasion for both used methods of tensile testing (strip method under 12 kPa and ball burst) are given in Fig. 4. There is a small difference in breaking forces before and after abrasion (strip method). For materials KF1 and KF3, the force slightly increases. Only for the material assigned as KF2, the decrease of force up to 5% for the abraded material is observed. The differences are more prominent when the results of the ball burst method are observed. After the abrasion process, an increase in breaking force is visible in all three tested samples. This can be explained by the fact that abrasion led to the formation of a more compact structure of the knitted fabric, whereby the protruding and separated fibers remained on the surface of the fabric and filled the voids within the structure.

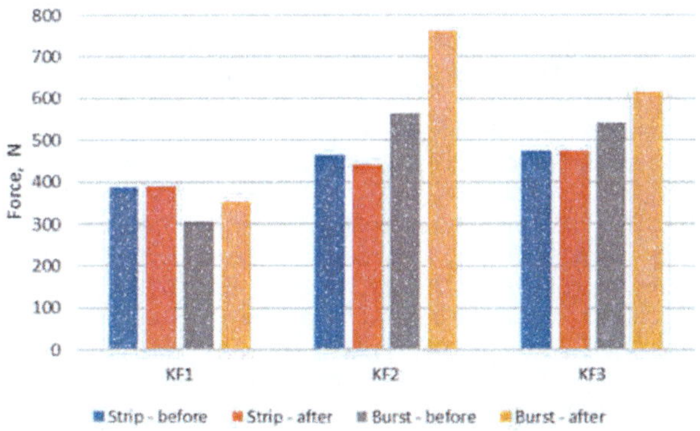

Fig. 4 Comparison of maximal forces before and after abrasion of materials

4 Conclusion

This research deals with the tendency of the material to pilling, and the action of different forces on the material itself after abrasion. The materials are intended for sportswear and as such should be extremely comfortable and functional. The performance properties of a product are significantly reduced after the abrasion process. The direct effect of abrasion under the action of mechanical friction reduces the quality of the product itself and leads to damage, peeling, fading, etc. According to the test results, it can be concluded that similar materials made of the same yarns show very different abrasion behaviour. Tensile properties and the appearance of the material surface themselves differ significantly. The increase in breaking force after abrasion can be explained by the formation of various protruding and detached fibres that have entered the structure of the knitted fabric and formed a much more compact structure. The results indicated that materials with lower surface mass demonstrate better resistance to piling (when measured without the pressure or under the pressure of 9 kPa), higher ability to elongate and higher resistance to the action of forces. Therefore, from the aspect observed, such materials are to be recommended for the use in the production of functional sportswear.

Acknowledgements HrZZ Croatian Science Foundation This work has been fully supported by Croatian Science Foundation under the project IP-2020-02-5041 „Textile Materials for Enhanced Comfort in Sports" – TEMPO.

References

1. Salopek Čubrić, I., Čubrić, G., Potočić Matković, V.M., Pavko, ČA.: The comfort of knitted fabrics: interaction of sportswear and athlete's body. Commun. Dev. Assembl. Text. Prod. **2**(1), 70–79 (2021)
2. Milun, K., Čubrić, G., Salopek Čubrić, I.: Evaluating the tactile comfort of sportswear fabrics. In: Ćorić, D., Šolić, S., Ivušić, F. (eds.) MATRIB 2021, HDTM - Hrvatsko Društvo za materijale i tribologiju, pp. 376–384 (2021)
3. Marmaralı, A.: Yuvarlak Örme Makineleri. Ege Üniversitesi Yayınları, İzmir (2008)
4. Bhat, Z.U.H., Sultana, D., Bashir, S.: Fibers for sportswear and influence of smart materials on sports performance. Res. Trends Phys. Educ. **3**, 49–64 (2019)
5. Özdil, N., Marmaralı, A., Kretzschmar, S.D.: Effect of yarn properties on thermal comfort of knitted fabrics. Int. J. Therm. Sci. **46**(12), 1318–1322 (2007)
6. Hansen, S., Atwood, K.B.: Polyester Fibers. Kirk-Othmer Encyclopedia of Chemical Technology, pp. 1–19 (2005)
7. Seventekin, N.: Kimyasal lifler, E.Ü. Tekstil ve Konfeksiyon Araştırma- Uygulama Merkezi Yayını, pp. 1–13 (2015)
8. Saleh, H.E.D.M.: Polyester. InTech, Rang-Du-Fliers (2012)
9. Jordeva, S., Ćortoševa, S., Mojsov, K., Žežova, S., Risteski, S., Kuzmanoska, V.D.: The influence of the structural characteristics of cotton and polyester knitted fabrics on thermo-physiological comfort. Adv. Technol. **6**(1), 88–92 (2017)

10. Ahmad, H.S., Jamshaid, H.: Development of thermo-physiologically comfortable knit structure for sports application. Tekstil Ve Konfeksiyon **29**(2), 105–112 (2019)
11. Kaloğlu, F., Önder, E., Özipek, B.: Influence of varying structural parameters on abrasion characteristics of 50/50 wool/polyester blended fabrics. Text. Res. J. **73**(11), 980–984 (2003)
12. Kothari, V.R.: Basics of Knitting. Views, pp. 20–22 (2010).
13. Hassan, M., Qashqary, K., Hassan, H.A., Shady, E., Alansary, M.: Influence of sportswear fabric properties on the health and performance of athletes. Fibres Text. Eastern Europe **20**(4), 82–88 (2012)
14. Manshahia, M., Das, A.: High active sportswear. Indian J. Fibre **39**, 441–449 (2012)
15. Susich, G.: Abrasion damage of textile fibers. Quartermaster Res. Dev. Lab. **24**(3), 210–228 (1954)
16. Özdil, N., Özçelik, G., Kayseri, Süpüren Mengüç, G.: Abrasion Resistance of Materials. Ege University, Textile Engineering Department, Izmir (2012)
17. Textor, T., Derksen, L., Bahners, T., Gutmann, J., Mayer-Gall, T.: Abrasion resistance of textiles: gaining insight into the damaging mechanisms of different test procedures. J. Eng. Fibers Fabr. **14**, 1–7 (2019)
18. Özgüney, A.T., Kretzschmar, S.D., Özçelik, G., Özerdem, A.: The comparison of cotton knitted fabric properties made of compact and conventional ring yarns before and after the printing process. Text. Res. J. **78**, 138–147 (2008)
19. Abrasion Resistance: Considerations for Textile Specifiers. Association for contract textiles (2011)
20. Manich, A.M., Castellar, M.D.D., Sauri, R.M., Miguel, R.A., Barella, A.: Abrasion kinetics of wool and blended fabric. Text. Res. J. **71**, 469–474 (2001)
21. Çelik, N., Değirmenci, Z., Kübra H., Kaynak.: Effect of nano-silicone softener on abrasion and pilling resistance and colorfastness of knitted fabrics. Tekstil Ve Konfeksiyon **20**(1), 41–47 (2009)
22. Gintis, D., Mead, E.J.: The mechanism of pilling. Text. Res. J. **29**(7), 578–586 (1959)
23. Okur A.: Textile and engineer, pp. 10–18 (1994)
24. Akter Smriti, A., Azharul Islam, M.D.: An exploration on pilling attitudes of cotton polyester blended single jersey knit fabric after mechanical singeing. Sci. Innovation **3**(1), 18–21 (2015)
25. ISO 12945–2:2000(en) Textiles — Determination of fabric propensity to surface fuzzing and to pilling—Part 2: Modified Martindale method

Virtual Reality Application in Maxillofacial and Plastic Surgery

Nikita Tulikov, Artem Smolin, Dmitry Shtennikov, Anna Lysenko, Elizaveta Ivanova, Igor Klimov, Andrei Mironov, and Aleksandr Volosiuk

Abstract The article describes a joint study of the Center for Usability and Mixed Reality (ITMO University) and the Department of Maxillofacial and Plastic Surgery (First Pavlov State Medical University) on applying Virtual Reality and Neurotechnology in medicine. The authors designed a VR application simulating the operating room at the Department of Maxillofacial and Plastic Surgery, to be used for the preoperative preparation of patients. The VR application was then tested in an experiment comparing the physiological parameters of two groups of patients - those who used the application and those who did not. The experiment results demonstrated the high efficiency of applying VR technologies in the preoperative preparation of patients.

Keywords virtual reality · medicine · 3D modeling · neurotechnology · surgery

---

N. Tulikov—Self-employed

N. Tulikov
Saint-Petersburg, Russia

A. Smolin · D. Shtennikov · I. Klimov · A. Mironov · A. Volosiuk (✉)
ITMO University, Saint-Petersburg, Russia
e-mail: aavolosiuk@gmail.com

A. Smolin
e-mail: smolin@itmo.ru

A. Lysenko · E. Ivanova
First Pavlov State Medical University, Saint-Petersburg, Russia

A. Volosiuk
St. Petersburg Electrotechnical University, Saint-Petersburg, Russia

I. Salopek Čubrić et al. (eds.), *Proceedings of the 9th International Ergonomics Conference*, Lecture Notes in Networks and Systems 701,
https://doi.org/10.1007/978-3-031-33986-8_37

1 Introduction

Healthcare professionals of various medical fields are increasingly using use Virtual Reality (VR). VR technologies are used in diagnostics, treatment processes, surgical interventions, as well as rehabilitation processes, and allow improving the efficiency of healthcare workers [1, 2].

One of the most important and, possibly, dominant areas of VR technologies application in medicine is education. Production of authentic educational simulations allows conducting full-fledged lessons with students of medical schools, as well as improving the qualification of already practicing medical staff. For example, Russian company Modum Lab in cooperation with Pirogov Russian National Research Medical University designed the Simulator for Ambulance Paramedics [3].

The Simulator for Ambulance Paramedics is an authentic simulator for the training of ambulance and emergency care in an ambulance vehicle. With a VR headset on, an ambulance paramedic can analyze the patient's condition and fulfill the necessary actions to stabilize the patient for further transportation to the hospital. Paramedics are able to use eleven diagnostic methods, and the simulator itself allows working out algorithms for providing emergency care according to six scenarios (Fig. 1):

1. Acute coronary syndrome with cardiogenic shock;
2. Acute coronary syndrome with pulmonary edema;
3. Anaphylactic shock;
4. Internal bleeding;
5. Respiratory and cardiac arrest;

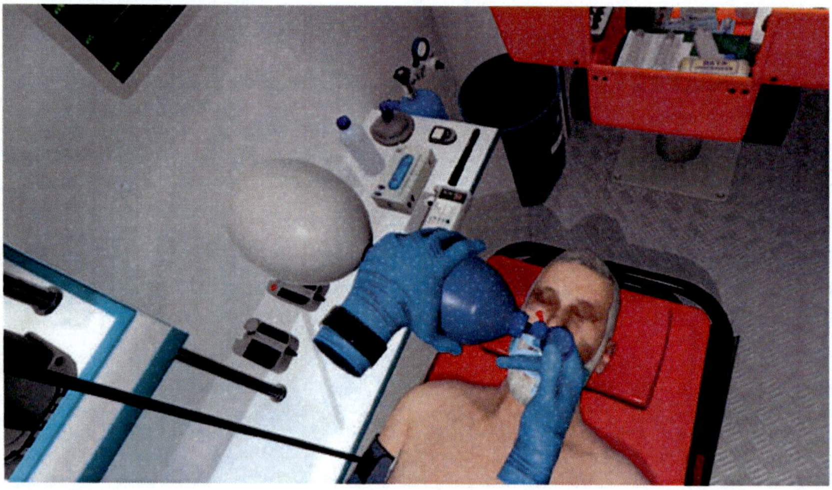

Fig. 1 VR simulator for ambulance paramedics

6. Syncope.

Another important example of VR technologies application in medicine is preoperative preparation, which allows working out the algorithms of varying complexity for operations without prejudice to the real patients.

Surgeons from the Great Ormond Street Hospital (GOSH) managed to fulfill a unique operation on the separation of Siamese twins with their skulls and blood vessels fused together, using VR technology and 3D printing [4]. The surgeons visualized anatomical features of the twins in virtual reality, which made it possible to work out the algorithm of their actions and thus prepare for the real operation (Fig. 2).

An important area of VR technologies application in medicine is the treatment of various phobias. Psychologists treat patients by placing them in a specific VR environment in a controlled manner. Various virtual locations, designed together with psychologists, help treat, for example, the following diseases: arachnophobia, claustrophobia, acrophobia, etc., achieving significant patient improvement.

A good example of such an application is the C2Phobia created by Canadian company C2Care. C2Phobia consists of more than seventy virtual locations aimed to provide treatment for various phobias and anxiety disorders [5]. The C2Phobia application has various setting options that allow regulating the interaction conditions depending on the patient's progress (Fig. 3).

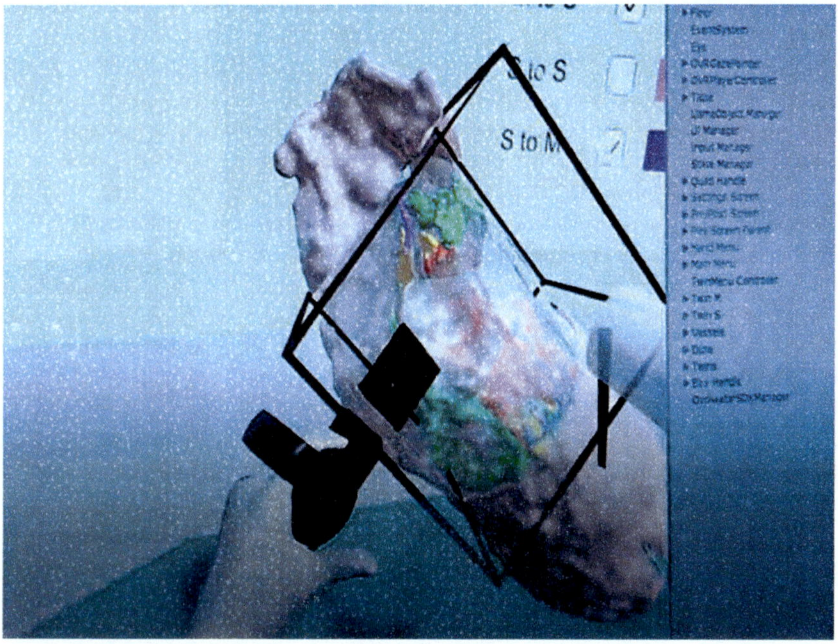

Fig. 2 VR simulation of the twins' anatomical features

Fig. 3 Acrophobia treatment using the C2Phobia application

The articles "Video Game With Biofeedback Helps Kids and Teens Regulate Stress and Anger" and "A "Proof of Concept" Randomized Controlled Trial of a Video Game Requiring Emotional Regulation to Augment Anger Control Training" discuss a VR application using the biofeedback to keep the player's heart rate low, thus helping the player to regulate the anger [6]. The authors of the studies conducted an experiment on a selection of forty patients at the Boston Children's Hospital. Based on a subsequent questionnaire, the authors demonstrated that the application users were able to easily control their emotional state, which clearly confirmed the positive effect of the VR application for controlling the emotional and thus functional state.

The article "Applications of virtual reality for pain management in burn-injured patients" discusses VR systems application for pain treatment in patients with burn injuries [7]. The authors state that immersion in virtual reality provides an intense form of cognitive distraction during short painful procedures. That has also been confirmed by previous studies, particularly, in reviews of immersive VR analgesia [8, 9]. The authors of the reviews suggest that while the pharmacological analgesics act directly on receptors, VR applications may demonstrate their pain relief-related benefits by altering the pain perception. The authors describe the effect as if the nerve centers were "distracted" by tasks related to the perception of additional audio-visual information and information on orientation in space during immersion in the VR application.

The most popular theory states that virtual reality competes effectively for a large proportion of the user's conscious attention, distracting the user's attention from a simultaneous painful medical procedure, and replacing it with more pleasant sensory input from the virtual experience. It is assumed that the descending inhibitory pathways in the central nervous system may play a role in this process. The exact neurophysiological mechanisms of this change in pain perception are not clearly defined and are the subject of ongoing research. However, mechanistic studies of VR analgesia in the controlled experimental pain conditions suggest that: (1) the magnitude of the analgesic effect depends on the user's sense of "presence" (immersion) in the virtual environment [10], (2) subjective VR analgesia is accompanied by a simultaneous decrease in pain-related brain activity in the cerebral cortex and brainstem [11], (3) VR analgesia is similar in magnitude and complements clinically relevant doses of concurrent analgesics [12].

Thus, the virtual reality environments application makes it possible to establish biological feedback with the user, which contributes to the correction of his psycho-emotional state and is able to distract it from external influences, competing for a limited amount of information storage and processing of the central nervous system.

The article "Manipulating presence influences the magnitude of virtual reality analgesia" [9] provides a comparison between immersion in a VR environment and a video game, used to distract attention for patients with burn injuries. Comparative analysis of the subjective pain and anxiety evaluation according to the 100-point visual analogue scale (VAS) shows significant improvements in all parameters during the use of virtual reality compared with the use of video games. (Fig. 4) [13].

Patients' fear to undergo surgery is one of the major issues at Maxillofacial and Plastic Surgery Department of First Pavlov State Medical University. The fear may lead up to severe emotional imbalance and loss of consciousness. For these

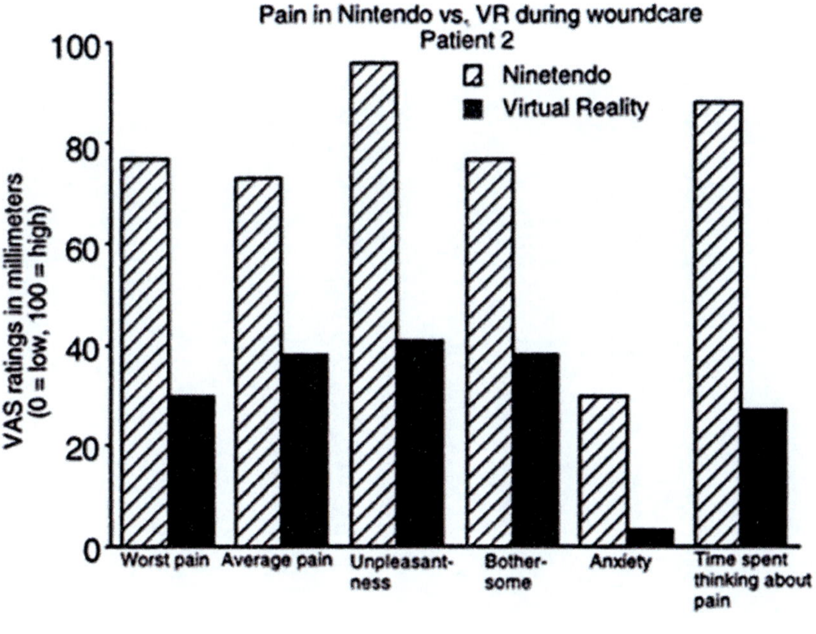

Fig. 4 Comparative study on the pain distraction using 2D video game and VR immersion

reasons, the department has a professional psychologist to carry out the preoperative preparation of patients.

The joint interdisciplinary research group of the Center of Usability and Mixed Reality (ITMO University) and the Department of Maxillofacial and Plastic Surgery (First Pavlov State Medical University) proposed the following hypothesis:

The preoperative preparation of patients will be more efficient when using a VR application for the patient's immersion in a virtual 3D environment that simulates a dental operating room.

To test this hypothesis, the authors designed and conducted experiments that will be describes in detail in the following paragraphs.

2 Materials and Methods

2.1 VR Application Design

To conduct the study the authors designed an authentic three-dimensional virtual location of the maxillofacial and plastic surgery department dental operating room. Virtual three-dimensional simulation of the operating room allowed to immerse the patient in the appropriate environment, thereby preparing him for the real operating room (Fig. 5).

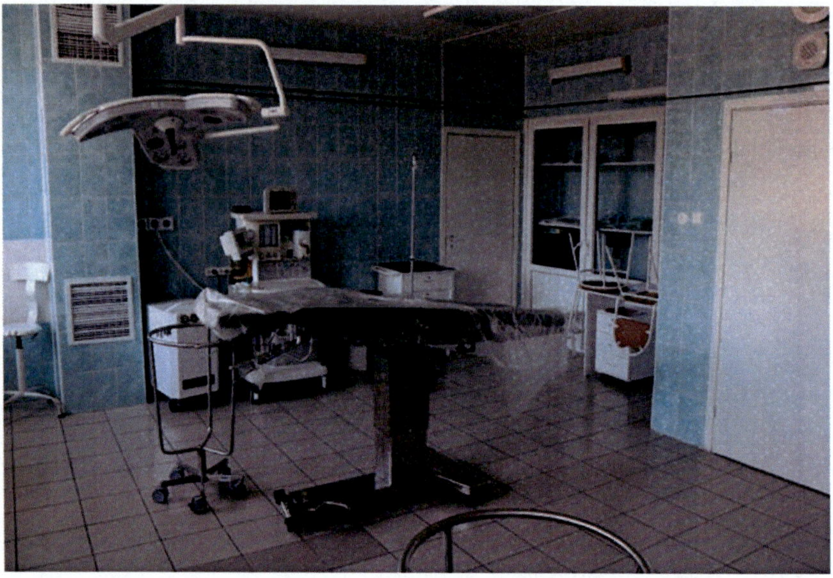

Fig. 5 Dental operating room of the Maxillofacial and Plastic Surgery Department

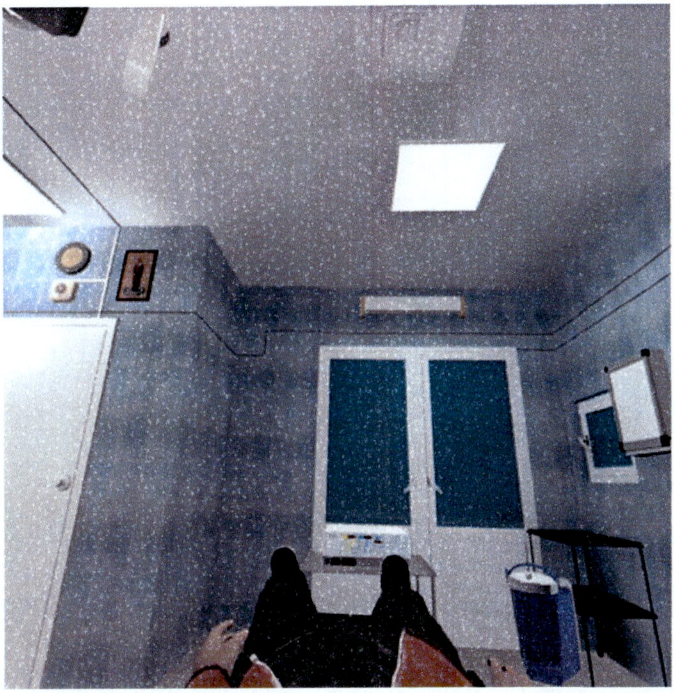

Fig. 6 The designed VR location as observed in the Samsung Gear VR headset

The authors used 3D graphics software Blender to model a three-dimensional space of the operating room. When the three-dimensional model of the operating room was ready, we exported it into the 3D graphics engine Unity. Unity allows producing a three-dimensional location for observing in a virtual reality headset. We chose the Samsung Gear VR headset was due to its convenience, as it is mobile thus allowing us to conduct the necessary research at First Pavlova State Medical University. Figure 6 shows a screenshot of the designed VR application for the Samsung Gear VR headset.

2.2 *Conducting an Experiment in Preoperative Patient Preparation*

We divided patients into two groups for the experiment: the first group had a session with a psychologist who used the designed VR application, the second group had a session with a psychologist without using the designed VR application. During the sessions, special equipment measured the most important stress indicators of the

Fig. 7 The process of conducting the experiment and obtaining the physiological data

patients - their heart rate registered by an optical plethysmograph and galvanic skin response (GSR) (Fig. 7).

To measure the physiological parameters, we used a software-hardware system designed by the Bitronics Lab Company [14]. Since we decided to record two physiological parameters of the patients, the program code for the pulse sensor was supplemented with the code for the GSR sensor, and the data from the second sensor were redirected to the second available signal output channel.

A focus group of twenty people participated in the preoperative preparation experiment. Two factors defined the number of subjects: a narrow target group (patients waiting for planned operations) and the context of the coronavirus pandemic (COVID-19). The patients were randomly assigned to one of the two groups: the one to undergo the preoperative preparation using the designed VR application, and the other to undergo the standard preoperative preparation procedure. Before and after the preoperative preparation procedure we measured and recorded the patients' physiological data. We used statistical analysis to process the recorded data comparing the physiological state of patients before and after the preoperative preparation procedure.

3 Results and Discussion

First, we recorded the patients' physiological data provided by the sensors at the psychologist's office. Then, the patients did or didn't use the designed VR application. The system recorded the heart rate (number of beats per minute) and GSR parameters (in conventional units).

Pulse beats could vary from 60 (low heart rate) to 150 (high heart rate). The norm/expectation for this parameter was 80 beats per minute.

The GSR varied from 0 to 5. The norm/expectation for this parameter was 2.5 units. The collected data are shown in Table 1.

At the first stage, we checked the correlation between the heart rate and GSR parameters. Sorting the data in ascending order, and using the formula for Spearman's rank correlation coefficient (CORREL function in Google Sheets), we found the correlation value to be -0.76.

At the second stage, we assessed the significance level of the obtained value. With a given sample size, according to the table of critical values, the *p*-value of 0.005 corresponds to the value of the coefficient 0.794, and for the *p*-value = 0.01 - 0.746. Therefore, the analysis demonstrates a statistically significant inverse relationship between the values of the samples at the level of 1%.

Scatter diagrams below show the dependence of the heart rate and GSR (Fig. 8), and the confidence interval (Fig. 9).

To prove the hypothesis that the designed VR application can effectively prepare a patient for surgery, the Mann–Whitney U-test was used for the heart rate data.

The data were ranked in ascending order for the heart rate (pulse) parameter. The sums of the ranks for the group that did not use the VR headset and for the one that did were 155 and 55, respectively.

Since the heart rate parameters in the second sample (when using a VR headset) were less than in the first one, we used the formula for the left-side hypothesis testing

Table 1 The results of experiment on two groups of respondents

	VR supplemented		standard procedure	
subject	pulse	GSR	pulse	GSR
1	89	2.43	81	2.59
2	97	2.42	80	2.58
3	103	2.38	73	2.54
4	134	2.42	78	2.67
5	115	2.35	79	2.50
6	123	2.35	81	2.57
7	145	2.39	80	2.59
8	114	2.41	72	2.56
9	151	2.22	73	2.53
10	107	2.24	67	2.54

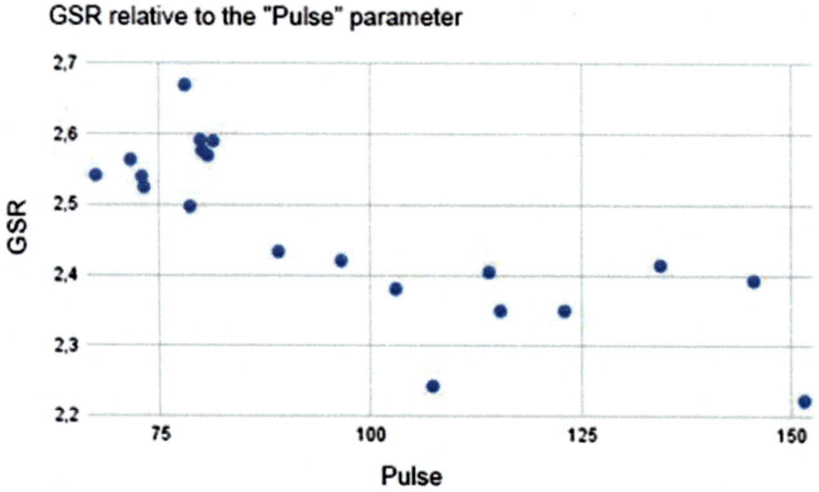

Fig. 8 Scatter diagram for the pulse and GSR dependence

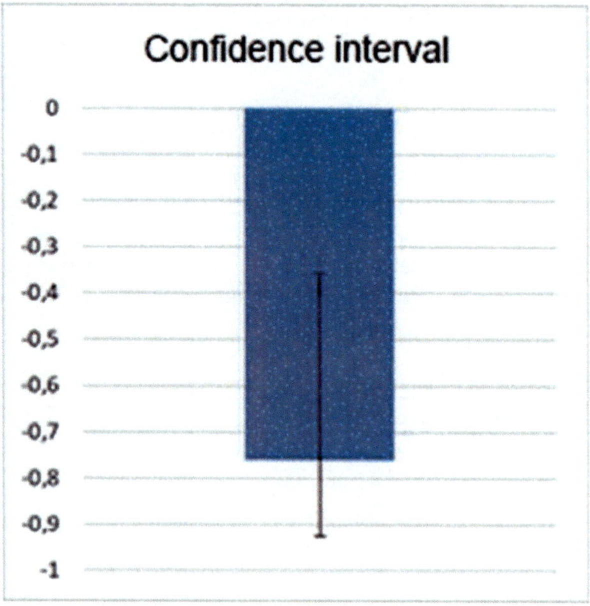

Fig. 9 Confidence interval for the pulse and GSR correlation

to calculate the U-criterion:

$$U = T_y - n \cdot (n + 1)/2, \tag{1}$$

where T_y– the sum of the second sample ranks; n – second sample size.

As a result, we obtained the final value $U = 0$.

Since the sizes of both samples are greater than eight, the distribution of the U-test can be considered as normal, therefore, instead of the U-test, significance can be calculated using the z-test, which has a standard normal distribution and fixed critical values for each significance level according to the formula:

$$z = \frac{\mathrm{U} - \frac{\mathrm{m \cdot n}}{2} + 0.5}{\sqrt{\frac{m \cdot n \cdot (m+n+1)}{12}}} \tag{2}$$

where z – z-criterion; m – first sample size; n – second sample size; + 0.5 is the continuity correction.

The calculation results for the heart rate: z-criterion = – 3.741, and p-value = 0.0091, as calculated in Google Sheets using the "NORMSDIST" function, which corresponds to a statistically significant result at the 1% significance level. The results demonstrate that the patients who used the VR headset to alleviate their psychophysiological state showed better results than the group of subjects who did not use the VR headset at the significance level of 1%.

To analyze the GSR data we used the Mann–Whitney U-test as well.

We ranked the data by the rise in GSR parameter. The sums of the ranks for both groups amounted to 155 and 55, respectively. Then, we used the Mann–Whitney U-test to calculate the U-value using the formula for the left-sided hypothesis testing (Eq. 1), and calculated the p-value with the z-test (Eq. 2).

The results for the GSR data calculations are: z-value = – 3.741, and p-value = 0.0091. This corresponds to a statistically significant result at the 1% significance level.

Thus, the study demonstrates that the use of a VR headset before dental procedures has a positive effect on patients. It helps patients to calm down, which shows in their heart rate and GSR parameters.

4 Conclusions

VR technologies in medicine present a promising tool that can be used both in research and in practice. VR application proved to be effective in the preoperative preparation of patients at the Department of Maxillofacial and Plastic Surgery of First Pavlova State Medical University. Statistical analysis of the data (pulse and GSR parameters) collected by the Bitronics Lab's hardware-software system showed that the efficiency of the preoperative preparation using the designed VR simulation was significantly higher than without it.

References

1. Advantages of virtual reality in medicine. https://www.vrs.org.uk/. Accessed 11 July 2022
2. Plotnikova, D., Volosiuk, A., Tikhonov, G., Tsynchenko, A., Luneva, A., Smolin, A.: VR application for vestibular system training (Pilot Study). In: Black, N.L., Neumann, W.P., Noy, I. (eds.) Proceedings of the 21st Congress of the International Ergonomics Association (IEA 2021). IEA 2021. Lecture Notes in Networks and Systems, vol. 223, pp. 552–558. Springer, Cham (2022). https://doi.org/10.1007/978-3-030-74614-8_68
3. How virtual reality technology helps paramedics to provide emergency care. https://vc.ru/. Accessed 11 July 2022
4. Separation of Siamese twins: a unique operation of British doctors. BBC News - Russian Service. https://www.bbc.com/russian. Accessed 11 July 2022
5. Psychological treatment by VR. https://psy.c2.care/en/. Accessed 11 July 2022
6. Ducharme P., et al.: A "proof of concept" randomized controlled trial of a video game requiring emotional regulation to augment anger control training. Front. Psychiatry (2021)
7. Sharar, S., Miller, W., Teeley, A., Soltani, M., Hoffman, H., Jensen, M., Patterson, D.: Applications of virtual reality for pain management in burn-injured patients. Expert Rev. Neurother. **8**(11), 1667–1674 (2008)
8. Gold, J.I., Belmont, K.A., Thomas, D.A.: The neurobiology of virtual reality pain attenuation. Cyberpsychol. Behav. Soc. Netw. **10**(4), 536–544 (2007)
9. Wismeijer, A.A.J., Vingerhoets, J.J.M.: The use of virtual reality and audiovisual eyeglass systems as adjunct analgesic techniques: a review of the literature. Ann. Behav. Med. **30**(3), 268–278 (2005)
10. Hoffman, H.G., Sharar, S.R., Coda, B., et al.: Manipulating presence influences the magnitude of virtual reality analgesia. Pain **111**(1–2), 162–168 (2004)
11. Hoffman, H.G., Richards, T.L., Coda, B., et al.: Modulation of thermal pain-related brain activity with virtual reality: evidence from fMRI. NeuroReport **15**(8), 1245–1248 (2004)
12. Hoffman, H.G., Richards, T.L., Van Oostrom, T., et al.: The analgesic effects of opioids and immersive virtual reality distraction: evidence from subjective and functional brain imaging assessments. Anesth. Analg. **105**(6), 1776–1783 (2007)
13. Hoffman, H.G., Doctor, J.N., Patterson, D.R., Carrougher, G.J., Furness, T.A., 3rd.: Virtual reality as an adjunctive pain control during burn wound care in adolescent patients. Pain **85**(1–2), 305–309 (2000)
14. Bitronics Lab company. https://bitronicslab.com/. Accessed 11 July 2022

Online Application for Fast Pure Tone Audiometry in Non-laboratory Conditions

Josipa Vujević and Kristian Jambrošić

Abstract Healthy hearing is very important in various professions, such as pilots, drivers, musicians, emergency services, etc. Sometimes it is difficult to regularly monitor hearing health with audiological examinations due to time or financial limitations. The same is true for psychoacoustic tests where it is important to receive information about the hearing health of the subjects. The importance of adapting sound reproduction systems to the individual hearing sensitivity is recognized by many smartphone manufacturers which offer testing of the user's hearing and adapting the sound reproduction using headphones to individual hearing. For the above reasons, an online hearing test application for pure tone audiometry was designed with the main goal to be accessible by any user with quality headphones in a quiet place, not just from a dedicated laboratory. The need for headphone calibration and equalization is further discussed. For precise results, the system requires calibration of subject headphones, but it is important to notice that the presented system and similar other audiometry applications can be used for early detection of hearing impairment if hearing is monitored at certain intervals, even if used in non-clinical surrounding and without a full calibration.

Keywords Pure Tone Audiometry · Hearing Monitoring · Hearing Impairment · Headphone Calibration

1 Introduction

The history of hearing tests begins in Greece all the way back in sixth century BC when they were done with the help of a tuning fork [1, 2]. David Edward Hughs had created a better solution for hearing testing than the tuning fork. This was a device

J. Vujević (✉) · K. Jambrošić
Faculty of Electrical Engineering and Computing, University of Zagreb, Zagreb, Croatia
e-mail: josipa.vujevic@fer.hr

K. Jambrošić
e-mail: kristian.jambrosic@fer.hr

I. Salopek Čubrić et al. (eds.), *Proceedings of the 9th International Ergonomics Conference*, Lecture Notes in Networks and Systems 701,
https://doi.org/10.1007/978-3-031-33986-8_38

called an audiometer. The physician had to alter sound frequencies on the audiometer that were reproduced over headphones to patients, and they indicated in return when they heard, or didn't hear the sounds by raised their hand. Those audiometers were only used in private practices by doctors and were not publicly available.

After World War I two physicists, H. Fletcher and R. L. Wenger, developed the first commercially available electric model of the audiometer, the 1-A, but there was no graphic data shown in the measurement report. Therefore, physicians could not document pathologic conditions of the ear in a proper way. That's when the audiogram was born. An audiogram is a chart where the hearing sensitivity is plotted as a curve for each measured ear, with frequencies listed along the abscissa axis and hearing sensitivity along the ordinate axis, Fig. 1.

Pure tone audiometry, as the name suggests, uses pure tones or other narrow-band stimuli as test signals to measure frequency-specific hearing thresholds. A pure tone audiometer is a device that changes, either manually or automatically, the level of a test signal in an adaptive process seeking the hearing threshold of the tested person.

Pure tone audiometry indicates what hearing thresholds (dB) are required to just be able to perceive a tone at different frequencies (Hz). A pure tone audiology threshold at a specific frequency is the decibel level at which a sound is perceived 50% of the time. The decibel scale used in pure tone audiometry is dB Hearing Level (dB HL). The dB HL intensity scale is based on normal human hearing with 0 dB HL representing the median threshold for ontologically normal young adults [2–4]. These combinations of frequencies and intensities are shown on the audiogram, Fig. 1.

To receive comparable results from the audiometry procedures, they are done according to the norm ISO 8253–1:2010, which in return considers the normed hearing threshold of the average human as defined in ISO 226:2003 [5, 6].

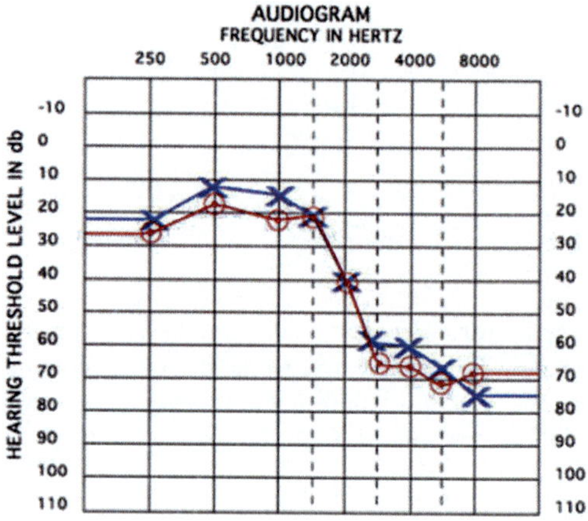

Fig. 1 An audiogram example for the left and right ear [3]

2 Development of Audiometric Application

The main goal of this paper is to present the design of an online hearing test application for pure tone audiometry that would make accessible the audiometry procedure to any user with quality headphones in a quiet place, not just in a dedicated laboratory. There are some other already existing online audiometer applications, but they have limited functionality and are mostly connected to some commercial companies.

2.1 Concept of the Audiometric Application

The application is envisioned as a web application. It consists of a general description of the test and instructions for calibration (currently in Croatian only). When the calibration is done, the user can enter the sound pressure level to which he calibrated his headphones and press the Test button, which then presents him with the test itself. This Introduction page is shown in Fig. 2.

The test is designed to work with stereo headphones because the sound played is a stereo sound with separate testing tones for the left and the right ear. The interactive tone reproduction interface for testing the left ear will play sound on the left earpiece, and the one for the right ear will play to the right. Sound is played by pressing the toggle round buttons, as seen in Fig. 3.

Below the test navigation interface is a table and a graph with the results. The results of the user's hearing sensitivity are dynamically filled in both the graph and table in real time as the test is proceeding. This can be seen in Fig. 4. During the testing process, the users can cancel the current results at any time and start from the

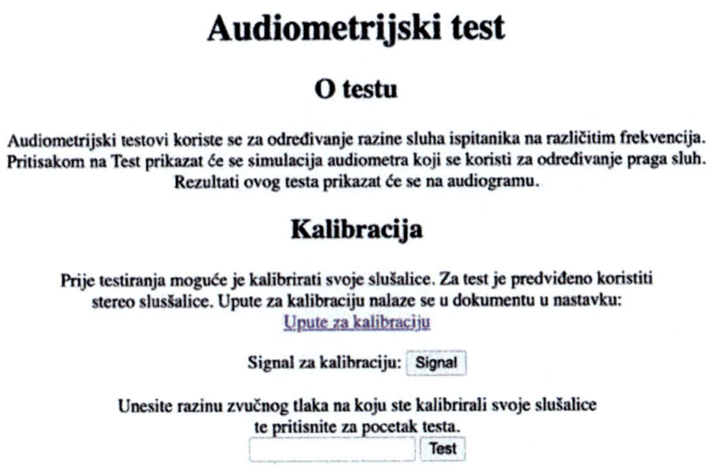

Fig. 2 Introduction page of the online audiometry application

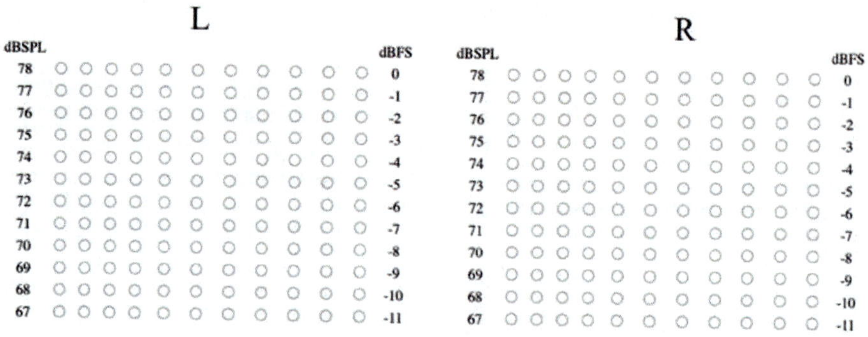

Fig. 3 Test buttons to navigate the reproduction of the test tones

beginning. When the testing is finished, the users can download the audiogram with the results to their computer, and, if necessary, forward them to the examiner by an e-mail.

2.2 *Used Technologies*

The presented application is fairly simple in design. It consists of two html documents for two mentioned pages: introductory page and test page. HTML, CSS and JavaScript were used on the client side. These pages are dynamically manipulated using the DOM (Document Object Model) and JavaScript. The Document Object Model is a programming interface for web documents. It represents a page so that programs can change the structure, style, and content of the document. DOM represents a document as nodes and objects; in this way coding languages can communicate with the page.

A web page is a document that can be displayed in a browser window or as an HTML source. In both cases, it is the same document, but the DOM view allows its manipulation. As an object-oriented web page representation, it can be modified by a scripting language such as JavaScript. The application was created in the NodeJS environment which is currently one of the most popular environments for executing JavaScript code.

Since the presented application is small and light, it is convenient that it runs in the browser without the need for connection to any kind of database. On the server side there is a simple server provided by the JS Express module.

The tests were also implemented using JavaScript, and the ChartJS JavaScript library was used for the graphical display of data [7]. For the deployment, Heroku was used, a platform that allows applications to be built and run entirely in the cloud [8].

Results

f	L (db SPL)	R (dB SPL)	L (dB HL)	R (dB HL)
125	22	14	0	-8
250				
500	18	7	14	3
750				
1000	13	3	11	1
1500	8		7	
2000				
3000	5		11	
4000	9		14	
6000	10		5	
8000	12		0	

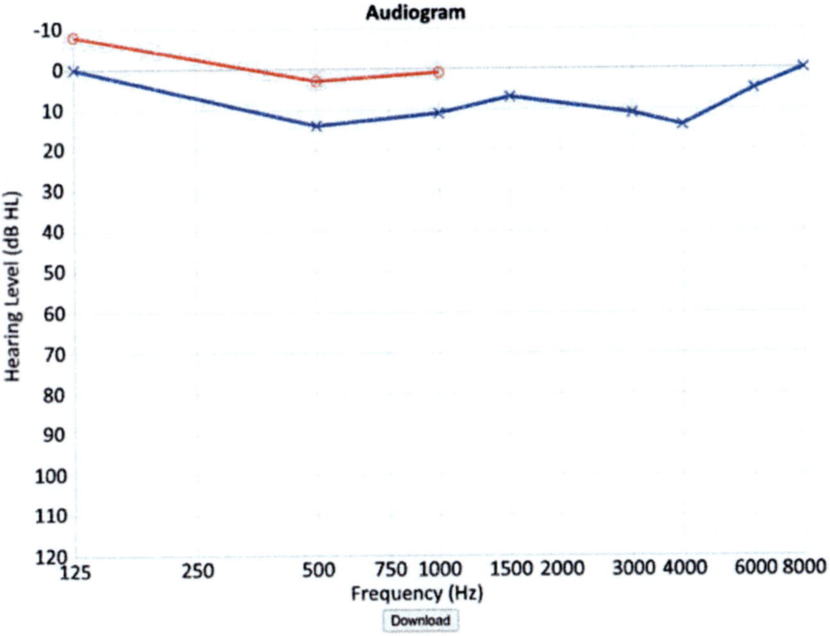

Fig. 4 Results of the hearing test, both in tabular and graphical form

2.3 *Test Phase*

The basic audiological assessment focuses on pure tone air conduction thresholds in the frequency range 0.25 - 8 kHz, according to the relevant norm ISO 8253–1:2010 [5]. The level of the test signal of the audiometer must be changed in the adaptive process, either manually or automatically.

The testing process itself is determined in [5] as well. For this application, the bracketing method was chosen, which is also described on the test page. It requires the.

test subject to start at a frequency of 1000 Hz and first reproduce a very quiet tone, then increases the level of the test tones until it becomes audible. The subject must remember at what level the tone was heard. Afterwards the fencing starts from above by repeating the process at the same frequency but starting from a tone level that is clearly audible and by decreasing it until it cannot be heard anymore. At the level when the tone is no longer audible, the subject switches to one step higher, that is, to the last one that was audible. At that moment, the level between the current one and the previous one must be remembered. This is the hearing threshold at that frequency and the last button should be left checked. Since the buttons are implemented as toggle buttons, the last button played will stay checked which is convenient. The process should be continued towards higher frequencies, and when the subject has finished with the higher frequencies, the procedure is returned to 1000 Hz and repeated by moving towards lower frequencies. The test should be done for both ears.

Since the sound played as a stereo signal, the user of the test doesn't have to think about whether the testing was done for the left or right ear. When the test is conducted for the right ear, the test signal will be played through the right earpiece and vice versa.

For reliable results, it is necessary to ensure that the masking effect of background noise does not increase the measured hearing thresholds. Otherwise, the test should be conducted in a sound isolated environment. It is highly recommended that closed back headphones are used and that they are properly calibrated.

3 Headphone Calibration Challenges

One of the main disadvantages of the proposed audiometry method is a lack of system calibration which would allow the adjustment of absolute sound pressure levels at the listeners ears. This topic was tackled before in several research studies, such as [9] and [10]. Some researchers argue that only Head and Torso Simulators (HATS) and Ear Simulators can provide a proper calibration of headphone sound reproduction levels so that they can be used in scientific research, medical diagnosis and all other applications.

Although this might be true, there are situations where one doesn't have access to a lab with calibrated HATS devices. A simplified method of calibration was proposed for the SATP project (Soundscape Attributes Translation Project) where many participants from all around the world had to listen to some binaural soundscape recordings in non-laboratory conditions [11]. The lack of calibrated headphones / listening units that provide sound reproduction over headphones with a defined, previously calibrated sound pressure level forced the project coordinators to think of a simple calibration method that can be used outside of laboratories with widely available equipment and devices.

Calibration must be done to identify the output voltage of the headphones amplifier that corresponds to a certain sound pressure level (SPL) at the output of the headphones. The required equipment includes: acomputer for playback of recorded

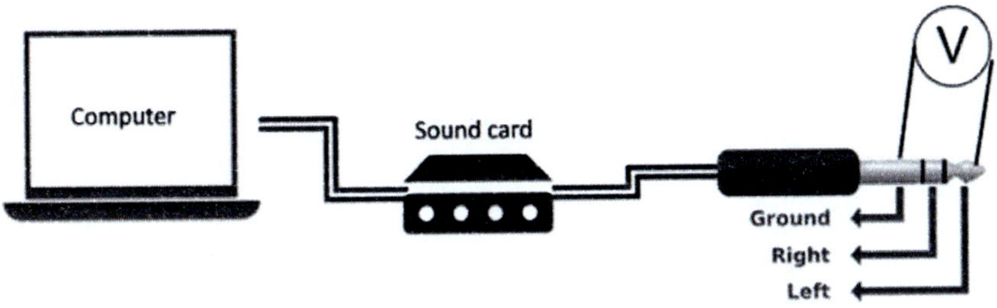

Fig. 5 The headphone calibration procedure setup

sound files, an external sound card with headphones output, with or without a headphones distribution amplifier (needed only if the sound card cannot achieve a high enough voltage level on its headphones output), a voltmeter/multimeter with AC RMS voltage measurement range below 1 V, and high quality headphones of the over-ear, open type. The calibration procedure can be seen in Fig. 5.

The headphones must be plugged into the headphones output of the sound card. If the calibration procedure shows that the headphones output does not provide a high enough voltage level to yield the proper sound pressure level, a headphones amplifier must be connected to the output of the sound card, and the headphones are then plugged into the amplifier.

The calibration is done by executing the following steps. First, the sensitivity of the headphones SPL_S must be known from the technical specifications (e.g. 103 dB (1 V_{RMS}) at 1 kHz). This means that the headphones will produce the sound pressure level of 103 dB at 1 kHz when the voltage of 1 V_{RMS} (RMS = Root Means Square) is applied to them.

The calibration signal SPL_C is a 1 kHz sine signal recorded at SPL 94 dB. Since we must achieve the same SPL with our headphones, the voltage that must be applied to the headphone jack can be calculated using the expression (1):

$$V_{RMS} = \frac{10^{\frac{SPL_C}{20}}}{10^{\frac{SPL_S}{20}}} = \frac{10^{\frac{94}{20}}}{10^{\frac{103}{20}}} = 0,355\text{V} = 355\text{mV} \tag{1}$$

If any other headphones are used, their corresponding sensitivity must be entered into the expression above.

Second, to tune the gain of the headphones output, first a balanced (stereo) 1/4″ TRS male to 1/4″ TRS male audio cable must be connected to the headphones output, Fig. 5. Next, a multimeter (AC measurement range) must be connected to the other end of the cable. The gain knob of the headphone output on the sound card must be adjusted until the displayed voltage at the voltmeter shows the value calculated in (1) (355 mV or whatever voltage you calculate for your model of headphones). After the gain is set, the gain knob mustn't be changed until the audiograms have been measured. Finally, the headphones are connected to the headphone output of the sound card, and the whole system is calibrated as good as possible, given the available equipment.

4 Conclusions

The creation of this application was motivated by the recent epidemiological situation since many research studies that use listening tests had to be done by test subjects from their homes and not from laboratories. The authors hope that this system will help anyone interested in testing their hearing in a simple way. It can also help researchers increase the amount of collected data for any kind of psychoacoustic research when it might be otherwise limited by time or space resources.

Moreover, the application can be also used for a continuous screening of hearing sensitivity over time by any professionals that relay on the health of their hearing. Even if the system is not calibrated, any evident drop of hearing sensitivity at certain frequencies over time might be an indication of hearing loss and an audiologist should be visited for confirming these findings.

The possible downside of this solution is the data storage itself. Current solution provides an audiogram and table with results, but the users are expected to copy paste the table if they want to share or store the results somewhere else to compare with other results or send them to the potential researcher. Nevertheless, when the tests are done correctly, it does seem intuitive to the user.

All potential users are invited to try out the application using the following link: https://audiogram-fer.herokuapp.com

References

1. Pulkki, V., Karjalainen, M.: Communication Acoustics: An introduction to Speech Audio and Psychoacoustics. Wiley, Finland (2015)
2. How long have audiograms been around?, Adirondack Audiology. https://adirondackaudiology.com/how-long-have-audiogramshearing-tests-been-around/. Accessed 29 Sept 2022
3. What is and audiogram? Work Health Professionals. https://www.workhealthonsite.com.au/what-is-an-audiogram/. Accessed 29 Sept 2022
4. Open access guide to audiology and hearing for otolaryngologists, Maggi Soer. http://www.entdev.uct.ac.za/guides/open-access-guide-to-audiology-and-hearing-aids-for-otolaryngologists/. Accessed 29 Sept 2022
5. ISO 8253–1:2010 Acoustics — Audiometric test methods — Part 1: Pure-tone air and bone conduction audiometry
6. ISO 226:2003 Acoustics — Normal equal-loudness-level contours
7. Chart.js, Chart.js. Poveznica. https://www.chartjs.org/docs/latest/. Accessed 1 June 2022
8. Heroku, HEROKU. https://www.heroku.com/what. Accessed 1 June 2022
9. Lam, B., et al.: Preliminary assessment of a cost-effective headphone calibration procedure for soundscape evaluations. In: Proceedings of the 28th International Congress on Sound and Vibration (ICSV28), Singapore, pp. 1–8 (2022)

10. Lam, B., Ooi, K., Ong, Z.-T., Watcharasupat, K.N., Wong, T., Gan, W.-S.: Assessment of a cost-effective headphone calibration procedure for soundscape evaluations. In: Proceedings of the 24th International Congress on Acoustics (ICA2022), Gyeongju, pp. 1–8 (2022)
11. Aletta, F., et al: Soundscape assessment: towards a validated translation of perceptual attributes in different languages. In: Proceedings of Inter-Noise 2020, Seoul, pp 1–10 (2020)